Environmental Chemistry for a Sustainable World

Volume 68

Series Editors
Eric Lichtfouse, Aix Marseille University, CNRS, IRD, INRA, Coll France, CEREGE, Aix en Provence, France
Jan Schwarzbauer, RWTH Aachen University, Aachen, Germany
Didier Robert, CNRS, European Laboratory for Catalysis and Surface Sciences, Saint-Avold, France

Environmental chemistry is a fast developing science aimed at deciphering fundamental mechanisms ruling the behaviour of pollutants in ecosystems. Applying this knowledge to current environmental issues leads to the remediation of environmental media, and to new, low energy, low emission, sustainable processes. The topics that would be covered in this series, but not limited to, are major achievements of environmental chemistry for sustainable development such as nanotech applications; biofuels, solar and alternative energies; pollutants in air, water, soil and food; greenhouse gases; radioactive pollutants; endocrine disruptors and other pharmaceuticals; pollutant archives; ecotoxicology and health risk; pollutant remediation; geoengineering; green chemistry; contributions bridging unexpectedly far disciplines such as environmental chemistry and social sciences; and participatory research with end-users.

The books series will encompass all scientific aspects of environmental chemistry through a multidisciplinary approach: Environmental Engineering/ Biotechnology, Waste Management/Waste Technology, Pollution, general, Atmospheric Protection/Air Quality Control/Air Pollution, Analytical Chemistry. **Other disciplines include:** Agriculture, Building Types and Functions, Climate Change, Ecosystems, Ecotoxicology, Geochemistry, Nanochemistry, Nanotechnology and Microengineering, Social Sciences. The aim of the series is to publish 2 to 4 book per year. **Audience:** Academic/Corporate/Hospital Libraries, Practitioners / Professionals, Scientists / Researchers, Lecturers/Tutors, Graduates, Type of books (edited volumes, monographs, proceedings, textbooks, etc.). **Edited volumes:** List of subject areas the series will cover: • Analytical chemistry, novel methods • Biofuels, alternative energies • Biogeochemistry • Carbon cycle and sequestration • Climate change, greenhouse gases • Ecotoxicology and risk assessment • Environmental chemistry and the society • Genomics and environmental chemistry • Geoengineering • Green chemistry • Health and environmental chemistry • Internet and environmental chemistry • Nanotechnologies • Novel concepts in environmental chemistry • Organic pollutants, endocrine disrupters • Participatory research with end-users • Pesticides • Pollution of water, soils, air and food • Radioactive pollutants • Remediation technologies • Waste treatment and recycling • Toxic metals

More information about this series at http://www.springer.com/series/11480

K. M. Gothandam • Ramachandran Srinivasan
Shivendu Ranjan • Nandita Dasgupta
Eric Lichtfouse
Editors

Environmental Biotechnology Volume 4

Springer

Editors
K. M. Gothandam
School of Bio Sciences and Technology
Vellore Institute of Technology
Vellore, Tamil Nadu, India

Ramachandran Srinivasan
Department of Genetic Engineering
SRM Institute of Science and Technology
Kattankulathur, Tamil Nadu, India

Shivendu Ranjan
Faculty of Engineering and the Built Environment
University of Johannesburg
Johannesburg, South Africa

Nandita Dasgupta
Department of Biotechnology
Institute of Engineering and Technology
Lucknow, Uttar Pradesh, India

Eric Lichtfouse
Aix Marseille University, CNRS, IRD, INRA, Coll France, CEREGE
Aix en Provence, France

ISSN 2213-7114 ISSN 2213-7122 (electronic)
Environmental Chemistry for a Sustainable World
ISBN 978-3-030-77794-4 ISBN 978-3-030-77795-1 (eBook)
https://doi.org/10.1007/978-3-030-77795-1

This Springer imprint is published by the registered company Springer Nature Switzerland AG
The registered company address is: Gewerbestrasse 11, 6330 Cham, Switzerland

Foreword

Since humans are a highly evolved species, it is their responsibility to maintain a proper ecosystem with good environmental quality for the well-being of every organism on Earth. Many environmental issues have emerged in the Anthropocene, such as pollution, soil degradation, global warming, climate change, ozone layer depletion, acid rain, ocean acidification, waste disposal, generating unsustainable waste, deforestation, and depletion of natural resources. Therefore, there is a need for advanced biotechnologies to clear pollutants and design goods and products that are fully recyclable in the context of the circular economy.

This book is the fourth volume on environmental biotechnology published in the series Environmental Chemistry for a Sustainable World. Chapters present the synthesis of bioplastics from various raw materials; the conversion of wastewater into valuable bioproducts by bacteria; removing phosphate in wastewater treatment plants; and agricultural waste biosorbents for heavy metal removal from wastewater (Fig. 1). This book also discusses recent advancement in pest control in rice; methods to measure the genotoxicity of soil samples; and the protective role of phytochemicals on acrylamide-induced toxicity in *Drosophila*. Mesoporous silica nanoparticles as nanocarriers for inhibiting breast cancer cells and biotechnological applications of marine fungal exopolysaccharides are also detailed.

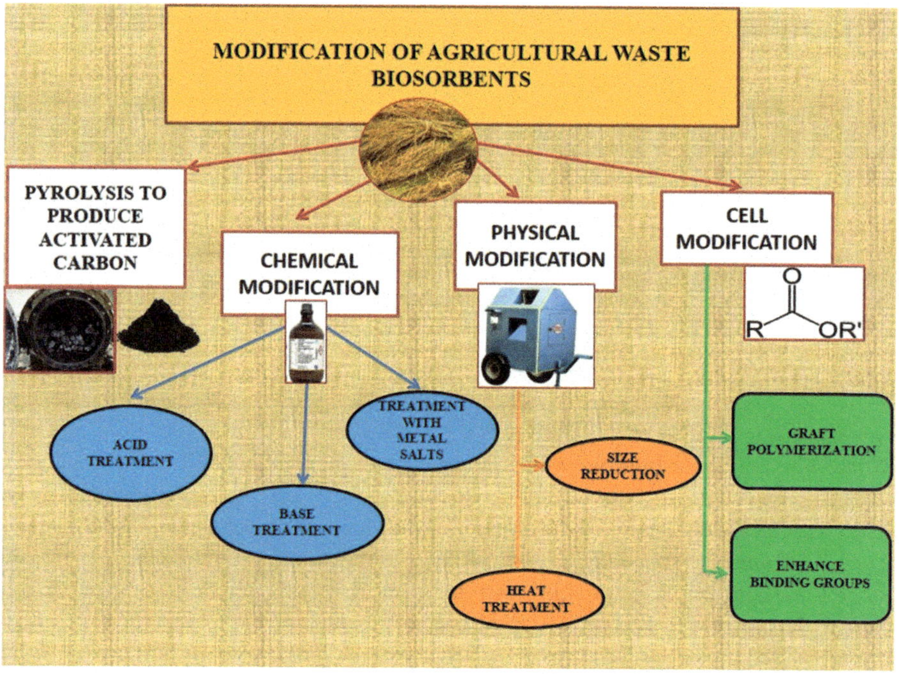

Fig. 1 Modification methods for producing effective agricultural waste biosorbents. (From Chap. 4)

Vellore, Tamil Nadu, India — K. M. Gothandam
Kattankulathur, Tamil Nadu, India — Ramachandran Srinivasan
Johannesburg, South Africa — Shivendu Ranjan
Lucknow, Uttar Pradesh, India — Nandita Dasgupta
Aix en Provence, France — Eric Lichtfouse

Contents

About the Editors

K. M. Gothandam is a professor of biotechnology, School of Bio Sciences and Technology, Vellore Institute of Technology, Vellore. He earned his Ph.D. from Bharathiar University, Coimbatore, and postdoctoral fellowship from School of Life Sciences and Biotechnology, Korea University, Seoul (2002–2007). He also served as HOD, Department of Biotechnology, and Dean, School of Bio Sciences and Technology (2016–2018). His research interest includes functional genomics, plant and microbial metabolites, cancer biology, and environmental biotechnology. He has published over 100 scientific research and review articles in international peer-reviewed journals. He also authored five book chapters and edited five books. He has guided 11 Ph.D. thesis and is now guiding 6 scholars. He is currently handling two funded projects funded by DBT and has completed five projects funded by DBT, DST, and CSIR.

Srinivasan Ramachandran is currently a national postdoctoral fellow in the Science and Education Research Board (SERB), Department of Genetic Engineering, SRM Institute of Science and Technology, Kattankulathur. He obtained his doctorate from Vellore Institute of Technology, Vellore. He did his postdoctoral research at the International Centre for Genetic Engineering and Biotechnology (ICGEB), New Delhi. His field of interest includes metabolic engineering of algae as well as photosynthetic organisms towards stress tolerance, and production of metabolites (pigments, lipids, carbohydrates, and proteins) and other high value-added products. He has published several scientific research articles in SCI-indexed and peer-reviewed journals and also refereed many peer-reviewed journals. He is also an active team member of SERB- and BIRAC-funded projects.

Shivendu Ranjan is senior research associate at the University of Johannesburg, South Africa. He is elected fellow of several scientific societies, e.g., Indian Chemical Society, Linnean Society (London), Bose Science Society, and Indian Engineering Teachers Association. His research interests include nanomedicine, biomaterials, and toxicology. He is associate editor of *Environmental Chemistry Letters* and editorial board member of several journals of international repute.

Nandita Dasgupta is working as an assistant professor in the Department of Biotechnology, Institute of Engineering and Technology, Lucknow, India. Her areas of interest include advanced materials fabrication and their applications in medicine, food, and biomedicine. She is the associate editor of *Environmental Chemistry Letters*. She has received several awards and recognitions from different national and international organizations. She is the elected fellow of the Linnean Society (London) and Bose Science Society.

Eric Lichtfouse is geochemist and professor of scientific writing at Aix-Marseille University, France, and visiting professor at Xi'an Jiaotong University, China. He has discovered temporal pools of molecular substances in soils, invented carbon-13 dating, and published the book *Scientific Writing for Impact factor Journals*. He is chief editor and founder of the journal *Environmental Chemistry Letters* and the book series Sustainable Agriculture Reviews and Environmental Chemistry for a Sustainable World. He has awards in analytical chemistry and scientific editing. He is World XTerra vice champion.

Chapter 1
Switching to Bioplastics for Sustaining our Environment

Priyanka, Durga Yadav, and Joydeep Dutta

Abstract The use of traditional plastics poses a major threat to the environment due to their non-biodegradability, which is also responsible for increasing human health risks. The majority of the traditional plastics are produced from fossil fuel-based non-renewable resources. It is estimated that more than 350 million tons of plastics are produced every year globally for versatile applications. In contrast, the production of bioplastics is still in the infancy stage. Further, the reservoir of fossil fuel is also getting exhausted over time. Due to these two main problems associated with traditional plastic, it is quite mandatory to look for new materials, eventually solving both purposes. From a social and economic viewpoint, unless an alternative is discovered, it is difficult to permanently escape from using these plastics because of the way they have ingrained in our daily lives. However, the environmental degradation caused by the accumulation of non-biodegradable plastic materials on our earth has been well documented. Several measures have also been taken to mitigate this damage but switching to purely bioplastics could essentially be a wise approach for sustaining our environment. Bioplastics are sourced from renewable raw materials found in nature, but it does not necessarily mean that all bio-based plastics are biodegradable. Therefore, a more thorough investigation is urgently required to identify bio-resourced plastics, which are eventually biodegradable. Moreover, keeping in mind the misconception of bioplastics, in this book chapter, we clearly defined all the terminologies related to plastics and bioplastics. Further, we reviewed in detail the most important raw materials, namely, polylactic acid, polyglycolic acid, polylactic-co-glycolic acid, and polyhydroxyalkanoates for the production of bioplastics in terms of their syntheses, properties, and degradation mechanisms. In addition, a futuristic sustainable model was also discussed as a probable solution to have a safer and greener environment.

Keywords Bioplastics · Bio-based plastics · Biodegradable · Polylactic acid · Polyglycolic acid · Polylactic-co-glycolic acid · Polyhydroxyalkanoates

---

Priyanka · D. Yadav · J. Dutta (✉)
Department of Chemistry, Amity School of Applied Sciences, Amity University Haryana, Manesar, Haryana, India

K. M. Gothandam et al. (eds.), *Environmental Biotechnology Volume 4*, Environmental Chemistry for a Sustainable World 68,
https://doi.org/10.1007/978-3-030-77795-1_1

1.1 Introduction

The way plastic has been ingrained in our day-to-day lives, it becomes challenging to think of something without it. Undoubtedly, it is quite amazing that plastics greatly dominates the entire world. Notably, all plastics are said to be polymers, but the reverse is not always true. Synthetic polymers have been produced by hundreds or thousands of single, small chemical units called monomers with the linkages of strong covalent chemical bonds. In the early twentieth century, bakelite was the first synthetic polymer developed by condensation reaction between phenol and formaldehyde. Still, the primary production of polymers started only in the 1950s. Since then, plastics materials have become an indispensable part of our modern lives due to their versatile designs, significantly low cost, too light weights, and appealing features, to name a few. Let's look around the world at this polymer age; we can easily see that the essential things are clothes, toothbrushes, food containers, cars, electronic devices, credit cards, etc. They are typically made from different types of plastics materials (Chamas et al. 2020).

The term *plastic* is etymologically derived from the Greek word '*plastikos*' and the Latin word 'plasticus' both meaning "able to be molded". Further, these words refer to the malleability or plasticity of materials during the manufacturing process, which means the materials can be easily cast, pressed, and extruded into various forms such as films, fibres, plates, tubes, bottles, and boxes, to mention a few (Millet et al. 2018). Based on their molecular structures, plastics are of two types: thermoplastics and thermosetting plastics. Polyethylene, polypropylene, polycarbonates, polystyrene, etc., are typical thermoplastics, whereas phenolic and acrylic resins, polyurethane, polyesters, silicone, etc., are the typical ones of thermosetting plastics. Thermoplastics can be reshaped or remoulded even after fabricating products upon heating, but thermosetting plastics cannot be reshaped or remoulded upon heating. This phenomenon gives us a clear message that once the products formed using thermosetting, plastics cannot be further processed through applying heat, instead these products get broken down. Traditional plastics are produced from non-renewable resources such as coal, oil, and natural gas. In 2015, the global production of fossil-based plastics had tremendously increased to 322 million tons from more than 16.1 million tons in 1964, this indicates that simultaneously the demand for plastics in our modern lives is also gradually increasing (Wei and Zimmermann 2017).

Since the 1940s, owing to their outstanding and remarkable properties, namely mechanical strength, lightweight, flexibility, and durability, synthetic plastics have found various applications ranging from household to spacecraft (Albuquerque and Malafaia 2017; Millet et al. 2018). In addition, traditional plastics are chemically resistant, highly durable, and also non-degradable. Despite their non-degradability, the use of conventional plastics is enormously increasing, considering their other outstanding properties to meet global needs. But that has nowadays become one of the major concerns for environmental pollution because of direct or indirect disposal of articles made of these plastics into the environment, which eventually do not get degraded rather accumulated in the form of a heap of garbages (Emadian et al. 2017). Thus, plastic waste management has become a significant challenge to keep our environment free from pollution. It is worth noting that about 93% of plastic waste is dumped into oceans and landfills, which consequently introduces adverse effects such as greenhouse gas emissions on the environment and drastically hampers the ecological balance.

Moreover, this non-degradable plastic waste continues to exist in the environment for a very long period because of its extremely inertness towards microorganisms. Therefore, it is responsible not only for causing severe health risks to humans but also for degrading the environment. Further, it is reported that toxic chemicals evolved from plastics get accumulated in living organisms and in the food chain, which are the potential causes of public health-related issues (Burgos et al. 2016). Hence, several measures have been taken to mitigate these plastic waste-related issues to sustain our environment from further degradation. Although cent percent replacement of plastics is not possible keeping in mind their huge demand worldwide and lack of suitable alternatives, the problem associated with the environmental degradation can be alleviated to a greater extent by switching to purely bioplastics as these are sourced from renewable natural resources such as biomass. Bioplastics' demand arises due to its 100% biodegradability, which does not cause any harmful effects on the environment.

Interestingly, it is worth noting that all bio-based plastics are not entirely biodegradable. Therefore, an urgent and thorough investigation is required to identify fully biodegradable bioplastics for increasing environmental sustainability, which on the other hand, can be able not only to serve various societal needs but also to boost up bioeconomy in the world. However, there is no exaggeration to say that bioplastics are eco-friendly. Still, bioplastics production may be affected due to several constraints such as high manufacturing cost, low mechanical strength, and the global food crisis, to name a few (Jain and Tiwari 2015; Yates and Barlow 2013). This chapter shall mainly discuss raw materials such as polylactic acid, polyglycolic acid, polyhydroxyalkanoates, etc., to be used for bioplastics production; and their syntheses, properties, and plausible degradation mechanisms.

1.2 Uncertainty of Life Without Plastics – A Myth

Although plastics' discovery had brought a revolutionary change, plastics' exploitation did not happen outside the military due to World War II. In 1950, plastic materials started to be developed on a large scale for domestic and commercial purposes. As a result, plastics production has tremendously increased from 0.5 million tonnes in 1950 to over 380 million tonnes in 2015 (Geyer et al. 2017). As of today, we find the use of plastics almost everywhere. Be it in transport, telecommunications, clothing, footwear, packaging materials. Particularly for packaging materials, these are conveniently used to promote the transport of a wide variety of food, drink and other goods because of their lightweight (Andrady and Neal 2009). Therefore, nowadays, it is quite challenging to imagine our lives without plastics as they have become an indispensable part of our lives.

On the other hand, the whole world appears to be wrapped in plastics due to their rapid evolution, which also causes our lives to become so colourful. So all of a sudden, if the government and policymakers strictly ban plastic materials considering the environmental issues associated with these materials, then probably there would be no certainty of life without plastics. This kind of myth people perceives in their minds. But actually, it is not the case. However, 100% of plastic banning is impossible unless there is an alternative with similar or nearly similar features to substitute plastics.

On the other hand, as the reservoir of fossil fuel for plastics production is getting exhausted, researchers have started finding a probable solution and have begun using renewable resources such as biomass to produce bioplastics. But due to some limitations, the production of bioplastics on a large scale is getting hindered. Therefore, we have reached a kind of dilemma that we would be deprived of using plastics one day if the production of bioplastics would not suffice! That is the kind of uncertainty of life without plastics and intriguingly may be without bioplastics. However, many efforts are being put from different angles throughout the world by academia, industries, and researchers to developing biodegradable bioplastics – promising substitutes of plastics from different renewable resources to offer us a pollution-free environment.

1.3 Worldwide Demand for Plastics

In the majority, fossil fuel-based feedstocks, especially petroleum-based feedstocks, are used to produce plastics. Approximately 4–5% of the total global oil and natural gas producing non-renewable resource is primarily used as feedstock for plastics production, and 3–4% of it is further consumed as energy for their manufacturing (Hopewell et al. 2009; Andrady 2015). With increasing the demand for plastics, its production is also growing. According to the data revealed on plastics production, i.e., 311 million tonnes in 2014, it is further expected to be doubled in the next two decades and to be almost quadrupled by 2050 (Ellen MacArthur Foundation 2017). China, North America, Europe, Africa, Japan, Latin America, and India are the major plastics producing and plastics consumption countries (Nkwachukwu et al. 2013; Ryberg et al. 2018). The countrywide contribution towards the global production of plastics is illustrated in Fig. 1.1. The compound annual growth rate of global production of plastics in 2015 was approximately 8.4%, which was roughly 2.5 times more than the compound annual growth rate in 1950 (PlasticsEurope 2006, 2016).

Usually, plastics are composed of 93% of polymer resins and 7% of additives by weight. More than thousands of polymeric raw materials are synthesized on a commercial scale to produce plastics (Rosato 2013). Most of India's market shares are mainly generated from some important polymers used as raw materials such as

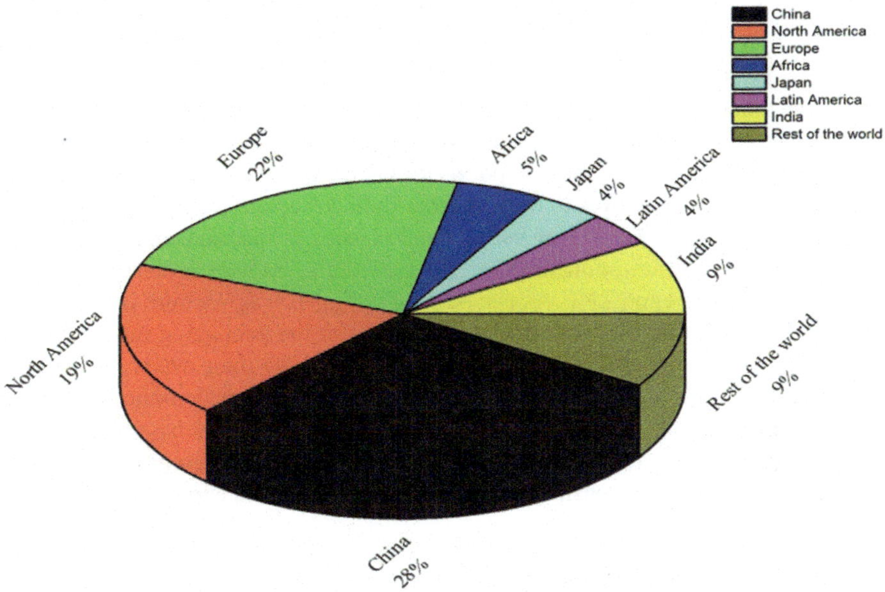

Fig. 1.1 The country-wise contribution towards total production of plastics in the world

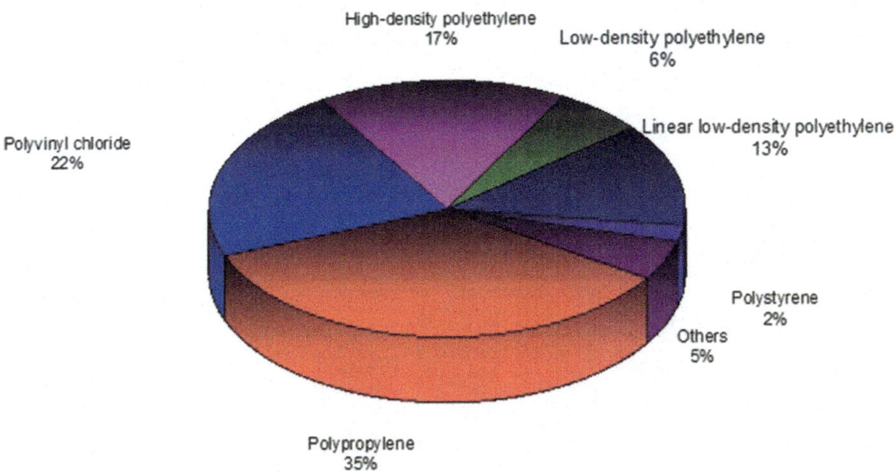

Fig. 1.2 Consumption of common polymeric raw materials for manufacturing plastics in India

polyethylene terephthalate, high-density polyethylene, low-density polyethylene, linear-low density polyethylene, polyvinyl chloride, polypropylene, polystyrene, and to name a few. The consumption of these raw materials for the production of their respective plastics in terms of percentage is shown in Fig. 1.2. Among them, polyethylene, including high-density polyethylene, low-density polyethylene, and linear-low density polyethylene, has the highest share up to 6% annually in the production of plastics by the Indian polymer industry than any other polymers (A report on Plastic Industry, FICCI 2017).

On the other hand, polypropylene has the second-largest share with a growth rate of 11% per annum, whereas the share from other polymers is up to 10% annually. The usage analysis of the consumption of plastics in different sectors in India is shown in Fig. 1.3. Here, it is to be noted that India's western part is the largest consumer of plastics, about 47% of India's total plastics production. In contrast, plastics' respective consumption in other parts of India, namely eastern, north, and south, are 9%, 23%, and 21%. As can be seen in Fig. 1.3, agriculture is the second-largest plastics consuming industry in India, next to the packaging industry. Thus, plastics production is also increasing substantially as per their needs (A report on Plastic Industry, FICCI 2017). On the one hand, due to the recalcitrance and inflexibility of the plastics, they have become so attractive that they have found a vast array of applications ranging from food transportation to packaging to sterile medical uses to building and construction to electrical appliances and electronics but on

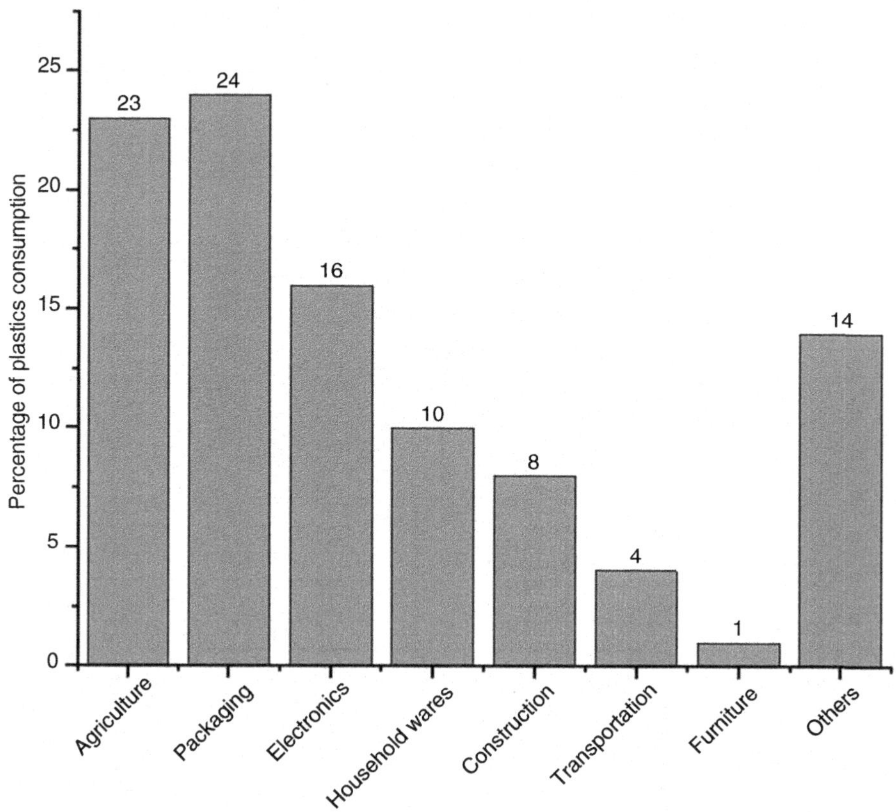

Fig. 1.3 Consumption of plastics in terms of percentage in different sectors in India

the other hand, their long-lived residence time has made them unsustainable towards the environment once they are discarded (Cole et al. 2011; Geyer et al. 2017). Notably, the packaging sector is accountable for holding the largest shares worldwide for the production of plastics. About half of the total plastics-based articles such as packaging objects, agricultural films, and other disposable items produced globally are discarded after a single application. In addition, waste electrical and electronic equipment, wherein approximately 30% of the mass electronic scrap is made of plastics, is the second-largest source for generating solid waste (Nkwachukwu et al. 2013).

Moreover, other plastic objects such as microbeads in cosmetics, food packaging, plastic bags, plastic bottles, and disposable cups, to mention a few (Anderson et al. 2016; Wikström et al. 2016; Martinho et al. 2017; Orset et al. 2017; Poortinga and Whitaker 2018) are also major concerns for environmental pollution because of their continuous accumulation on earth, which in turn, do not get degraded. Of them, plastic bags have been consumed mostly in every sector. Thus, the ever-increasing demand for plastics ironically has become a significant threat to the world's economy because these have contributed to the rise in the concentration of atmospheric carbon dioxide over the past two centuries, which in turn, also affect the world's climate. The massive use of fossil-based materials for plastics production is also an ethical problem because more than 90% of these materials are used up to meet the global energy demand. Above all, considering the various merits and demerits of plastics, only 18% of total plastics waste is recycled; 24% is incinerated, and the remainder, i.e., 58%, undergoes anaerobic decomposition over a prolonged period after landfilling or dumping into the natural environment.

On the other hand, it is estimated that 13% of plastic waste is generated from total plastic production in the United States, i.e., 35.4 million tonnes in 2017. The landfilling rate for these plastics waste is increased up to 75%. According to the current growth rate of plastic production, the assembled plastic waste in landfills and the natural environment is portended to reach approximately 12,000 million tonnes worldwide by 2050 (Zheng and Suh 2019). Despite all the issues such as human health risk, environmental pollution and so on associated with non-degradable plastics, the global demand for plastics is steadily increasing.

1.4 Classification of Plastics

Plastic plays a vital role in our society; it becomes an indispensable part of our daily lives. There is hardly any sector that remains untouched by the use of plastics. They find a wide array of applications ranging from biomedical to packaging to communication technology to automobiles. Different types of plastics are available in the market, according to their characteristics, these are used for specific applications. Classification of plastics is made based on several parameters. It is worth mentioning that the source of raw materials is a vital criterion to classify plastics. In addition, while classifying the plastics, one should also keep in mind that many commercial bloomberg data point formulations combine materials from various sources to reduce cost and enhance performance in terms of biodegradability. So depending on several ingredients present in the formulations, the classification of plastics can be customized.

Moreover, plastics can be intrinsically biodegradable or can be inert towards biotic components. Furthermore, biodegradability is an essential factor of plastics, which actually decides their final fate, and obviously, the plastics' biodegradability depends on the ingredients used for their manufacturing. Nevertheless, plastics can broadly be classified into two heads: (a) fossil fuel-based plastics and (b) renewable resources-based plastics.

1.4.1 Fossil Fuel-Based Plastics

Today, about 50% of packaging products are comprised mostly of plastics, which are produced from fossil fuels. Further, fossil fuel-based plastics can be sub-divided into three parts, namely non-degradable plastics, oxo-degradable plastics, and biodegradable plastics.

1.4.1.1 Non-degradable Plastics

Basically, the plastics that are predominantly made from crude oils and not decomposed by microorganisms when disposed into the environment after their use are termed non-degradable plastics. Due to their non-degradability, they cause adverse effects on the environment because they do not readily get decomposed by the microbial activity, resulting in the formation of heaps of garbage in plastic waste (Mohee and Unmar 2007). Further, due to their vast applications in the last several decades, they have been significant concerns for related environmental issues - exposure of the ecosystem to hazardous pollutants and evolution of harmful greenhouse gases, mainly methane and carbon dioxide, due to landfilling of these non-degradable plastics (Leja and Lewandowicz 2010; Crowley et al. 2003). Further, on burning, they also release carbon dioxide, which contributes to global warming. That is why many countries have banned non-degradable plastics for food packaging purposes. Polyethylene, polypropylene, polyvinyl chloride, etc., are the most common types of non-degradable commodity plastics, which are extensively used to manufacture several daily usage products plastics bags.

1.4.1.2 Oxo-Degradable Plastics

Oxo-degradable plastics falling under fossil-fuel-based plastics are usually prepared from common polymers, such as polyethylene. These plastics also contain special additives as catalysts, such as metal salts of carboxylic acid and dithiocarbonates, which, in turn, are also responsible for accelerating these plastics' degradation. In this regard, the most exploited additives are very common transition metals, such as manganese, iron, nickel, and cobalt, to name a few (Wiles 2005). The additives used in oxo-degradable plastics are entirely different from the ones used in petroleum-based plastics. Oxo-degradable plastics are delineated as non-compostable by the international standards organization, namely EN13432 and ASTM 6400 (ASTM 2012). Biodegradation of these plastics can only occur after they have fragmented into smaller pieces due to additives, and the remaining materials fragmented into microplastics do not get degraded easily. In essence, the degradation rate of oxo-degradable plastics is prolonged as compared to compostable plastics, which are to be discussed later. However, it is advised that these plastics should not be included with the mainstream plastic waste undergoing composting because oxo-degradable plastics do not get completely degraded except their recyclate containing oxo-degradable additives, which are responsible for causing them more vulnerable to degradation. Moreover, the fragments left out during the overall composting process badly influence the compost's quality and marketability. According to the life cycle analysis of oxo-degradable plastics, incineration is the best disposal option for recovering energy. Further, in case of non-availability of the incineration facility with energy recovery, landfilling would be the next suitable and viable option. Thomas et al. (2010) demonstrated the effect of additives contained in oxo-degradable plastics on the environment. They showed that incorporating the additives into fossil fuel-based plastics accelerated the degradation rate, followed by producing harmful residues that are not good enough for the environment. Notably, oxo-biodegradable plastics used for packaging, such as carrier bags, are environmentally friendly and a solution to plastics pollution (Chiellini et al. 2007).

1.4.1.3 Biodegradable Plastics

Another class of fossil-fuel-based plastics is biodegradable plastics. As its name implies, these types of plastics can be degraded easily into water, inorganic compounds, carbon dioxide or methane and biomass within a specific time frame by the natural action of microorganisms such as bacteria, fungi, algae (Greene 2007). Here, it is to be noted that all fossil fuel-based plastics are not biodegradable. Some of them are compostable. Even though the source of origin is a non-renewable fossil fuel, these plastics are made by combining partly petrochemical carbon and partly biogenic (renewable) carbon derived from biomass. For almost the last two decades, to decrease the direct environmental impact of conventional plastics, these have been blended with some naturally occurring compounds to make them biodegradable to a certain extent (Anderson et al. 1998; Gross and Kalra 2002; Zhang et al. 2000; Demirbas 2007). If we look around the world, the use of 30% biodegradable polymers per annum is steadily increasing (Leja and Lewandowicz 2010). These types of plastics can undergo either aerobic or anaerobic biodegradation. Complete biodegradation can be stated as the degradation of plastics altogether into biomass, whereas mineralization can be explained as the full conversion of original polymer into gas and minerals. Further, the biodegradation of plastics occurs in two phases. Firstly, microorganisms attack the polymer chain to convert it into simpler molecules through bond cleavage, and secondly, natural biodegradation of the plastics occurs. These activities are strictly governed by endogenous factors such as crystallinity, molecular flexibility, molecular weight; as well as by exogenous factors such as pH, temperature, and humidity, oxygen availability, and presence of an enzyme, which are collectively or individually responsible for changing the pathway of the biodegradation process (Kale et al. 2007). Because of their biodegradability characteristics, from a waste management perspective, these plastics offer various value propositions, namely short-term or single-use applications such as bags for food packaging, disposable tableware, collection of organic waste or plants pot in agricultural and horticultural sectors, and mulch films, etc. But still, there are some problems associated with designing biodegradable plastics in terms of optimizing their physical, chemical, and mechanical properties and the extent of their biodegradable characteristics (Zee 1995; Fritz et al. 1994). A few examples of petrochemical-based biodegradable plastics such as polybutylene succinate, polycaprolactone, polyglycolic acid, and copolymers of polyvinyl alcohol and polybutylene succinate terephthalate are prepared from their respective monomers, which are obtained during petrochemical refining processes and consequently, the said polymers exhibit necessary biodegradability (Clarinval and Halleux 2005). Among other petroleum-based biodegradable polymers, polyvinyl alcohol, polyvinyl chloride, acylated starch-based plastics are used for the preparation of agricultural films (Kyrikou and Briassoulis 2007).

1.4.2 Renewable Resources-Based Plastics

The adverse impact of conventional plastics and increasing petroleum price and their future shortage pushes us to develop renewable resources-based plastics. Preparing plastics from renewable feedstocks is a critical mandate for sustaining our environment. Due to the abundant biomass availability as one of the natural renewable resources, it can easily be deployed as a precursor for plastics production. Utilization of renewable bioresources to prepare bio-based plastics, especially those used in great quantity, would be a wise approach to sustaining the environment (Philp et al. 2013; Reddy et al. 2013). Biomass yields the same basic chemical intermediate needed to prepare novel biobased plastics, which are identical to petroleum-based conventional plastics. These plastics offer many environmental advantages over fossil-fuel-based plastics, such as lower carbon dioxide gas emissions than traditional plastics. Manufacturing plastics from renewable resources can be proved to be a promising solution to prevent environmental pollution, which is caused by fossil fuel-based plastics waste. Notably, all bio-derived plastics are not necessarily meant to be biodegradable. Thus, renewable resources-based plastics are classified into two heads: (a) compostable plastics and (b) biodegradable bioplastics.

1.4.2.1 Compostable Plastics

Compostable plastics can be termed bioplastics because their raw materials are derived from renewable biomass through extraction, followed by various chemical conversion. These plastics can be degraded through biological activities to be executed by microorganisms while composting, which does not leave any different, visible and toxic residue (Riggle 1998). As one of the most common raw materials, corn starch is used to prepare compostable plastics. Polylactic acid is known as a typical compostable plastic, which is derived from corn starch. Composting is one of the most viable options for waste management of degradable plastics (Greene 2007). It involves two steps. In step 1, an abiotic process such as photochemical or thermal treatment is done, and in step 2, biotic process is adopted to degrade the polymer under specified conditions. Thus, it results in the formation of low-molecular-weight species. Nevertheless, the remaining reside should be completely decomposed by the microorganisms. Otherwise, it can cause several risks to our health and environment (Narayan 2006a, b).

1.4.2.2 Biodegradable Bioplastics

A bioplastic is a type of manmade polymer in which raw material is obtained from living organisms, generally from plants. Although these renewable feedstocks used to produce bio-based plastics possess identical chemistry to traditional fossil fuel-based plastics, there is still a significant difference between them depending on their source of origin. Edible food crops such as soybean and corn; non-edible food crops such as switchgrass, or agricultural waste as biomass can be used to produce bio-based plastics. Polyhydroxyalkanoates, polylactic acid, starch blends, etc., are suitable examples of biodegradable bioplastics.

Undoubtedly, the classification of plastics plays a very crucial role. Therefore, for better understanding, it is schematically shown in Fig. 1.4.

1.5 Differentiation Between Compostable, Biodegradable and Bio-based Plastics

All the terms, namely compostable, biodegradable, and bio-based, are interchangeably used due to lack of proper awareness. But there are significant differences between them; each term has its own distinct meaning with respect to its source of origin. A bio-based material is obtained from renewable plant and animal feedstocks. For example, starch, corn, sugarcane, and other sources, including potatoes, algae, mycelium, and food waste, are used as feedstocks to produce bio-based plastics. Although biodegradability and compostability refer to the package's end-of-life, it does not necessarily mean that all bio-based plastics are biodegradable rather,

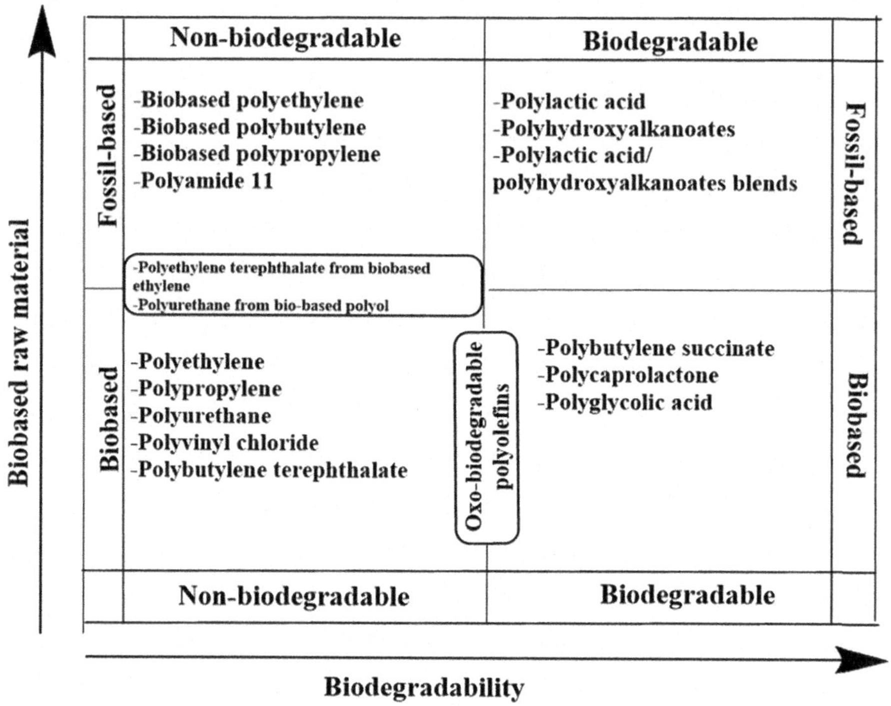

Fig. 1.4 Classification of plastics with suitable examples depending on the various source of origin

it can be said that all biodegradable plastics are compostable, but all compostable plastics are not biodegradable. It is quite noteworthy to mention that compostable plastics can be produced from raw materials obtained from renewable biomass such as starchy plants, and they can get degraded with the help of microorganisms within a sufficiently short period under the specified conditions of a composting operation. Therefore, compostable plastics are said to be synthetic biodegradable bio-based plastics based on their source of origin. These are still not fully biodegradable, whereas biodegradable plastics are fully degradable irrespective of their basis of origin. Hence, it can further be simplified that some bio-based plastics are biodegradable and some of them are not.

1.6 Understanding the Concept of Bioplastics

Although the differences between compostable, biodegradable and bio-based plastics have been discussed in the foregoing section, it is still not enough to gain complete knowledge, especially about bioplastics. Therefore, in this section, a special effort has been made to clarify the bioplastics, which is schematically presented in Fig. 1.5. The concept of bioeconomy defines the use of renewable natural resources

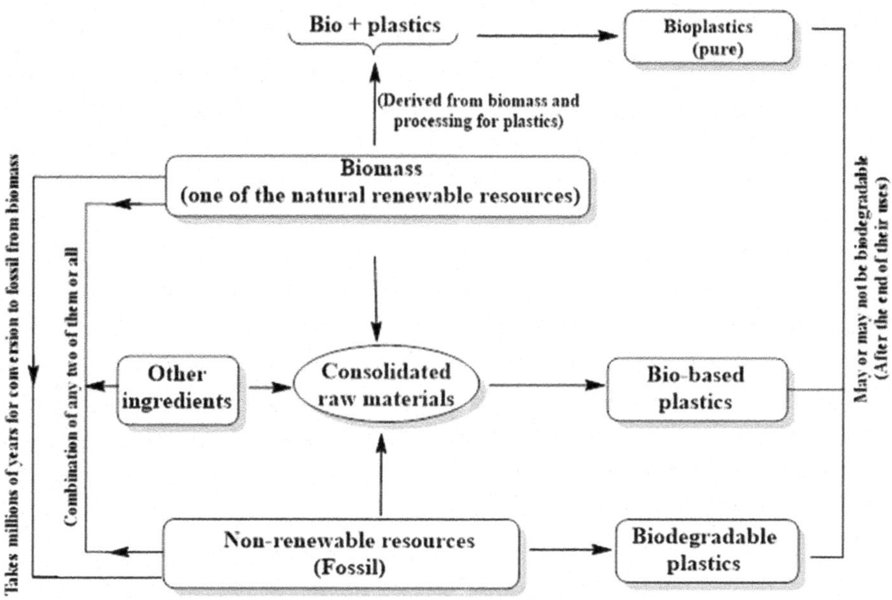

Fig. 1.5 Understanding the concept of bioplastics, bio-based plastics, and biodegradable plastics

for the industry and plants for an integrated revolution in the economy and society. It is replacing fossil resources and reducing greenhouse gas emissions and the potential benefit for almost all stakeholders for creating new jobs, especially in rural and coastal areas, without any risks. We shall also collectively promote our environment towards sustainability (European Bioplastics 2016). Nowadays, most plastics are based on fossil resources. To avoid environmental changes and reduce the employment of limited fossil resources, plant-based renewable resources have received considerable attention from researchers to produce bioplastics in the last several decades.

The word bioplastic itself defines that it is derived from biomass, which is further used for the production of plastics and then it is named bioplastics. In other words, it can be said that bioplastics are typically produced from naturally-occurring polymers, which are extracted chemically or biologically from either plants or animal-based renewable resources. They can be polysaccharides, proteins, lignin, or nucleic acids. However, there is a clear distinction between bioplastics and bio-based plastics. For instance, polylactic acid is a bio-based plastic, whereas polyhydroxyalkanoate is a bioplastic.

Further, it can be a little elaborated to make it more comprehensive. Polylactic acid-based plastics are the most important bio-based plastics because their raw materials are synthesized through polymerization of their corresponding lactic acid monomers, which in turn, are produced from starchy plants through several operations, whereas polyhydroxyalkanoates are bioplastics as their corresponding raw materials are found naturally (Kabasci 2014; Raza et al. 2018). Moreover, some other bio-based plastics that originated from renewable resources are bio- polyethylene terephthalate, bio-polyethylene, and bio-polypropylene, which have similar characteristics forefather, but they are not biodegradable. On the other hand, polyglycolic acid, polycaprolactone, polybutylene succinate-co-terephthalate, polybutylene adipate-co-terephthalate, and polyvinyl alcohol are generated from petroleum-based resources with certain degrees of biodegradability (Song et al. 2009). All these above-mentioned bio-based plastics have been depicted in Fig. 1.4.

1.7 A Dire Need for Switching to Bioplastics

Undoubtedly, it is very hard to imagine a world without plastics as they play a very crucial role in our daily lives. As a consequence of their versatility and low-cost, they find tremendous applications in our modern lives. As plastics are produced from fossil fuel-based resources, so these are non-degradable and cause environmental pollution. More than 350 million tons of plastics are manufactured globally and used for diversified applications (Banks et al. 2016; Statista 2018). The packaging industry uses a massive amount of plastics produced globally. It is to be mentioned that 14% of plastic packaging is accumulated for recycling, 14% is gone to incineration, 40% is landfilled, and the remaining 32% is dumped illegally into the environment (Ellen MacArthur Foundation 2017). In addition, building and construction sectors are also accountable for contributing to environmental problems during the process (Emadian et al. 2017).

Moreover, the recovery of energy from plastic wastes increases the amount of CO_2 emissions. Thus, for increasing environmental sustainability, the use of non-degradable plastics waste has to be drastically reduced. In this context, there is a dire need to switch to bioplastics production due to their biodegradability. Bioplastics are considered promising alternatives to traditional plastics. Starch plays a crucial role in the development of bio-based plastics. It is used to develop polyhydroxyalkanoates where starch-derived sugar is utilized as a substrate for microorganisms, facilitating producing polyhydroxyalkanoates (Mekonnen et al. 2013). But not all bio-based plastics are biodegradable. Notably, about 60% of bio-based plastics made today are not biodegradable.

Nevertheless, bioplastics have several advantages over traditional plastics. However, the use of bioplastics dramatically reduces carbon dioxide gas emissions. Besides, the extent of biodegradability is relatively much more significant compared to conventional plastics. Therefore, in recent years, considerable attention has been paid to bioplastics production to replace fossil fuel-based plastics partially.

1.8 Raw Materials for Manufacturing Biodegradable Bioplastics and their Plausible Degradation Mechanisms

1.8.1 Polylactic Acid

Polylactic acid is an environmentally friendly biodegradable aliphatic polyester made from lactic acid monomers, as shown in Fig.1.6.

Lactic acid is extracted from 100% renewable resources such as corn starch, followed by its polycondensation polymerization to obtain polylactic acid. The latter offers several exciting and remarkable properties: high mechanical strength, biocompatibility, biodegradability, ease of flexibility, etc. In addition, it is inexpensive and also possesses a low impact strength. Polylactic acid is a versatile semicrystalline plastic widely used for commodity applications. The excellent balance of polylactic acid's physical and mechanical properties makes it a viable candidate for the replacement of petroleum-based thermoplastics for various industrial and biomedical applications.

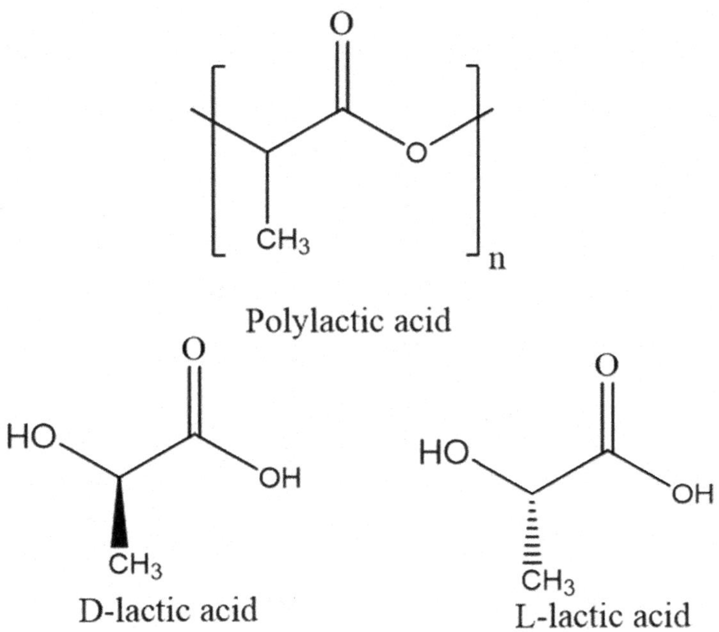

Fig. 1.6 The structural formulae of polylactic acid and its isomeric monomers

1.8.1.1 Synthesis of Polylactic Acid

Polylactic acid is prepared by polymerizing lactic acid, which is obtained through fermentation of starch into glucose, followed by conversion to lactic acid (Okada 2002; Jem et al. 2010). Lactic acid as a bioproduct is mostly found in L- configuration and can be polymerized into polyL-lactic acid. If D and L-configurations of lactic acid monomers are subjected to polymerization, then polyD, L-lactic acid is obtained. The characteristic properties of polyD, L-lactic acid exclusively depend on the feeding ratio of stereoisomers content taken to carry out polymerization reaction (Garlotta 2001).

Lactic acid is also converted into lactide and further to polylactide. Therefore, polymers obtained from lactic acid and a cyclic dimer of lactic acid are designated as polylactic acid and polylactide, respectively. As shown in Fig. 1.7, there are

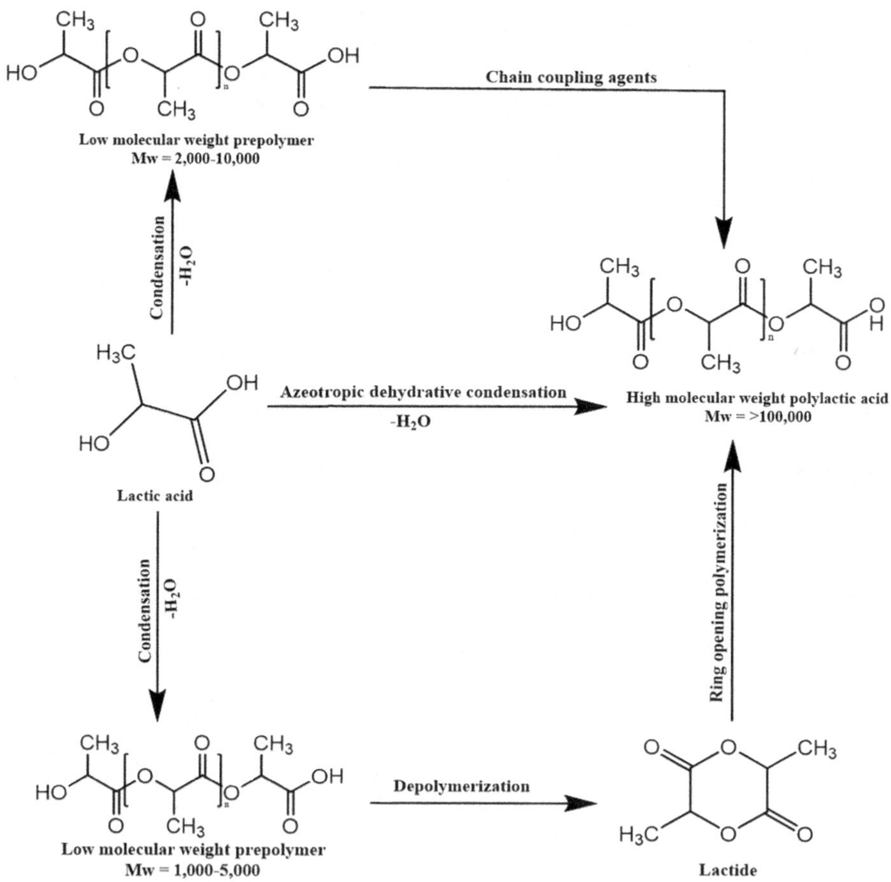

Fig. 1.7 Various polymerization techniques for polylactic acid production. The polymer obtained from lactic acid is known as polylactic acid, and the same obtained from the cyclic dimer of lactic acid is known as polylactide. (Adapted from Lunt 1998)

mainly two well-known processes: direct polycondensation and ring-opening polymerization, through which polylactic acid can be produced. Both methods are almost similar, but for producing a high-molecular-weight polymer, ring-opening polymerization is favoured. In addition, the manufacturing process of polylactic acid and its products is depicted in Fig. 1.8.

Further, during synthesis, coupling agents are added to increase the molecular weight of polylactic acid prepared from lactic acid itself. Moreover, lactide, a cyclic dimer of lactic acid can follow either cationic or anionic ring-opening polymerization depending on the selection of initiator of interest to yield high molecular weight polymer.

1.8.1.2 Properties of Polylactic Acid

Physicochemical Properties

As polylactic acid is biodegradable and biocompatible, it can readily undergo a hydrolysis reaction. The stereochemistry of its monomers significantly affects its chemical properties. Accordingly, polyD, L-lactide is amorphous and does not possess a melting point, whereas polyL-lactide and polyD-lactide obtained from the respective L-lactide D-lactide are homocrystalline (Tsuji et al. 2005). Various solvents, namely dioxane, dichloroacetic acid, dichloromethane, chloroform, acetonitrile, 1,1,2-trichloroethane, etc., used for the dissolution of polylactic acid. In addition, there are several solvents such as acetone, toluene, tetrahydrofuran, and ethylbenzene, wherein it can be dissolved while heating to boiling temperature. Moreover, lactic acid is not soluble in water, alkanes, and a few selective alcohols (Nampoothiri et al. 2010).

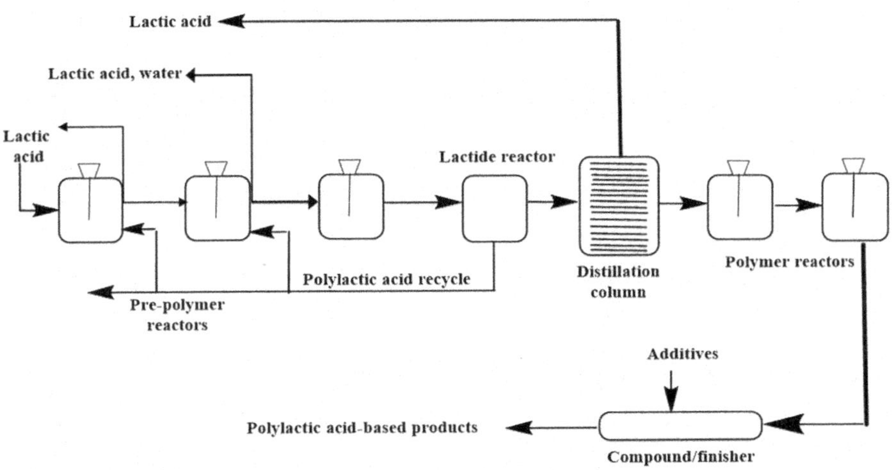

Fig. 1.8 A flow chart for manufacturing polylactic acid and its products. (Adapted from Dutta and Dutta 2006)

Thermal Properties

The thermal properties of polylactide are strictly dependent on the content of its stereoregular isomers, which greatly affect the mechanical properties and crystallinity of their corresponding polymers. PolyL-lactide and polyD-lactide are semicrystalline polymers with a melting point of 180 °C, whereas polyD, L-lactide is an amorphous material whose glass transition temperature falls in the range of 50–57 °C (Drumright et al. 2000). Due to its excellent heat sealability, polylactide finds suitable packaging right from food packaging to medical and pharmaceutical applications.

Optical Properties

Due to two different stereoisomers, D and L forms of polylactic acid are optically active, whereas the polymer of their racemic mixture is optically inactive.

Mechanical Properties

Mechanical properties of polylactic acid mainly depend on its molecular weight apart from stereochemistry, crystallinity, and internal molecular arrangement within the polymer. Therefore, as per the need of its end application, polylactic acid can be easily tuned by changing the parameters mentioned above to make it soft and stiff. Interestingly, the higher the stereochemical purity of polylactic acid, the higher is the tensile strength, but the lower is the impact strength. Without compromising the biodegradability, a lot of work on chemical modification of polylactic acid is done to enhance its mechanical properties. It is reported that the elongation at break of polylactic acid is less than 10%. Usually, plasticizers are used to increase its ductility and decrease its glass transition temperature. Martin and Averrouss (2001) reported the use of several plasticizers, such as polyethylene glycol, not only to plasticize polylactic acid but also to reduce its modulus ranging from 28% to 65%. Further, they showed that plasticizers' addition also increased elongation at break more than 200%. Furthermore, polylactic acid can be blended with other polymers such as polyhydroxyalkanoates, polypropylene carbonate, etc., to generate new materials with improved properties. For instance, mixing polylactic acid with polycaprolactone increases the former's toughness by increasing its elongation at break.

1.8.1.3 Degradation Mechanism of Polylactic Acid

Hydrolytic Degradation of Polylactic Acid

In hydrolytic degradation, polylactic acid undergoes hydrolysis to yield lactic acid, and this can be performed either in an acidic or alkaline medium. As polylactic acid belongs to aliphatic polyester groups, these are degraded in water through hydrolysis, followed by fragmenting firstly into lactic acid oligomers and eventually into lactic acid (Tsuji 2010). It is an internal autocatalytic process in which the carboxylic end group of L-lactic acid gets affected during hydrolytic degradation of polylactic acid, and degradation proceeds randomly along the polymeric backbone (Henton et al. 2005; Paul et al. 2005; Zhou and Xanthos 2008; Tsuji and Ikada 2000). Consequently, the polymeric chain scission occurs in amorphous regions, which, in turn, increases the crystallinity of the polymer (Hakkarainen 2002). Further, the accelerated study of hydrolytic degradation of polylactic acid is carried out in both alkaline and acidic solutions or buffer solutions, even at higher temperatures, and in the presence of enzymes to understand their hydrolytic effects on polylactic acid (Tsuji et al. 2004).

Fukushima et al. (2011) studied the accelerated hydrolytic degradation of polylactic acid-clay-based nanocomposite films at different temperatures. Briefly, a series of polylactic acid-clay based nanocomposites films were first prepared with the individual addition of 5% modified montemorillonite or flurohectorite and unmodified sepiolite and then each nanocomposite film was placed in phosphate-buffered saline solution at 37 °C. Further, the same operation was also carried out at 58 °C. The results showed that the individual presence of modified montemorillonite and unmodified sepiolite in the prepared nanocomposite films retarded the degradation of polylactic acid-based nanocomposite films at 37 °C. This phenomenon showed that crystallinity of the nanocomposite films increased due to the addition of clays, which in turn, also enhanced their water uptake efficiencies and thus, results in the reduction of the availability of water for hydrolysis of individual polymer matrices but at 58 °C, just an opposite phenomenon was observed. In other words, it can be said that at the latter temperature, the nanocomposite film degrades at a much faster rate compared to the same at 37 °C. It might be due to microscopic change in structural behaviour within the nanocomposite film while heating at a relatively higher temperature, which further facilitates water molecules to get absorbed into the core of polymer matrices for rapid hydrolysis. It was also found that the addition of modified fluorohectorite not only enhanced the polymer crystallinity but also accelerated the hydrolytic degradation of polylactic acid. Further, the addition of nanoclays did not markedly change the hydrolytic degradation pattern of polylactic acid at a higher temperature. Moreover, at this temperature, all the studied materials achieved a similar trend of molecular weight reduction because of the ease of access of water molecules into the polymer matrices, which caused rapid degradation of the nanocomposite films without allowing the interference of clays responsible for the crystallinity of the said films. In another study, it was found that the pH of the solution also affected the hydrolytic degradation of polylactic acid. Both at low and high pH, the rate of degradation got increased (Göpferich 1996). Jong et al. (2001) studied the influence of media pH on polylactic acid's cleavage mechanism. The hydrolytic degradation mechanisms of polylactic acid both in acidic and alkaline media are shown in Fig. 1.9.

In acidic medium, degradation proceeds through chain end scission to give rise to lactic acid, whereas in alkaline medium, the same occurs through back-biting so

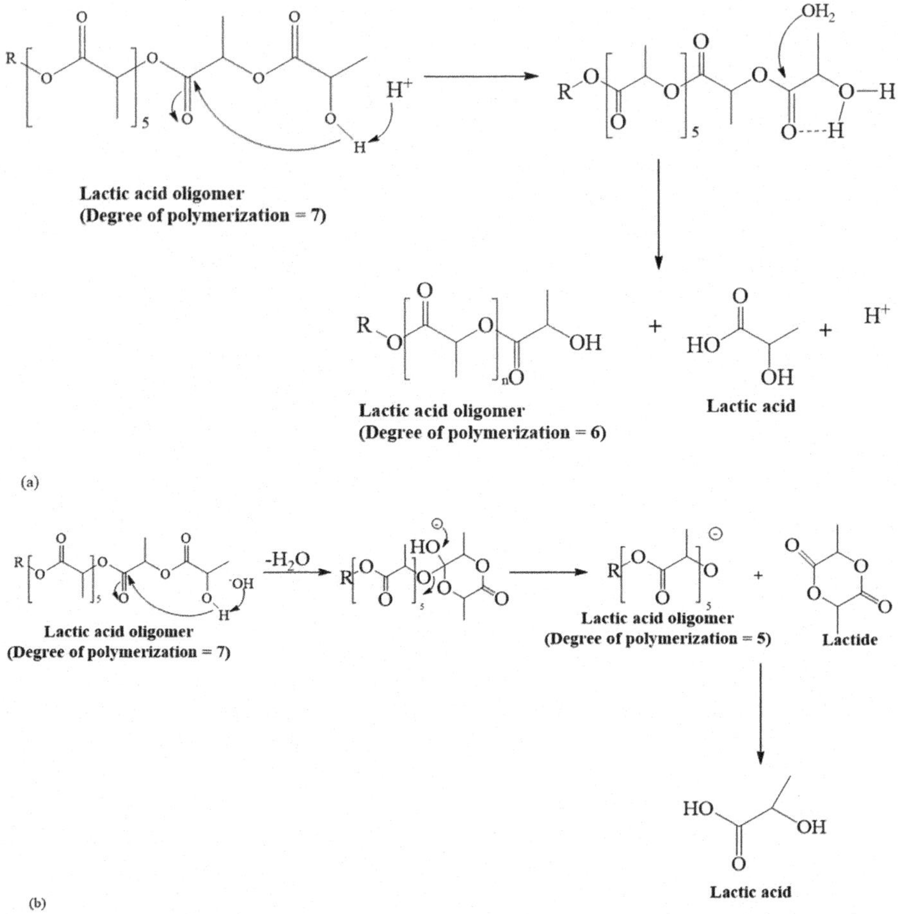

Fig. 1.9 Acid catalysed (**a**) and base catalysed (**b**) chemical degradation mechanisms of polylactic acid) via hydrolysis. (Adapted from Jong et al. 2001)

as to produce lactide, a cyclic dimer of lactic acid, which subsequently undergoes hydrolyzation resulting in the formation of lactic acid.

Thermal Degradation

It is reported in the literature that thermal degradation of polylactic acid is a non-radical chain scission with intramolecular transesterification. In addition, various researches suggested that polylactic acid starts degrading during thermal processing, resulting in reducing its molecular weight, which eventually decides the materials' final fates. Many mechanisms have already been reported for the thermal degradation of polylactic acid, such as Zipper-like depolymerization (Gupta and

Deshmukh 1982a) in the presence of catalysts through transesterification, thermo oxidative degradation (McNeil and Leiper 1985; Zhang et al. 1992). In a study, Nicolae et al. (2008) demonstrated an experiment to conduct thermal degradation of polylactic acid in an inert atmosphere. Therefore, polylactic acid was subjected to both isothermal and dynamic heating regimes under already specified conditions. Finally, they reported a concerted process resulting in the weight reduction in the range of 0–30% with an activation energy of 21–23 kJ/mol. In another study, it was reported that polylactic acid underwent thermo-oxidative degradation through a concerted mechanism with an activation energy of 22–28 kcal/mol resulting in lactide formation as a decomposition product (Gupta and Deshmukh 1982a, b). Still, there is a dilemma in the cleavage of bonds in a chain of polylactic acid during thermal degradation. Nicolae et al. (2008) claimed that carbonyl carbon-oxygen bond was more efficient than carbonyl carbon-carbon linkage as claimed by Gupta and Deshmukh (1982b) for undergoing thermal degradation.

Environmental Degradation of Polylactic Acid

We need to pay more attention to the environmental degradation of polylactic acid as compared to others already discussed because of its large scale production, i.e., more than 100,000 tonnes every year and the majority of which is used especially for the development of commodity products as well as packaging materials (Sin et al. 2013). Most of the commodity products based on polylactic acid are discarded into the environment after their applications. Therefore, it is quite essential to understand the environmental degradation mechanism of polylactic acid because in recent years, it has been deployed to replace polyethylene-based commodity products to a greater extent. Polylactic acid can be degraded chemically at elevated temperature through hydrolysis followed by microbial degradation, which is a well-known method for its degradation under specified compositing conditions. Only the degradation mechanism of polylactic acid through chemical hydrolysis is represented in Fig. 1.10. It starts decomposing into carbon dioxide and methane during aerobic and anaerobic degradation (Lunt 1998). Moreover, it has also been reported that polylactic acid's chemical degradation does not get influenced with or without microorganisms. Still, in this case, only polymeric chain scission occurs at ester linkage through hydrolysis abiotically (Agarwal et al. 1998). Hence, there is no role of microorganisms in terms of influencing the rate of degradation. Polylactic acid follows both aerobic and anaerobic degradation pathways, governed by bacteria and fungi' assimilation with and without oxygen. However, polylactic acid's degradation rate is significantly impacted by its inbuilt materials characteristics, namely crystallinity, stereo complexity, molecular weight, and other external influences such as sunlight, oxygen, and moisture.

Polylactic acid is a compostable polymer, but it can be degraded under a particular set of conditions. Notably, the exact role of microorganisms is not yet deciphered during the environmental degradation of polylactic acid. Karamanlioglu and Robson (2013) conducted a simulation study on polylactic acid coupons obtained from

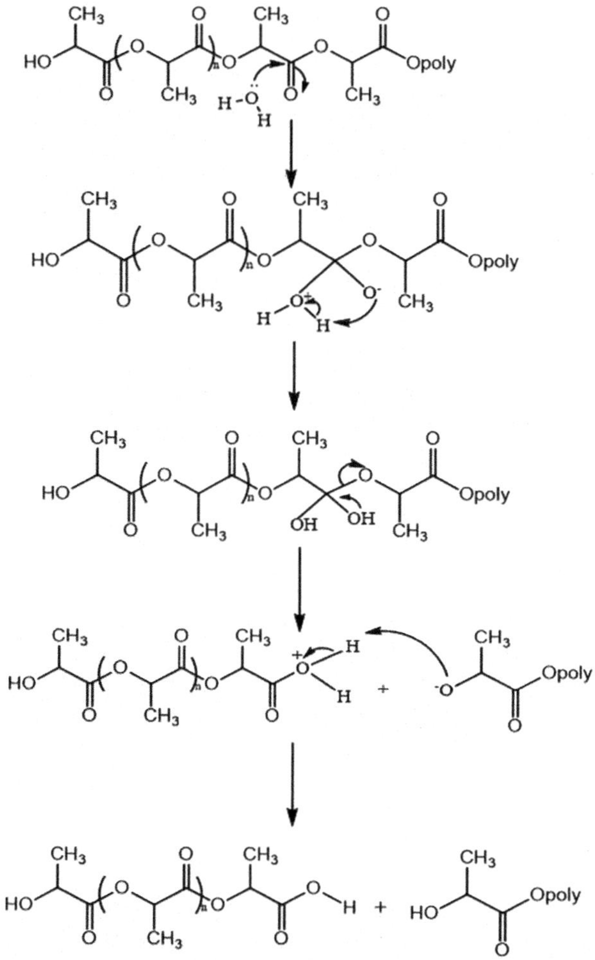

Fig. 1.10 Environmental degradation through chemical hydrolysis of polylactic acid. (Adapted from Lunt 1998)

commercial packaging by changing various biotic and abiotic to determine microorganisms' role and the environmental factors during environmental degradation. Succinctly, two different set of polylactic acid coupons with dimensions of 7 × 3 × 0.02 cm and 2 × 2 × 0.02 cm were cut, and surfaces of all the coupons were sterilized using 70% (v/v) ethanol. After that, all these coupons were allowed to dry at ambient temperature and buried at approximately 7 cm depth in compost or soil in rectangular plastic boxes. Finally, all the said boxes were incubated at different temperatures, namely 25°, 37°, 45°, 50°, and 55 °C.

Further, polylactic acid coupons with dimensions of 7 × 3 × 0.02 cm and 2 × 2 × 0.02 cm were recovered for measurements of tensile strength and molecular weight, respectively. They showed no significant weight reduction in polylactic acid

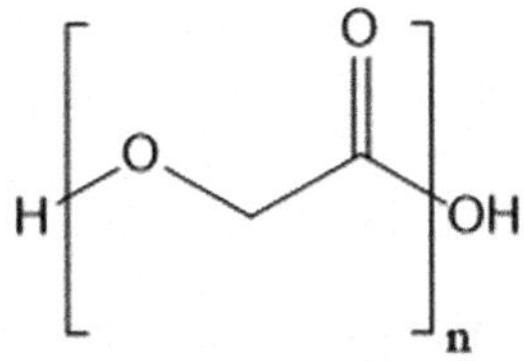

Fig. 1.11 Chemical structure of polyglycolic acid

coupons both at 25 and 37 °C. In contrast, at elevated temperatures, microorganisms' presence drastically enhanced their degradation rates, which were far better than chemical hydrolysis. Further, it was found that microorganisms were far active in compost rather than in soil at elevated temperatures. Therefore, further research is urgently needed to identify microorganisms powerful in degrading polylactic acid faster.

1.8.2 Polyglycolic Acid

Polyglycolic acid is a linear thermoplastic polyester. It is also biocompatible and biodegradable, like polylactic acid. It is mainly prepared through the polymerization of glycolic acid shown in Fig. 1.11.

Due to its biodegradability, it can be easily degraded into non-toxic degradation products (Marin et al. 2013). It also possesses good mechanical stability (Moon et al. 2004). Because of its suitable mechanical strength, biodegradability, and biocompatibility, polyglycolic acid finds a wide variety of applications ranging from absorbable sutures to tissue engineering to drug delivery to disposable medical devices (Lu et al. 2014; Schmidt et al. 2014; Takahashi et al. 2000; Gaudin et al. 2016).

1.8.2.1 Synthesis of Polyglycolic Acid

Usually, direct polycondensation and azeotropic condensation polymerization synthesize low molecular weight polyglycolic acid using glycolic acid as a monomer. On the other hand, ring-opening polymerization is used to synthesize high molecular weight polyglycolide using a cyclic dimer of glycolic acid.

Direct Polycondensation Polymerization

As direct polycondensation polymerization of polyglycolic acid yields low molecular weight, this method is not widely used. Polyglycolic acid can be synthesized via either solid-state polycondensation polymerization or solution polymerization. Briefly, glycolic acid is heated in the temperature range of 175–185 °C under

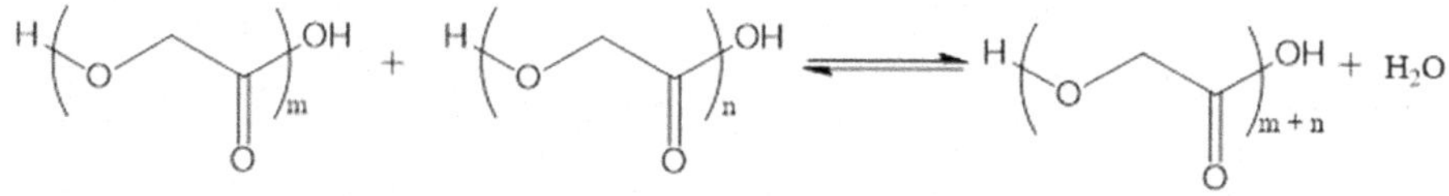

Fig. 1.12a Esterification reaction equilibrium during polycondensation polymerization of glycolic acid. (Adapted from Singh and Tiwari 2010)

atmospheric pressure to remove water from the system. Afterward, the system's pressure is lowered to 150 mm Hg without changing the system temperature and allowed to continue for another 2 h. Thus, it results in the formation of low molecular weight polyglycolic acid. Two different types of chemical equilibrium occur during direct polycondensation: (a) esterification reaction equilibrium leading to the removal of water (Fig. 1.12a) and (b) depolymerization equilibrium, which leads to depolymerize polymeric chain followed by the formation of glycolide (Fig. 1.12b). Polyglycolic acid obtained through direct polycondensation possesses inferior mechanical properties, but these properties can be effectively used for drug delivery purposes.

Azeotropic Distillation

It is an effective polymerization technique to synthesize high molecular weight polyglycolic acid. It involves the use of highly active catalysts and low boiling organic solvents, which can be dried and recycled back to carry out the further reaction. The water is removed azeotropically to shift the chemical equilibrium towards the forward direction yielding high molecular weight polyglycolic acid. The reaction temperature should be set in such a way so that the polymer formed does not get depolymerized.

Ring-Opening Polymerization

Before synthesizing polyglycolic acid using ring-opening polymerization, glycolide is prepared from glycolic acid by forming polyglycolic acid oligomers (Lu et al. 2008). But this method is very expensive to produce polyglycolide with a high molecular weight because of additional purification process to obtain highly pure polyglycolic acid, which itself is also a very complex process (Shen and Yang 2013). It is worth noting that high molecular weight polyglycolic acid cannot be synthesized via bulk polymerization in the melt/solid-state because it requires a higher temperature than the melting temperature of polyglycolic acid, which in turn exhibits poor thermal stability of the polymer (Fig. 1.13).

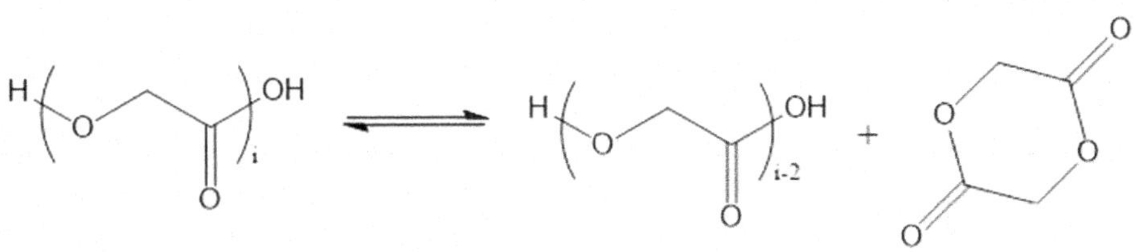

Fig. 1.12b Depolymerization equilibrium during polycondensation polymerization of glycolic acid. (Adapted from Singh and Tiwari 2010)

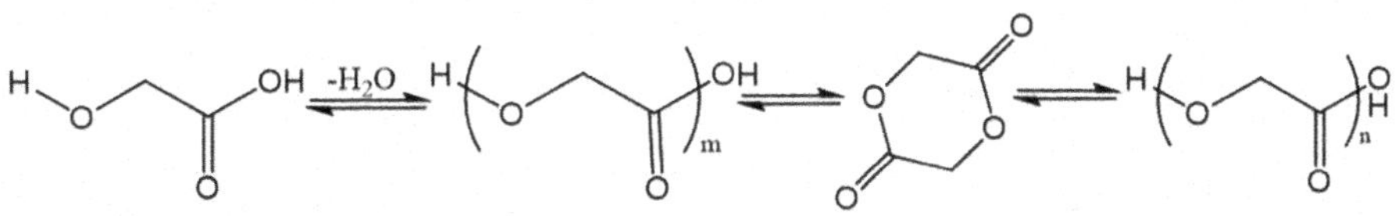

Fig. 1.13 Synthesis of polyglycolic acid with a high molecular weight using ring-opening polymerisation. (Adapted from Singh and Tiwari 2010)

1.8.2.2 Properties of Polyglycolic Acid

Polyglycolic acid exhibits remarkable properties such as biodegradability, biocompatibility, and high tensile strength, to name a few (Moon et al. 2004). It is also structurally similar to polylactic acid and does not contain pendant methyl groups in its polymer backbone. Thus, it is more hydrophilic than the latter and possesses high crystallinity, which is 45–55%. Further, due to its high crystallinity, it has a high melting point, i.e., 220–225 °C. Furthermore, polyglycolic acid's glass transition temperature is reported to be in the range between 35 and 40 °C (Shalaby et al. 1994). It shows low solubility towards conventional organic solvents but is soluble in some special solvents such as hexafluoroisopropanol, which is quite expensive (Shalaby et al. 1994). It possesses some free volume due to its chemical structure, which results in the high density of polyglycolic acid, leading to increased gas barrier performance (Kazuyuki Yamane et al. 2014). After all, the by-products obtained through degradation are non-toxic (Marin et al. 2013).

1.8.2.3 Degradation of Polyglycolic Acid

It is noteworthy to mention that polyglycolic acid degrades faster than polylactic acid. Chu 1981, demonstrated the hydrolytic degradation of polyglycolic acid using sterilized sutures in size 2–0. Briefly, the sutures with a length of over 43 cm were stored in a dessicator, which was filled with a mixture of phosphorous pentoxide (P_2O_5) and calcium sulphate ($CaSO_4$) before immersing into distilled water. Then 0.191313 g of the dried sutures were dipped into 30 mL of distilled water contained in glassware. After that, the glassware was sealed and placed in a septical hood to prevent contamination before experimenting. Taking this precaution is that the presence of bacteria reduces the rate of degradation in vitro and *in vivo*. Each suture specimen was separately immersed in the medium and placed in an incubator maintaining the temperature at 37 ± 1 °C for a definite period.

Every 7, 14, 21, 28, 49, 60, and 90 days, the suture specimen was taken out from the incubator for further examination. This study found that a drastic decline in tensile strength occurred between 7 and 14 days, and a similar pattern was also observed in the case of strain during this said period. Further, the percentage of the degree of degradation of polyglycolic acid over 49, 60, and 90 days were 42, 56, and 70%, respectively. Moreover, no tensile strength was observed at 49 days of immersion instead, more than 50% of the original suture remained undegraded. On the other hand, the degree of crystallinity increased initially, followed by a gradual reduction with immersion time. This phenomenon can be explained as follows: during immersion, water molecules entered into the polymer through diffusion occupy the voids in the amorphous region, whereas the crystalline region remains highly packed and densely ordered and thus, do not allow penetration of even little or no water. Therefore, the amorphous region of the polymer starts degrading hydrolytically faster than the crystalline region because of the degradation of tie-chain segments into fragments, which in turn reduces the probability of entanglement of long polymer chains residing in this amorphous region and also allows them to reorganise in an ordered state from a distorted state to increase the crystallinity of the polymer further. Degradation starts in the crystalline region once the entire amorphous region is removed by hydrolysis, which leads to a further decrease in crystallinity. Moreover, it can be said that the degradation of polyglycolic acid is due to its molecular weight and degree of crystallinity.

Furthermore, it is reported that polyglycolic acid takes several weeks to months to get completely degraded (Domb et al. 1997). Braunecker et al. (2004) demonstrated the effect of molecular weight and porosity on polyglycolic acid degradation. They showed that the rate of degradation of polyglycolic acid increased with decreasing its molecular weight. Additionally, it is reported that the process used to prepare polyglycolic acid can also have a significant impact on degradation. Nevertheless, polyglycolic acid also undergoes thermal degradation, which starts at 240 °C and causes 50% weight reduction at 346 °C (Chujo et al. 1967). Ginde and Gupta (1987) showed that the rate of degradation of polyglycolic acid pellets and fibers got affected during heating and might be due to a change in the molecular chain arrangement. Apart from this, during the hydrolytic degradation of polyglycolic acid, non-toxic products such as glycolic acid, water, and carbon dioxide are randomly produced. The same phenomenon can also be observed enzymatically through the polyglycolide chain cession mechanism (Gunatillake and Adhikari 2003).

1.8.3 Polylactic-Co-Glycolic Acid

Polylactic-co-glycolic acid is a linear synthetic copolymer. It is made of lactic acid and glycolic acid. By changing these constituent monomers' ratios, different grades of polylactic-co-glycolic acid copolymers can be synthesized. Further, it is to be noted that its physical and chemical properties exclusively depend on the individual constituent monomer. It finds immense potential to be used as a drug delivery vehicle as well as tissue engineering application.

1.8.3.1 Synthesis of Polylactic-Co-Glycolic Acid

Polylactic-co-glycolic acid can be obtained by the direct polymerization of lactic acid and glycolic acid. It is usually synthesized by mixing 75% of lactic acid and 25% glycolic acid, followed by subsequent polymerization (Zhou et al. 2004). Various methods can process it, and the physicochemical properties of polylactic-co-glycolic acid depend on feeding ratios of lactic acid and glycolic acid. Polylactic-co-glycolic acid with a low molecular weight is obtained using polycondensation reaction of lactic acid and glycolic acid at above 120 °C followed by removal of water (Wang et al. 2006). On the other hand, polylactic-co-glycolic acid with high molecular weight is synthesized by ring-opening polymerization of lactide and glycolide in the presence of catalysts such as stannous octanoate at elevated temperatures in the range of 130–220 °C. The reaction rate can be controlled by adding a chain regulator containing a free hydroxyl group (Kricheldorf et al. 1992).

Moreover, polylactic-co-glycolic acid is also prepared by the enzymatic ring-opening using lipase enzyme under mild conditions. Still, it requires a longer reaction time, but it yields a low molecular weight polymer (Duval et al. 2014). It is evident from the literature that polylactic-co-glycolic acid is degraded quickly by enzymatic ring-opening than by normal ring opening. Therefore, enzymatic degradation is favourable.

1.8.3.2 Physicochemical Properties of Polylactic-Co-Glycolic Acid

A distinct advantage of using polylactic acid-co-glycolic acid copolymers over polylactic acid and polyglycolic acid is that these copolymers are easily soluble in a wide array of conventional organic solvents such as chloroform, dichloromethane, acetone, ethyl acetate, and tetrahydrofuran, to name a few (Makadia and Siegel 2011). Polylactic acid-co-glycolic acid can be designed into distinct shapes and sizes. Further, polylactic acid-co-glycolic acid's physicochemical properties depend on various factors such as the ratio of lactic acid and glycolic acid, reaction time, catalyst concentration, and temperature (Houchin and Topp 2009). Moreover, it is to be noted that crystallinity and the degradability of polylactic acid-co-glycolic acid exclusively depend on the ratio of lactic acid and glycolic acid.

1.8.3.3 Degradation of Polylactic Acid-Co-Glycolic Acid

Polylactic acid-co-glycolic acid is a precious material due to its biocompatibility and degradability. It gets hydrolytically degraded into lactic acid and glycolic acid monomers. The hydrolytic degradation of polylactic acid-co-glycolic acid involves four steps: (i) hydration allowing water to enter into amorphous region leads to reduction in glass transition temperature of the copolymer, which is associated with the Van der Waals forces as well as hydrogen bonds; (ii) degradation is initiated with the involvement of breaking down ester linkages, which causes molecular weight reduction; (iii) autocatalysis occurs during degradation, which further leads to a weight reduction by breaking the backbone of covalent groups with loss of purity; (iv) solubilization permits chain fragmentation through dissolution in the aqueous environment (Engineer et al. 2011).

Lactic acid and glycolic acid are produced as by-products during degradation. The degradation rate depends on several factors, such as molecular weight, ratios of lactic acid and glycolic acid, stereoisomerism, and chain ends chemical composition (Avgoustakis 2005). Chemical composition and stereoisomerism are the major factors for the degradation of polylactic-co-glycolic acid. It is reported that if the concentration of lactic acid is higher than that of glycolic acid in the polylactic-co-glycolic acid, then it exhibits less hydrophilicity, which in turn, shows poor water absorption capacity leading to slower degradation. It is worth noting that polylactic-co-glycolic acid-containing equal proportions of lactic acid and glycolic acid shows faster degradation that occurs within 2 months. The degradation rate of polylactic-co-glycolic acid increases with increasing the hydrophillicity, but an opposite phenomenon is observed while increasing the crystallinity. D and L forms of lactic acid are mostly exploited for the synthesis of polylactic-co-glycolic acid, and the degradation rate is accelerated when an adequate amount of water gets absorbed in the amorphous region. Further, polylactic-co-glycolic acid degradation behavior is strongly affected by the shape of the material and also depends on water accessibility.

Moreover, the environmental pH is also an essential factor to influence the degradation rate during the autocatalysis (Wu and Wang 2001). Additionally, the degradation of polylactic-co-glycolic acid is greatly influenced by alkaline and strongly acidic media. The degradation rate is increased with assembling the carboxylic end groups during hydrolysis of the ester linkages. However, it is to be mentioned that polylactic-co-glycolic acid is stable in the dry state, and hydrolytically it degrades into small molecules through the cleavage of ester linkages and finally gets converted into lactic acid and glycolic acid monomers (Lao et al. 2011). As far as degradation of polylactic-co-glycolic acid is concerned, only the amorphous nature of this polymer is accountable for rapid hydrolysis because of diffusion of water molecules throughout the bulk matrix, leading to complete degradation (Bagheri et al. 2017).

1.8.4 Polyhydroxyalkanoates

Polyhydroxyalkanoates have gained considerable attention from researchers as one of the promising biomaterials of the future due to their physicochemical properties, such as biodegradability and biocompatibility. Polyhydroxyalkanoates belonging to the polyester family can be synthesized by the various microorganisms which use the excess of carbon as an energy source for their metabolism apart from using nutrient elements such as nitrogen, magnesium, phosphorus, and oxygen in a limited way during growth in the environment (Anderson and Dawes 1990). The formation of this kind of polymer mainly depends on the type of bacteria used for their growth (Mitra 2014). Lemoigne invented Polyhydroxyalkanoates in 1925, and initially, *Bacillus megaterium* was used for its synthesis. Polyhydroxyalkanoates have mainly 3-hydroxyalkanoates as building blocks of monomers. Usually, it is represented as poly3-hydroxybutyrate, which is popularly known as polyhydroxybutyrate.

Based on the number of carbon atoms present in the monomer, broadly, polyhydroxyalkanoates can be categorized into two heads: short-chain length consists of 3–5 carbon atoms, whereas medium-chain-length consists of 3–14 carbon atoms. Depending on the type of microorganisms during synthesis, different polyhydroxyalkanoates such as poly4-hydroxybutyrate, poly3-hydroxyvalerate, poly3-hydroxyhexanoate, poly3-hydroxyoctanoate, etc. can be produced. The availability of polyhydroxybutyrate from renewable resources has received researchers' tremendous attention to developing novel synthetic biodegradable materials for various purposes, especially for bioplastics production.

1.8.4.1 Synthesis of Polyhydroxyalkanoates

Polyhydroxyalkanoates are the member of intracellular biopolymers, which are biosynthesised from renewable feedstocks by bacterial fermentation of sugar, lipids, organic acids, and gases like methane and carbon dioxide. The various synthesis routes for the production of polyhydroxyalkanoates are shown in Fig. 1.14 (Haddouche et al. 2011). Various factors such as types of microorganisms, substrate, media, purification process, flow rate, pH, temperature, pressure, oxygen supply, etc., are having a direct impact on the molecular weight of polyhydroxyalkanoates (Verlinden et al. 2007). Principally, biological methods are adopted to synthesize polyhydroxyalkanoates by using a broad spectrum of microorganisms such as *E. coli, P. putida, Bacillus megaterium, Methylobacterium rhodesianum, Alcaligenes eutrophus, M. extorquens,* and *Sphaerotilus natans,* to name a few under specified conditions. Depending on the saturated, unsaturated, straight, cyclic or branched side chain of monomer units, homo and copolymers of polyhydroxyalkanoates can be obtained. This functional side chain is aliphatic and can modify chemically or enzymatically to synthesize new polymers (Kim and Lenz 2001). Here, we shall mainly focus on the synthesis of polyhydroxybutyrate. To synthesize

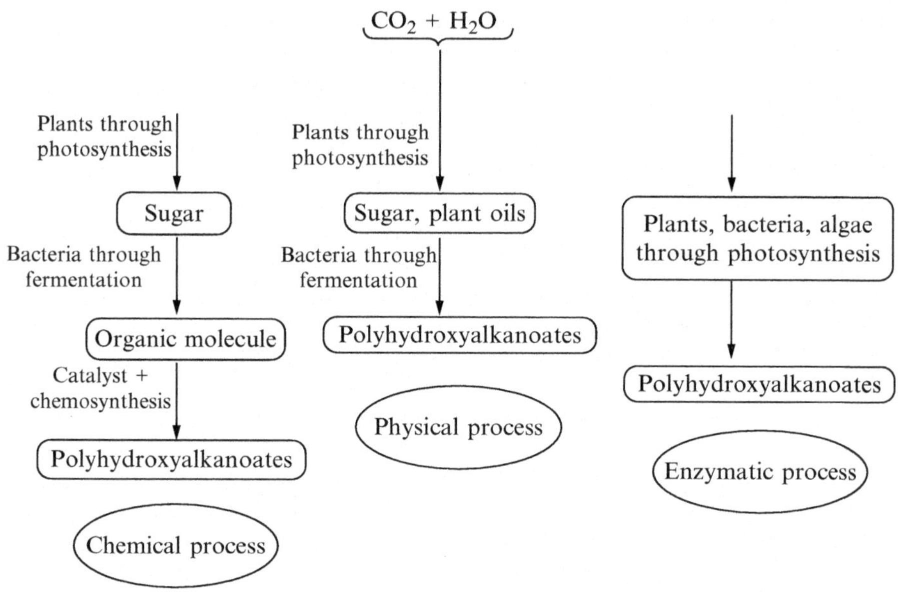

Fig. 1.14 Various synthesis routes for the production of polyhydroxyalkanoates. (Adapted from Asther 2016)

polyhydroxybutyrate from simple carbon sources, chemolithotrophic bacteria such as *Ralstonia eutropha, Cupriavidus metallidurans, Alcaligenes latus* is used. In addition, some gram-positive bacteria such as *Bacillus spp., B. subtilis, Bacillus thuringiensis, and B. cereus* and some gram-negative bacteria such as *Alcaligenes eutrophus*, *Pseudomonas spp.* are also used for the production of polyhydroxybutyrate. Notably, among gram-negative bacteria, *Alcaligenes eutrophus* is efficient in producing approximately 80% of poly3-hydroxybutyrate homopolymer with respect to its original cell dry weight (Fukui and Doi 1998). Further, it is to be noted that the said bacteria can be used to produce poly3-hydroxybutyrate-co-3-hydroxyvalerate, an industrially important copolymer, while adding propionic acid to the glucose feedstocks (Poltronieri and Kumar 2019).

Polyhydroxybutyrates can also be produced from agricultural waste, wherein glucose is used as a feedstock. Briefly, two fermentation steps are required to produce polyhydroxybutyrates. The first step is pre-fermentation, wherein bacterial cells are permitted to grow at 30 °C aerobically. After 24 h, the cells are found to increase at a rate of approximately 20 g/L. On the other hand, in the second step, during fermentation, the cells are further allowed to grow for another 40–80 h to obtain discrete granules with 0.3–1 μm diameters of cells as storage materials. Finally, about 80% of polyhydroxybutyrate is obtained from the original cells' dry weight during the fermentation process (Bugnicourt et al. 2014).

In another study, Leong et al. (2016) reported the production of polyhydroxybutyrate using three different carbon sources, namely soybean oil, cooking oil, and refined glycerol. Succinctly, a 5 L of the bioreactor was separately charged with

3.5 L of yeast extract with a concentration of 2 g/L and an individual carbon source of definite concentration. Then *Cupriavidus necator* as a hydrogen-oxidizing bacterium was cultivated at 30 °C and a pH of 7 for 72 h in this bioreactor. By controlling the stirring speed, the concentration of dissolved oxygen and airflow rate was maintained at 50% and 1 vvm, respectively. The flow chart for manufacturing polyhydroxybutyrate from the said bacterium is shown in Fig. 1.15.

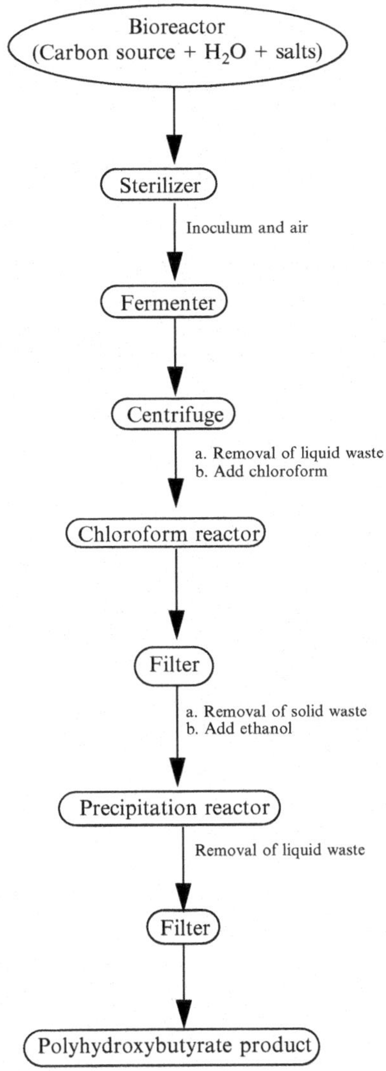

Fig. 1.15 Flow chart for manufacturing polyhydroxybutyrate. (Adapted from Leong et al. 2016)

1.8.4.2 Properties of Polyhydroxyalkanoates

In terms of physicochemical and mechanical properties, polyhydroxyalkanoates are similar to conventional polymers such as polyethylene and polypropylene. Interestingly, polyhydroxyalkanoates are not soluble in water, but they are completely biodegradable and possess good mechanical properties along with long-term stability. It exhibits better oxygen barrier properties than other biopolymers. The average molecular weight of polyhydroxyalkanoates varies from 2000 to 4000 Dalton, and correspondingly, the polydispersity index is around 2. Polyhydroxybutyrate has D-3 hydroxybutyric acid as its monomer unit and contains ester groups, making it more advantageous to target microbial enzymes frequently. It is semi-crystalline (Ray 2017). The respective tensile strength and Young's modulus of polyhydroxybutyrate are about 45 MPa and 3.5 GPa, which are closely similar to polypropylene. But the elongation at break for the latter one is 400%, which is 80 times higher than that of polyhydroxybutyrate (Penkhrue et al. 2020). Polyhydroxybutyrate does not possess any chain branching during the processing. It exhibits thermoplasticity and biodegradability properties for commercial usage. It is more advantageous than polypropylene in terms of food packaging applications because of its excellent water vapour permeability and oxygen barrier properties. It is chemically resistant to be attacked by either acids or alkalis. On the contrary, high crystallinity, low thermal stability, and the high melting point of polyhydroxybutyrate have restricted its broader applicability. To overcome these limitations, polyhydroxybutyrate is blended with other polymers, and further, plasticizers are also used not only to reduce its brittleness but also to improve its degradation rate by lowering the process temperature (Wang and Mao 2012).

1.8.4.3 Degradation Mechanism of Polyhydroxyalkanoates

As polyhydroxyalkanoates are synthesized from natural renewable resources through microbial fermentation, they can be greatly exploited for bioplastics production due to their excellent biodegradability. Usually, degradation is a structural or morphological change within the polymer when exposed to heat, light, etc. In addition, the same phenomenon may also occur during biological degradation. Photodegradation takes place under UV irradiation, thermal degradation occurs once the materials are treated at elevated temperatures, and biological degradation happens through the active participation of microorganisms (Sreedevi et al. 2014). As shown in Fig. 1.16, polyhydroxyalkanoates or polyhydroxybutyrates granules undergo enzymatic degradation and get converted into their corresponding monomeric units using suitable enzymes such as intracellular or extracellular polyhydroxyalkanoate depolymerase, which in turn, depends on the types of granules present in the bacterial cells. Polyhydroxyalkanoates undergo intracellular degradation through hydrolysis with intracellular polyhydroxyalkanoate depolymerase as an enzyme secreted by the bacteria once they feel stressed due to limited carbon and energy sources. On the other hand, once polyhydroxyalkanoates are exposed to the

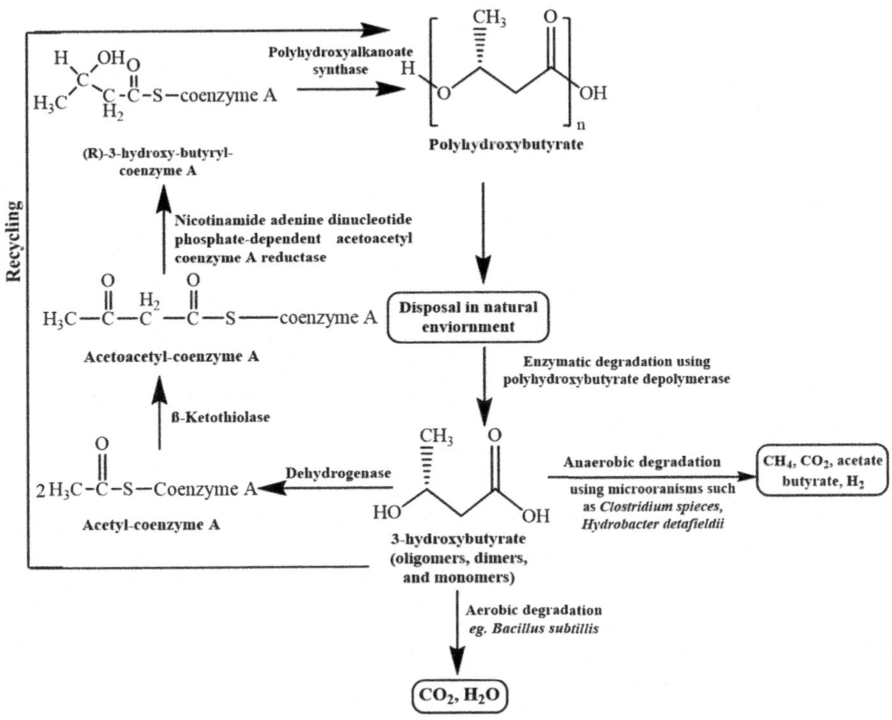

Fig. 1.16 A schematic representation of enzymatic degradation of polyhydroxybutyrate in the natural environment in the presence of microorganisms and its recycling process

environment, they undergo extracellular degradation due to the secretion of extracellular polyhydroxyalkanoate depolymerase enzyme by bacteria. However, the intracellular polyhydroxyalkanoate depolymerase is supposed to have different mechanisms due to its amorphous nature compared to the extracellular polyhydroxyalkanoate depolymerase (Volova et al. 2017). Polyhydroxybutyrate may undergo microbial biodegradation either aerobically or anaerobically. It is degraded into water (H_2O) and carbon dioxide (CO_2) in a moist warm aerobic environment, whereas in the anaerobic environment, it is broken down into methane (CH_4), hydrogen sulphide (H_2S), CO_2 and H_2O.

Of all the above-mentioned raw materials, polyhydroxyalkanoates are the most promising ones, which can be successfully used to produce bioplastics due to their outstanding biodegradability. On the other hand, polylactic-co-glycolic acid also shows better biodegradability than polylactic acid and polyglycolic acid. But unfortunately, it is not yet explored for industrial applications, especially for bioplastics production. Instead, its usage has been well-documented for biomedical applications. A summary of all the above-mentioned raw materials, their syntheses, and their plausible degradation mechanisms are discussed in Table 1.1.

Table 1.1 A snapshot of raw materials with various information for the production of bioplastics and their degradation techniques

SN	Raw materials for bioplastics production	Chemical structure of raw materials	Source of origin	Method of synthesis	Type of degradation mechanism	Time required for degradation	Reference
1.	Polylactic acid	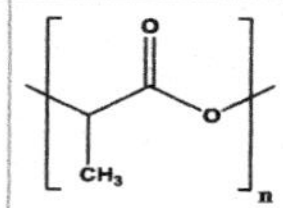	Lactic acid from sugar, corn, beets, cellulose, and other polysaccharides	(a) Direct polycondensation of lactic acid (b) Azeotropic dehydrative mechanism of lactic acid (c) Ring opening polymerization of lactide	(a) Hydrolytic and enzymatic degradation. Ester groups can be hydrolytically degraded in presence of water to yield lactic acid (b) Thermal degradation mainly caused by intra-molecular transesterification reaction followed by cyclic oligomers of lactic acid and lactide. It is a concerted process accompanied by 0–30% of its weight loss and found to have activation energy of 21–23 KJ/mol. (c) Environmental degradation involves mainly two steps: (i) chemical hydrolysis (ii) microbial degradation	6–12 month	Tsuji (2010), Kopinke et al. (1996), Wachsen et al. (1997), Nicolae et al. (2008), and Lunt (1998)

SN	Raw materials for bioplastics production	Chemical structure of raw materials	Source of origin	Method of synthesis	Type of degradation mechanism	Time required for degradation	Reference
2.	Polyglycolic acid	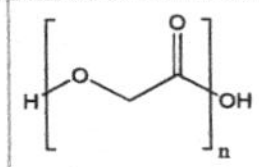	Synthetic glycolic acid and glycolide	(a) Direct polycondensation of synthetic glycolic acid (b) Azeotropic condensation polymerization of glycolic acid (c) Ring-opening polymerization of glycolide	(a) Thermal degradation occurs at 240 °C and causes 50% weight loss at 346 °C. (b) Hydrolytic degradation accompanied by two competing process, hydrolysis and induced crystallization.	>24 month	Chujo et al. (1967) and Chu (1981)
3.	Polylactic-co-glycolic acid	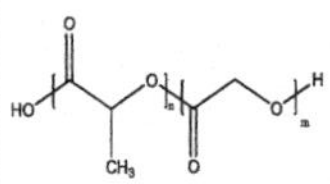	Lactic acid and glycolic acid	(a) Polycondensation process with lactic acid and glycolic acid (b) Ring opening polymerization of lactic acid and glycolic acid	Hydrolytic degradation i.e. hydrolysis of ester groups of polylactic-co-glycolic acid to produce lactic acid and glycolic acid monomers. It involves four steps: (a) hydration (b) initial degradation (c) constant degradation (d) solubilization	5–6 month	Engineer et al. (2011)

(continued)

Table 1.1 (continued)

SN	Raw materials for bioplastics production	Chemical structure of raw materials	Source of origin	Method of synthesis	Type of degradation mechanism	Time required for degradation	Reference
4.	Polyhydroxyalkanoates/ Polyhydroxybutyrates	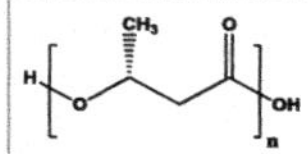	Sugar, lipids, organic acids or gases like methane and carbon dioxide	(a) Chemical process (b) Physical process (c) Enzymatic process	Enzymatic degradation of polyhydroxybutyrate seems to occur through two steps: (a) Adsorption (b) Hydrolysis It can also follow either aerobic or anaerobic degradation mechanism. Aerobic degradation produces H_2O and CO_2 whereas anaerobic degradation produces CH_4, H_2S, CO_2, and H_2O as end products	3–12 month	Lee and Choi (1999) and Roohi and Kuddus (2018)

1.9 Future of Bioplastics

The exploitation of traditional plastics virtually in every industrial sector has become a significant concern for environmental pollution due to their non-degradability. Therefore, we can increase the use of bioplastics to a certain extent for several applications, especially for the packaging, by minimizing traditional plastic usage. On the other hand, fossil-fuel-based resources are getting exhausted over time, leading researchers, academicians, and industrialists to explore various possibilities to find an alternative to replace traditional plastics for biodegradable production bioplastics. Over several decades, researchers are aggressively working towards developing bioplastics from raw materials obtained from natural renewable resources. Hence, as to make them fully biodegradable, which in turn, will not be the cause of environmental pollution, although it is quite known that 100% replacement of petroleum-based plastics is not possible because of relatively small scale production of bioplastics and cost of bioplastics production is very high. However, the demand for bioplastics is overgrowing, and these are being produced from renewable resources such as crops. In the recent past, considerable interest has been paid to agricultural wastes to reduce the fear of food scarcity. Bioplastics production is similar to bioenergy and biofuels that are strongly influenced by the technical and political aspects. Polylactic acid, polyhydroxyalkanoates, and polybutylene succinate are promising biodegradable bioplastics. It is expected that they will soon capture the worldwide market by replacing non-degradable plastics for various applications. There are two main critical points for growing the interest in bioplastics: the first one is biological feedstocks for the development of products, and the second one is complete biodegradation, whether environmentally or in a biological system. According to international standard agencies for biodegradation, a material is designated biodegradable if the material gets completely degraded in the aquatic environment and soil within 56 days and two years, respectively (Narancic et al. 2018).

The bioplastics products should be designed by considering the potential end-of-life option. Further, to improve mechanical properties and modify in a desirable shape for a specific application, bioplastic is blended with various additives or biopolymers, which holds the greatest promise to unlock their potential cost-effectiveness. Moreover, it also needs to be communicated with the public of the societies and the end-users about utilizing bioplastics products for sustainable development.

From the authors' perspective, a futuristic sustainable model shown in Fig. 1.17 for the production of bioplastics on a large scale can be developed, keeping in mind the food scarcity because of edible plants and crops etc., for the said purpose. In this regard, we can think of cultivating some other genetically modified plants which would not fall in the category of edible foods and, on the other hand, could also absorb atmospheric carbon dioxide in an adequate amount for producing food as an energy storage material during the photosynthesis process, which further can be used for bioplastics production. As a consequence of that, we would be able to avail

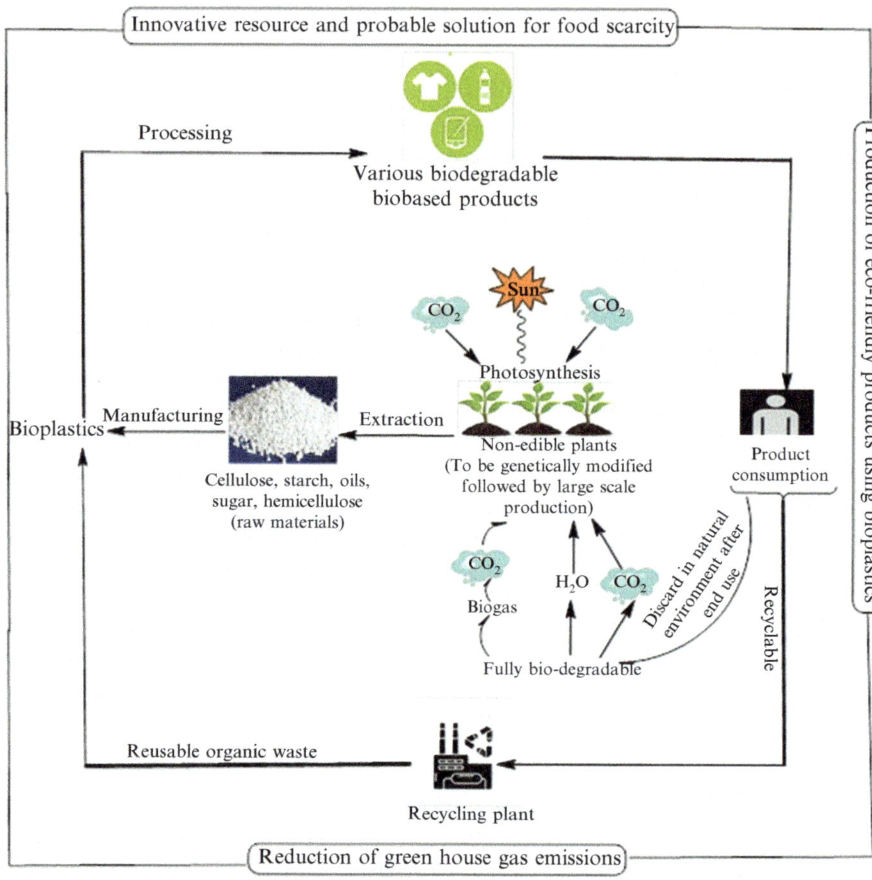

Fig. 1.17 A futuristic sustainable model for bioplastics production to avail triple benefits

triple benefits – relief from fear of food crisis, production of eco-friendly bioplastics from those genetically modified non-edible plants and their related biodegradable products; and tremendous reduction of carbon dioxide gas emissions into the atmosphere, which in turn, would offer us a sustainable and greener environment.

1.10 Conclusions

Plastics offer an essential role in our daily lives, but unfortunately, due to their non-degradability, they have become one of the major concerns for environmental pollution. Plastics are solely produced from limited fossil-fuel based resources. On the other hand, random disposal of plastics waste has considerably become the cause of producing toxic pollutants in the form of gas and liquid into the environment, which poses a severe threat to public health and nature. Despite their adverse effects on the environment, truly, it is very difficult to replace the plastics from our lives permanently, but the related environmental issues can be mitigated to a greater extent by introducing bioplastics for their huge potentials to sustain the environment. Bioplastics are produced from natural renewable resources, and relentlessly, we can use these sources to produce bioplastics as, unlike petroleum-based resources, these resources never get exhausted. Notably, once we talk about plastics/bioplastics, classification of them is very important, which clearly defines the actual nature of the plastics of interest. Moreover, the biodegradation of bioplastics is a crucial factor that would essentially solve all the problems associated with traditional plastics because of their biodegradability in the natural environment. On the other hand, recycling of bioplastics is relatively expensive because of the non-utility of the conventional processes used for traditional plastics. But considering the enormous potentials of bioplastics, these problems can be overcome with time. Undoubtedly, the bioplastics based on polylactic acid, polyglycolic acid, polylactic-co-glycolic acid, polyhydroxyalkanoates would be the most powerful competitors to defeat petrochemical-based plastics in the near future. In this chapter, a detail of various raw materials for the production of bioplastics and their plausible degradation mechanisms were discussed. However, 100% replacement of traditional bioplastics is not possible, but the introduction of the above-mentioned bioplastics to a great extent can be a probable solution in the context of environmental pollution. Further, a futuristic sustainable model was also discussed in this context, which could offer us a safer and greener environment upon successful execution. Finally, we must search for new materials for large-scale bioplastics production to increase their market share globally to compete with traditional plastics.

References

Agarwal M, Koelling KW, Chalmers JJ (1998) Characterization of the degradation of polylactic acid polymer in a solid substrate environment. Biotechnol Prog 14:517–526. https://doi.org/10.1021/bp980015p

Albuquerque PBS, Malafaia CB (2017) Perspectives on the production, structural characteristics and potential applications of bioplastics derived from polyhydroxyalkanoates. Int J Biol Macromol 107:615–625. https://doi.org/10.1016/j.ijbiomac.2017.09.026

Anderson AJ, Dawes A (1990) Occurrence, metabolism, metabolic role, and industrial uses of bacterial polyhydroxyalkanoates. Microbiol Rev 54:450–472. PMID: 2087222; PMCID: PMC372789

Anderson JM, Hiltner A, Wiggins MJ, Schuber MA, Collier TO, Kao WJ, Mathur AB (1998) Recent advances in biomedical polyurethane biostability and biodegradation. Polym Int 46:163–171. https://doi.org/10.1002/(SICI)1097-0126(199807)46:3%3C163::AID-PI972%3E3.0.CO;2-9

Anderson AG, Grose J, Pahl S, Thompson RC, Wyles KJ (2016) Microplastics in personal care products: exploring perceptions of environmentalists, beauticians and students. Mar Pollut Bull 113:454–460. https://doi.org/10.1016/j.marpolbul.2016.10.048

Andrady AL (2015) Plastics and environmental sustainability: fact and fiction. John Wiley and Sons, Hoboken, New Jersy pp 352

Andrady AL, Neal MA (2009) Applications and societal benefits of plastics. Phil Trans R Soc B 364:1977–1984. https://doi.org/10.1098/rstb.2008.0304

Asther SA (2016) Intoduction to bioplastics engineering. William Andrew Chapter 5:81–151. https://doi.org/10.1016/c2014-0-04010-5
Avgoustakis K (2005) Polylactic-co-glycolic acid (PLGA). Encyclopedia of biomaterials and biomedical engineering, second edition – four volume set, (scheme 2), pp 2259–2269. https://doi.org/10.1081/E-EBBE-120013950
Bagheri AR, Laforsch C, Greiner A, Agarwal S (2017) Fate of so-called biodegradable polymers in seawater and freshwater. Global Chall 1:1–5. https://doi.org/10.1002/gch2.201700048
Banks I, Smet MDD, Linder M (2016) Towards a new plastics economy. Retrieved from https://www.chemistryworld.com/opinion/towards-anewplasticseconomy/1010279.article
Braunecker J, Baba M, Milroy GE, Cameron RE (2004) The effects of molecular weight and porosity on the degradation and drug release from polyglycolide. Int J Pharm 282:19–34. https://doi.org/10.1016/j.ijpharm.2003.08.020
Bugnicourt E, Cinelli P, Lazzeri A, Alvarez VA (2014) Polyhydroxyalkanoate (PHA): review of synthesis, characteristics, processing and potential applications in packaging. Express Polym Lett 8:791–808. https://doi.org/10.3144/expresspolymlett.2014.82
Burgos N, Valdés A, Jiménez A (2016) Valorization of agricultural wastes for the production of protein-based biopolymers. J Renew Mater 4:165–177. https://doi.org/10.7569/JRM.2016.634108
Chamas A, Moon H, Zheng J, Qiu Y, Tabassum T, Jang JH, Omar MMA, Scott SL, Suh S (2020) Degradation rates of plastics in the environment. ACS Sustain Chem Eng 8:3494–3511. https://doi.org/10.1021/acssuschemeng.9b06635
Chiellini E, Corti A, D'Antone S (2007) Oxo-biodegradable full carbon backbone polymers—biodegradation behaviour of thermally oxidized polyethylene in an aqueous medium. Polym Degrad Stab 92:1378–1383. https://doi.org/10.1016/j.polymdegradstab.2007.03.007
Chu CC (1981) Hydrolytic degradation of polyglycolic acid: tensile strength and crystallinity study. J Appl Polym Sci 26:1727–1734. https://doi.org/10.1002/app.1981.070260527
Chujo K, Kobayashi H, Suzuki J, Tokuhara S, Tanabe M (1967) Ring-opening polymerization of glycolide. Die Makromolekulare Chemie 100:262–266. https://doi.org/10.1002/macp.1967.021000128
Clarinval AM, Halleux J (2005) Classification of biodegradable polymers, in: Smith R (ed) Biodegradable polymers for industrial applications. Cambridge, Chapter pp 3–31
Cole M, Lindeque P, Halsband C, Galloway TS (2011) Microplastics as contaminants in the marine environment: a review. Mar Pollut Bull 62:2588–2597. https://doi.org/10.1016/j.marpolbul.2011.09.025
Crowley D, Staines A, Collins CB, Bracken J, Bruen M, Fry J, Hrymak V, Malone D, Magette B, Ryan MF, Thunhurst C (2003) Health and environmental effects of landfilling and incineration of waste – a literature review. In: reports, ed. paper 3
Demirbas A (2007) Biodegradable plastics from renewable resources. Energy Sources Part A 29:419–424. https://doi.org/10.1080/009083190965820
Domb AJ, Kost J, Wiseman DM (1997) Handbook of biodegradable polymers. Harwood Academic, Amsterdam, pp 87–97
Drumright RE, Gruber PR, Henton DE (2000) Polylactic acid technology. Adv Mater 12:1841–1846. https://doi.org/10.1002/1521-4095(200012)12:23<1841::AIDADMA1841>3.0.CO;2-E
Dutta J, Dutta PK (2006) Lactic acid and lactic acid based industries: ecofirendly in nature. Everyman's Sci XLI 2:103–109
Duval C, Nouvel C, Six JL (2014) Is bismuth subsalicylate an effective nontoxic catalyst for plga synthesis? J Polym Sci Part A 52:1130–1138. https://doi.org/10.1002/pola.27096
Ellen MacArthur Foundation (2017) The new plastics economy: Catalysing action. Ellen MacArthue Foundation and World Economic Forum
Emadian SM, Onay TT, Demirel B (2017) Biodegradation of bioplastics in natural environments. Waste Manag 59:526–536. https://doi.org/10.1016/j.wasman.2016.10.006
Engineer C, Parikh J, Raval A (2011) Review on hydrolytic degradation behavior of biodegradable polymers from controlled drug delivery system. Trends Biomater Artif Organs 25:79–85

European Bioplastics (2016) Bioplastics market data 2016. Retrived from https://www.epa.gov/facts-and282figures-about-materials-waste-and-recycling/advancing-sustainable-materials-management

Fritz HG, Seidenstuker T, Bolz U, Juza M, Schroeter J, Enders J (1994) Study on production of thermoplastics and fibers based mainly on biological materials. Agro Ind Res Division:392

Fukui T, Doi Y (1998) Efficient production of polyhydroxyalkanoates from plant oils by Alcaligenes eutrophus and its recombinant strain. Appl Microbiol Biotechnol 49:333–336. https://doi.org/10.1007/s002530051178

Fukushima K, Tabuani D, Dottori M, Armentano I, Kenny JM, Camino G (2011) Effect of temperature and nanoparticle type on hydrolytic degradation of poly(lactic acid) nanocomposites. Polym Degrad Stabil 96:2120–2129. https://doi.org/10.1016/j.polymdegradstab.2011.09.018

Garlotta D (2001) A literature review of poly(lactic acid). J Polym Environ 9:63–84. https://doi.org/10.1023/A:1020200822435

Gaudin R, Knipfer C, Henningsen A, Smeets R, Heiland M, Hadlock T (2016) Approaches to peripheral nerve repair: generations of biomaterial conduits yielding to replacing autologous nerve grafts in craniomaxillofacial surgery. Biomed Res Int 2016:1–18. https://doi.org/10.1155/2016/3856262

Geyer R, Jambeck JR, Law KL (2017) Production, use, and fate of all plastics ever made. Sci Adv 3:e1700782. https://doi.org/10.1126/sciadv.1700782

Ginde RM, Gupta RK (1987) In vitro chemical degradation of poly (glycolic acid) pellets and fibers. J Appl Polym Sci 33:2411–2429. https://doi.org/10.1002/app.1987.070330712

Göpferich A (1996) Mechanisms of polymer degradation and erosion. Biomater Res 17:103–114. https://doi.org/10.1016/0142-9612(96)85755-3

Greene J (2007) Biodegradation of compostable plastics in green yard-waste compost environment. J Polym Environ 15:269–273. https://doi.org/10.1007/s10924-007-0068-1

Gross RA, Kalra B (2002) Biodegradable polymers for the environment. Green Chem 297:803–807. https://doi.org/10.1126/science.297.5582.803

Gunatillake PA, Adhikari R (2003) Biodegradable synthetic polymers for tissue engineering. Eur Cell Mater 5:1–16. https://doi.org/10.22203/ecm.v005a01

Gupta MC, Deshmukh VG (1982a) Thermal oxidative degradation of poly-lactic acid part I: activation energy of thermal degradation in air. Collo Polym Sci 260:308–311. https://doi.org/10.1007/BF01447969

Gupta MC, Deshmukh VG (1982b) Thermal oxidative degradation of poly-lactic acid part II: molecular weight and electronic spectra during isothermal heating. Coll Polym Sci 260:514–517. https://doi.org/10.1007/BF01452999

Haddouche R, Poirier Y, Delessert S, Sabirova J, Pagot Y, Neuveglise C, Nicaud JM (2011) Engineering polyhydroxyalkanoate content and monomer composition in the oleaginous yeast Yarrowia lipolytica by modifying the ß-oxidation multi- functional protein. Appl Microbiol Biotechnol 91:1327–1340. https://doi.org/10.1007/s00253-011-3331-2

Hakkarainen M (2002) Aliphatic polyesters: abiotic and biotic degradation and degradation products. Polymer 157:113–138. https://doi.org/10.1007/3-540-45734-8_4

Henton DE, Gruber P, Lunt J, Randall J (2005) Polylactic acid technology. In: Mohanty AK, Misra M, Drzal LT (eds) Natural fibers, biopolymers, and biocomposites. Taylor & Francis, Boca Raton, pp 527–577

Hopewell J, Dvorak R, Kosior E (2009) Plastics recycling: challenges and opportunities. Philos Trans R Soc Lond Ser B Biol Sci 364:2115–2126. https://doi.org/10.1098/rstb.2008.0311

Houchin ML, Topp EM (2009) Physical properties of plga films during polymer degradation. J Appl Polym Sci 114:2848–2854. https://doi.org/10.1002/app.30813

Jain R, Tiwari A (2015) Biosynthesis of planet friendly bioplastics using renewable carbon source. J Environ Health Sci Eng 13:11. https://doi.org/10.1186/s40201-015-0165-3

Jem KJ, Por ZFVD, Sd V (2010) Microbial lactic acid, its polymer poly(lactic acid), and their industrial applications. In: Chen G-Q (ed) Plastics from Bacteria: natural functions and applications, Microbiology Monographs, vol 14, pp 323–345

Jong SJD, Arias ER, Rijkers DTS, Nostrum CFV, Bosch JJKVD (2001) New insights into the hydrolytic degradation of poly(lactic acid): participation of the alcohol terminus. Polymer 42:2795–2802. https://doi.org/10.1016/S0032-3861(00)00646-7

Kabasci S (2014) Bio-based plastics: materials and applications John Wiley and Sons Ltd Handbook Chapter 7.3.2

Kale G, Kijchavengkul T, Auras R, Rubino M, Selke S, Singh SP (2007) Compostability of bioplastic packaging materials: an overview. J Macromol Biosci 7:255–277. https://doi.org/10.1002/mabi.200600168

Karamanlioglu M, Robson GD (2013) The influence of biotic and abiotic factors on the rate of degradation of poly(lactic) acid (PLA) coupons buried in compost and soil. Polym Degrad Stabil 2013:1–9. https://doi.org/10.1016/j.polymdegradstab.2013.07.004

Kim YB, Lenz RW (2001) Polyesters from microorganisms. Adv Biochem Eng Biot 71:51–79. https://doi.org/10.1007/3-540-40021-4_2

Kopinke FD, Remmler M, Mackenzie K, Moder M, Wachsen O (1996) Thermal decomposition of biodegradable polyesters-II. Poly(lactic acid). Polym Degrad Stab 53:329–342. https://doi.org/10.1016/0141-3910(96)00102-4

Kricheldorf HR, Boettcher C, Tonnes KU (1992) Polylactones: polymerization of racemic and mes dl-lactide with various organotin catalysts stereochemical aspects. Polymer 33:2817–2824. https://doi.org/10.1016/0032-3861(92)90459-A

Kyrikou J, Briassoulis D (2007) Biodegradation of agricultural plastic films: a critical review. J Polym Environ 15:125–150. https://doi.org/10.1007/s10924-007-0053-8

Lao LL, Peppas NA, Boey FYC, Venkatraman SS (2011) Modeling of drug release from bulk-degrading polymers. Int J Pharm 418:28–41. https://doi.org/10.1016/j.ijpharm.2010.12.020

Lee SY, Choi JI (1999) Production and degradation of polyhydroxyalkanoates in waste environment. Waste Manag 19:133–139. https://doi.org/10.1016/S0956-053X(99)00005-7)10.1016/S0956-053X(99)00005-7)

Leja K, Lewandowicz G (2010) Polymer biodegradation and biodegradable polymers - a review. Pol J Environ Stud 19:255–266

Leong YK, Show PL, Lin HC, Chang CK, Loh HS, Lan JCW, Ling TC (2016) Preliminary integrated economic and environmental analysis of polyhydroxyalkanoates (PHAs) biosynthesis. Bioresour Bioprocess 3:42. https://doi.org/10.1186/s40643-016-0120-x

Lu DD, Yuan JC, Li H, Lei ZQ (2008) Synthesis and characterization of a series of biodegradable and biocompatible PEG-supported poly(lactic-ran-glycolic acid) amphiphilic barbell-like copolymers. J Polym Sci 46:3802–3812. https://doi.org/10.1002/pola.22729

Lu Y, Schmidt C, Beuermann S (2014) Fast synthesis of high-molecular weight polyglycolide using diphenyl bismuth bromide as catalyst. Macromol Chem Phys 216:395–399. https://doi.org/10.1002/macp.201400474

Lunt J (1998) Large-scale production, properties and commercial applications of polylactic acid polymers. Polym Degrad Stabil 59:145–152. https://doi.org/10.1016/s0141-3910(97)00148-1

Makadia HK, Siegel SJ (2011) Poly lactic-co-glycolic acid (PLGA) as bio- degradable controlled drug delivery carrier. Polymer (Basel) 3:1377–1397. https://doi.org/10.3390/polym3031377

Marin E, Briceño MI, George CC (2013) Critical evaluation of biodegradable polymers used in nanodrugs. Int J Nanomedicine 8:3071–3091. https://doi.org/10.2147/IJN.S47186

Martin O, Averrous L (2001) Poly(lactic acid): plasticization and properties of biodegradable multiphase systems. Polymer 42:6209–6219. https://doi.org/10.1016/S0032-3861(01)00086-6

Martinho G, Balaia N, Pires A (2017) The Portuguese plastic carrier bag tax: the effects on consumers' behavior. Waste Manag 61:3–12. https://doi.org/10.1016/j.wasman.2017.01.023

Mcneill IC, Leiper HA (1985) Degradation studies of some polyesters and polycarbonates. 2. Polylactide: degradation under isothermal conditions, thermal degradation mechanism and photolysis of the polymer. Polym Degrad Stab 11:309–326. https://doi.org/10.1016/0141-3910(85)90035-7

Mekonnen T, Mussone P, Khalil H, Bressler D (2013) Progress in bio-based plastics and plasticizing modifications. J Mater Chem A 1:13379–13398. https://doi.org/10.1039/C3TA12555F

Millet H, Vangheluwe P, Block C, Sevenster A, Garica L, Romanos A (2018) The nature of plastics and their societal usage. In: Plastics and the environment, pp 1–20. https://doi.org/10.1039/9781788013314-00001
Mitra BC (2014) Environment friendly composite materials: biocomposites and green composites. Defence Sci J 64:244–261. https://doi.org/10.14429/dsj.64.7323
Mohee R, Unmar G (2007) Determining biodegradability of plastic materials under controlled and natural composting environments. Waste Manag 27:1486–1498. https://doi.org/10.1016/j.wasman.2006.07.023
Moon S, Deguchi K, Miyamoto M, Kimura Y (2004) Synthesis of polyglactin by melt/solid polycondensation of glycolic/L-lactic acids. Polym Int 53:254–258. https://doi.org/10.1002/pi.1335
Nampoothiri KM, Nair NR, John RP (2010) An overview of the recent developments in polylactide (PLA) research. Bioresour Technol 101:84938–84501. https://doi.org/10.1016/j.biortech.2010.05.092
Narancic T, Verstichel S, Chaganti SR, Morales-Gamez L, Kenny ST, De Wilde B, Padamati RB, Connor KEO (2018) Biodegradable plastic blends create new possibilities for end-of-life management of plastics but they are not a panacea for plastic pollution. Environ Sci Technol 52:10441–10452. https://doi.org/10.1021/acs.est.8b02963
Narayan R (2006a) Biobased and biodegradable polymer materials: rationale, drivers, and technology exemplars. Am Chem Soc Symp Ser 939:282–306. https://doi.org/10.1021/bk-2006-0939.ch018
Narayan R (2006b) Rationale, drivers, standards, and technologyfor biobased materials. In: Graziani M, Fornasiero P (eds) Renewable resources andrenewable energy. CRC Press, Taylor & Francis Group, Boca Raton
Nicolae CA, Grigorescu MA, Gabor RA (2008) An investigation of thermal degradation of poly(lactic acid). Eng Lett 16:568
Nkwachukwu OI, Chima CH, Ikenna AO, Albert L (2013) Focus on potential environmental issues on plastic world towards a sustainable plastic recycling in developing countries. Int J Ind Chem 4:1–13. https://doi.org/10.1186/2228-5547-4-34
Okada M (2002) Chemical syntheses of biodegradable polymers. Prog Polym Sci 27:87–133. https://doi.org/10.1016/S0079-6700(01)00039-9
Orset C, Barret N, Lemaire A (2017) How consumers of plastic water bottles are responding to environmental policies? Waste Manag 61:13–27. https://doi.org/10.1016/j.wasman.2016.12.034
Paul MA, Delcourt C, Alexandre M, Degée P, Monteverde F, Dubois P (2005) Polylactide/montmorillonite nanocomposites: study of the hydrolytic degradation. Polym Degrad Stab 87:535–542. https://doi.org/10.1016/j.polymdegradstab.2004.10.011
Penkhrue W, Jendrossek D, Khanongnuch C, Pathomaree W, Aizawa T, Behrens RL, Lumyong S (2020) Response surface method for polyhydroxybutyrate (PHB) bioplastic accumulation in Bacillus drentensis BP17 using pineapple peel. PLoS One 15:1–21. https://doi.org/10.1371/journal.pone.0230443
Philp JC, Ritchie RJ, Guy K (2013) Biobased plastics in a bioeconomy. Trends Biotechnol 31:65–67. https://doi.org/10.1016/j.tibtech.2012.11.009
PlasticsEurope (2006) The compelling facts about plastics: an analysis of plastic production, Demand and Recovery in Europe
PlasticsEurope (2016) Plastics-the facts 2016: an analysis of European plastics production, Demand and Waste Data
Poltronieri P, Kumar P (2019) Polyhydroxyalkanoates (PHAs) in industrial applications. Handbook of Ecomaterials Cham Springer, 2843–2872. doi:https://doi.org/10.1007/978-3-319-68255-6_70
Poortinga W, Whitaker L (2018) Promoting the use of reusable coffee cups through environmental messaging, the provision of alternatives and financial incentives. Sustainability 10:1–12. https://doi.org/10.3390/su10030873

Potential of Plastics Industry in Northern India with Special Focus on Plasticulture and Food Processing – 2017: A Report on Plastic Industry FICCI 34. Retrieved from http://ficci.in/spdocument/20396/Knowledge-Paper-ps.pdf

Ray D (2017) Biocomposites for high-performance applications: current barriers and future needs towards industrial development. Else Sci 2017

Raza ZA, Abid S, Banat IM (2018) Polyhydroxyalkanoates: characteristics, production, recent developments and applications. Int Biodeterior Biodegradation 126:45–46. https://doi.org/10.1016/j.ibiod.2017.10.001

Reddy MM, Vivekanandhan S, Misra M, Bhatia SK, Mohanty AK (2013) Biobased plastics and bionanocomposites: current status and future opportunities. Prog Polym Sci 38:1653–1689. https://doi.org/10.1016/j.progpolymsci.2013.05.006

Riggle D (1998) Moving towards consensus on degradable plastics. Biocycle 39:64–70

Roohi ZMR, Kuddus M (2018) PHB (poly-β-hydroxybutyrate) and its enzymatic degradation. Polym Advan Technol 29:30–40. https://doi.org/10.1002/pat.4126

Rosato D (2013) Designing with plastics and composites: a handbook. Springer, Boston

Ryberg MW, Laurent A, Hauschild M, Tonda E, Averous S, Yan C, Hasegawa K (2018) Mapping of global plastics value chain 96. Retrieved from https://gefmarineplastics.org/files/2018mappingofglobalplasticsvaluechainandhotspots-finalversionr181023.pdf

Schmidt C, Behl M, Lendlein A, Beuermann S (2014) Synthesis of high molecular weight polyglycolide in supercritical carbon dioxide. RSC Adv 4:35099–35105. https://doi.org/10.1038/s41428-019-0182-7

Shalaby SW, Johnson RA, Deng M (1994) Degradation of polymeric materials. Biomed Polym 3:1–34

Shen K, Yang S (2013) Preparation of high-molecular-weight poly(glycolic acid) by direct melt polycondensation from glycolic acid. Adv Mater Res 821–822:1023–1026. https://doi.org/10.4028/www.scientific.net/AMR.821-822.1023

Sin L, Rahmat A, Rahman W (2013) Degradation and stability of poly(lactic Acid). 247–299. https://doi.org/10.1016/b978-1-4377-4459-0.00007-x

Singh V, Tiwari M (2010) Structure-processing-property relationship of poly(glycolic acid) for drug delivery systems 1: synthesis and catalysis. Int J Polym Sci 2010:1–23. https://doi.org/10.1155/2010/652719

Song JH, Murphy RJ, Narayan R, Davies GBH (2009) Biodegradable and compostable alternatives to conventional plastics. Phil Trans R Soc Lond B Biol Sci 364:2127–2139. https://doi.org/10.1098/rstb.2008.0289

Sreedevi S, Unni KN, Sajith S, Priji P, Josh MS, Benjamin S (2014) Bioplastics: advances in polyhydroxybutyrate research. Adv Polym Sci, Springer https://doi.org/10.1007/12_2014_297

Statista (2018) Plastics industry – Statistics & facts. Retrieved from https://www.statista.com/topics/5266/plastics-industry/

Takahashi K, Taniguchi I, Miyamoto M, Kimura Y (2000) Melt/solid polycondensation of glycolic acid to obtain high-molecular-weight poly(glycolic acid). Polym 41:8725–8728. https://doi.org/10.1016/S0032-3861(00)00282-2

Thomas NL, Clarke J, McLauchlin AR, Patrick SG (2010) Assessing the environmental impacts of oxodegradable plastics across their life cycle. EV0422. Report to the Department for Environment, Food and Rural Affairs. Loughborough University, Loughborough, England

Tsuji H (2010) Poly(lactic acid): synthesis, structures, properties, processing, and applications. Hydro Degrad:343–381. https://doi.org/10.1002/9780470649848.ch21

Tsuji H, Ikada Y (2000) Properties and morphology of poly (L-lactide): effects of structural parameters on long-term hydrolysis of poly (L-lactide) in phosphate buffered solution. Polym Degrad Stab 77:1452–1464. https://doi.org/10.1002/1097-4628(20000815)77:7<1452::AIDAPP7>3.0.CO;2-S

Tsuji H, Ikarashi K, Fukuda N (2004) Poly (L-lactide): XII, formation, growth, and morphology of crystalline residues as extended-chain crystallites through hydrolysis of poly(L-lactide) films in phosphate-buffered solution. Polym Degrad Stab 84:515–523. https://doi.org/10.1016/j.polymdegradstab.2004.01.010

Tsuji H, Echizen Y, Saha SK, Nishimura Y (2005) Photodegradation of poly(L-lactic acid): effects of photosensitizer. Macromol Mater Eng 290:1192–1203. https://doi.org/10.1002/mame.200500278

Verlinden RA, Hill DJ, Kenward MA, Williams CD, Radecka I (2007) Bacterial synthesis of biodegradable polyhydroxyalkanoates. J Appl Microbiol 102:1437–1449. https://doi.org/10.1111/j.1365-2672.2007.03335.x

Volova TG, Prudnikova SV, Vinogradova ON, Syrvacheva DA, Shishatskaya EI (2017) Microbial degradation of Polyhydroxyalkanoates with different chemical compositions and their biodegradability. Microb Ecol 73:353–367. https://doi.org/10.1007/s00248-016-0852-3

Wachsen O, Reichert KH, Kruger RP, Much H, Schulz G (1997) Thermal decomposition of biodegradable polyesters-III. Studies on the mechanisms of thermal degradation of oligo-L-lactide using SEC, LACCC and MALDI-TOF-MS. Polym Degrada Stab 55:225–231. https://doi.org/10.1016/S0141-3910(96)00127-9

Wang J, Mao Q (2012) Methodology based on the PVT behavior of polymer for injection molding. Adv Polym Technol 32:474–485. https://doi.org/10.1002/adv.21294

Wang ZY, Zhao YM, Wang F, Wang J (2006) Syntheses of poly(lactic acid-co-glycolic acid) serial biodegradable polymer materials via direct melt polycondensation and their characterization. J Appl Polym Sci 99:244–252. https://doi.org/10.1002/app.22468

Wei R, Zimmermann W (2017) Biocatalysis as a green route for recycling the recalcitrant plastic polyethylene terephthalate. Microb Biotechnol 10:1302–1307. https://doi.org/10.1111/1751-7915.12714

Wikström F, Williams H, Venkatesh G (2016) The influence of packaging attributes on recycling and food waste behaviour–an environmental comparison of two packaging alternatives. J Clean Prod 137:895–902. https://doi.org/10.1016/j.jclepro.2016.07.097

Wiles DM (2005) Oxo-biodegradable polyolefins in packaging. In: Smith R (ed) Biodegradable polymers for industrial applications. Woodhead Publishing, Cambridge, pp 57–76

Wu XS, Wang N (2001) Synthesis, characterization, biodegradation, and drug delivery application of biodegradable lactic/glycolic acid polymers. Part II: Biodegradation J Biomater Sci Polym Ed 12:21–34. https://doi.org/10.1163/156856201744425

Yamane K, Sato H, Ichikawa Y, Sunagawa K, Shigaki Y (2014) Development of an industrial production technology for high-molecular-weight polyglycolic acid. Polym J 46:769–775. https://doi.org/10.1038/pj.2014.69

Yates MR, Barlow CY (2013) Life cycle assessments of biodegradable, commercial biopolymers-a critical review. Resour Conserv Recycl 78:54–66. https://doi.org/10.1016/j.resconrec.2013.06.010

Zee MVD (1995) Structure-biodegradability relationships of polymeric materials. 1. Effect of degree of oxidation on biodegradability of carbohydrate polymers. J Environ Polym Degr 3:235–242. https://doi.org/10.1007/BF02068678

Zhang X, Wyss UP, Pichora D, Goosen MFA (1992) An investigation of the synthesis and thermal stability of poly(dl -lactide). Polym Bull 27:623–629

Zhang JY, Beckman EJ, Piesco NP, Agarwal S (2000) A new peptide-based urethane polymer: synthesis, biodegradation, and potential to support cell growth in vitro. Biomater Res 21:1247–1258. https://doi.org/10.1016/s0142-9612(00)00005-3

Zheng J, Suh S (2019) Strategies to reduce the global carbon footprint of plastics. Nat Clim Chang 9:374–378. https://doi.org/10.1038/s41558-019-0459-z

Zhou Q, Xanthos M (2008) Nanoclay and crystallinity effects on the hydrolytic degradation of polylactides. Polym Degrad Stab 93:1450–1459. https://doi.org/10.1016/j.polymdegradstab.2008.05.014

Zhou SB, Deng X, Li X, Jia W, Liu L (2004) Synthesis and characterization of biodegradable low molecular weight aliphatic polyesters and their use in protein-delivery systems. J Appl Polym Sci 91:1848–1856. https://doi.org/10.1002/app.13385

Chapter 2
Bioenergy Production from Wastewater Resources Using *Clostridium* Species

Rajathirajan Siva Dharshini, Ramachandran Srinivasan, and Mohandass Ramya

Abstract Water management is a major issue globally; on the other hand, the accumulation of wastewater streams and their treatment strategies are hard to accomplish even in growing technologies. The discharge of wastewater is majorly unavoidable. Prominent wastewater disposals are agriculture wastewater, septic tanks, household kitchen wastewater, industrial disposal, and sewage sludge. Increased population also increases the waste disposal and recovery of the economy from waste is the smart approach to build the nation. European nations have already initiated waste treatment with beneficial outlets. Wastewater was the unique source for bioenergy production by presenting mixed polysaccharides, lipids, microbial loads, etc. Microbial populations differ according to the composition of wastewater. Microbial metabolism also helps hand in hand to generate energy from wastewater. Nutrient-rich wastewater is also advantageous for the growth of the industrial microbial population. Even many techniques like hybrid systems and microbial approaches are practiced, but the recovery rate is still limited, and the gain of benefits is countable. *Clostridium*, an industrial fermentative mesophile, was the candidate microorganism to convert the waste biomass into energy resources. It has an advantageous feature that it can sustain in anaerobic conditions. Microbial Fuel cells and biofuel production using *Clostridium* species are renowned examples of using waste biomass for bioenergy production. Hence, wastewater treatment directed towards energy recovery was the best choice to improve the economy and encourage an eco-friendly environment. It will balance both energy conservation and a pollution-free environment. This review discusses the forthcoming techniques for the bioenergy production derived from wastewater and their treatment using *clostridium* species.

Keywords *Clostridium* · Bioenergy · Wastewater · Effluent · Glycerol · Hydrogen · Acetone-butanol-ethanol

R. S. Dharshini · R. Srinivasan · M. Ramya (✉)
Molecular Genetics Laboratory, Department of Genetic Engineering, SRM Institute of Science and Technology, Kattankulathur, Tamil Nadu, India

K. M. Gothandam et al. (eds.), *Environmental Biotechnology Volume 4*, Environmental Chemistry for a Sustainable World 68,
https://doi.org/10.1007/978-3-030-77795-1_2

2.1 Introduction

Energy crisis and waste management are the two major problems prevailing in society. Increased pollution by the usage of fossil fuels can be resolved by bioenergy production. This can be achieved by the two-in-one process of using waste as the biomass for the production of bioenergy. Wastewater can be used as the microbes' growth medium to assimilate the substrate and releases bioenergy as a byproduct. These approaches are economical as well as reduction of waste accumulation (Prakash et al. 2018).

Different types of industries use large volumes of water, three fourth among them are discharged as wastewater. All those wastewaters are released into the water resources, and it mainly contributes to water pollution. Wastewater is an abundant pollutant all over the world. It constitutes many contaminants and toxic components, which are discharged by industries and sanitary facilities. These constituents are highly destructive to the environment, and they can cause health issues even among living organisms. The World Health Organization stated that 30% of human disease and 40% of morbidity are majorly due to wastewater.

Wastewater from industries is mostly composed of high amounts of sulfite, and its discharges affect the environment by corrosion, bad smell, and contamination with public water supply (Blázquez et al. 2019). Chemical wastewater treatments are costly, energy-consuming, and produce sludge to a great extent. Biological methods of wastewater treatment are specific regarding the nature of the wastewater in its nutrient removal, low sludge production, and reduced cost (Bertolino et al. 2012). Conventional methods for wastewater treatment were the microbial reduction of toxic substances and oxidation.

The emergence of resources from wastewater streams was currently focused on due to the shortage of energy resources (Mansouri et al. 2017) and the petroleum routes. Next-generation energy production was broadly focused on the waste streams as a substrate because of their carbon enrichment. It facilitates the production of chemicals and materials instead of petrochemical processes, which encourages the circular economy, industrial waste as a substrate for energy generation (Baleta et al. 2019).

On the other hand, microbial conversion of bioenergy production. Microbes play an essential role in wastewater treatment and bioenergy production. The selection of microbes was the main precaution to get effective and beneficial productivity. *Clostridium* was a highly preferable microorganism among all microbes because of its industrial applications and substrate assimilation. Diversified species of *Clostridium* can metabolize several substrates into valuable products (Liberato et al. 2019). The *Clostridium* species can easily assimilate enriched mixed substrates in the wastewater. In this chapter, literature data explains the role of wastewater in bioenergy production using *Clostridium* species.

2.2 Types of Wastewater and Its Composition

Wastewater was varied in its composition, processing, and nature. Major wastewater resources are municipal wastewater, food industry wastewater, agro-industry wastewater, manufacturing industry wastewater, distillery industry wastewater, and anaerobic digester effluents of various wastewater treatment units. The processes involved in wastewater treatment are represented in Fig. 2.1. Wastewater's composition was variable as per the industries and processing methods. Different types of wastewater are widely used for the production of bioenergy and value-added products. Bioenergy production from wastewater yielded zero cost consumption for substrate, and it was profitable in the same manner. During ethanol production, the leftover waste stream was known as stillage to generate 20 L for each liter of ethanol. It comprises the chemical oxygen demand of 30–91 g/L based on the type of feedstock. Glycerol was the byproduct of ethanol fermentation. Stillage usually contained glycerol in the range of 5.1–24.6 g/L (Ahn et al. 2011). The thin stillage was composed of sugar compounds such as glucose, arabinose, glucan, xylose, xylan, arabinan and glycerol and fibers, proteins, minerals, and fats (Kim et al. 2008). The presence of polysaccharides and fat content increased the application of bioenergy through fermentation.

Brewery wastewater was abundant because of its continuous processing and marketing. Beer took fifth place in the list of consumable beverages (Maiti et al. 2016). While beer production, wastewater was released in two stages, such as brewing and packaging. Beverage industrial wastewater was a prominent source for bioenergy production, which contains high Chemical Oxygen Demand, Biochemical Oxygen Demand, and organic content of sugars, soluble starch, ethanol, and volatile fatty acids (Amenorfenyo et al. 2019). Sugar-enriched beverages, including fruit

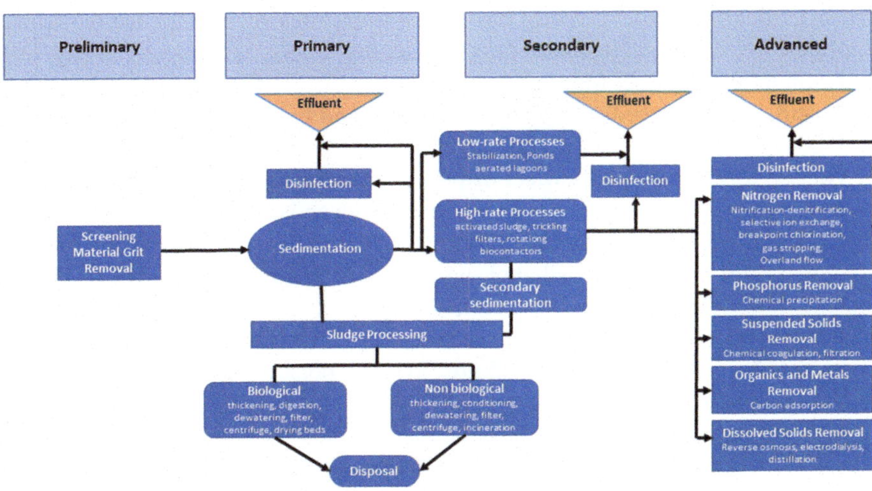

Fig. 2.1 Different processes involved in wastewater treatment

juices, syrups, soft drinks, and sports drinks, act as the carbon source for acetone butanol and ethanol production. Beverage industries used to discharge 0.1% of the waste stream every year. These waste streams include process wastewaters, seizures of goods, and stocks after expiration (Raganati et al. 2015). The process of bottle washing discharged 50% of the wastewater; caustic soda and sugar are the major constituents of these waste streams. The increased sugar content of these industrial wastewaters was an advantage for fermentation during bioenergy production.

Food industrial wastewater was varied according to their processing feedstocks and resources. Food industrial effluents are different from municipal wastewater, which constitutes biodegradable matter, suspended solids, Chemical Oxygen demand and Biochemical Oxygen demand (Amin 2019). Starch-rich wastewater was the common waste stream among food industries. Cassava processing was major for starch and flour production and wastewater disposal and around 7 m^3 per kg of cassava root. Cassava wastewater has a high content of carbohydrates and a Chemical Oxygen demand level of 5–15 g/L and Biochemical Oxygen Demand at the range of 20–50 g/L (Cappelletti et al. 2011). Tapioca starch was derived from cassava (*Manihot utilissima*). Tapioca starch wastewater generated in the range of 40–60 m^3 from one ton of cassava that contains carbohydrates, glucose (425–1850 mg/L), and nitrogen (97–182 mg/L) (Setyawaty et al. 2011). Corn steep liquor was another form of starchy wastewater discharged during the production of corn starch. It constitutes organic nitrogen, small peptides, vitamins, trace elements, and lactic acid (Loy and Lundy 2018). Depending on the composition, bioenergy production was increased from food waste.

Nowadays, biodiesel industries are widely used for bioenergy production. Even though it was an eco-friendly approach, it was also used to produce a large amount of wastewater. Glycerol was the main byproduct of the biodiesel industry. While transesterification, which generates 10% of raw oil by mass. Several technologies are there to convert glycerol into value-added products (Dams et al. 2016). After transesterification, crude glycerol has lots of impurities such as methanol, sulfate, chlorides, free fatty acid residues, fatty acid methyl esters, and soaps. The high fatty acid content of biodiesel industries is used to produce different chemicals such as citric acid, 1,3- propanediol, hydrogen, butanol, and polyhydroxyalkanoates (Yang et al. 2012).

Vinasse was residual wastewater disposed of during first-generation ethanol production. Vinasse was the major source of bioenergy and can be compared with ethanol's second-generation (Mazuchi 2018). In 2012–2013, 10–14 L of vinasse was generated with each liter of ethanol, which comprises a large amount of organic content, low pH, and nutrients (Ferraz Júnior et al. 2015). The composition of vinasse depends on the molasses and processing conditions. Vinasse has a high Chemical Oxygen Demand of 22–45 g/L, low pH, potassium, sulfate, other phenolic compounds, and melanoidins.

In this case, sugar industries have no exception, which used large amounts of water for their operations and processing. As same as the production of wastewater also to a great extent. One ton of sugar needs 20 m^3 freshwater, which discharges wastewater twice its amount in addition to the initial water used (Justo et al. 2016).

It was characterized by the high Biochemical Oxygen Demand, Chemical Oxygen Demand and suspended solids with carbohydrates, nutrients, oil, grease, sulfates, and heavy metals (Kushwaha 2015). Sugar industrial wastewater was broadly used as a source for ethanol and biogas production. Molasses are the viscous liquid produced during the sugar production process, consisting of 50% total sugars (sucrose, glucose, and fructose), suspended colloids, heavy metals, vitamins, and nitrogenous compounds. Molasses was one of the desirable substrates for ethanol and butanol production (Li et al. 2013a).

Palm oil was the prominent edible oil, and its production was about 33% in the world. Water consumption of palm oil industries was started from the processing of palm fruits itself, which leads to the production of a large amount of waste effluent called palm oil mill effluent. Usually, 10 liters of palm oil mill effluent was produced in the generation of one liter of oil produced. Palm oil mill effluent was an abundant pollutant with decreased pH composed of organic and free fatty acids. Typically, it contains total solids (40,500–75,000 mg/L), oil, grease (2000–8300 mg/L), suspended solids (18,000–47,000 mg/L), nitrogen (400–800 mg/L) and ash content (3000–42,000 mg/L). Palm oil mill effluent has a Biochemical Oxygen Demand in the range of 25,000–54,000 mg/L and Chemical Oxygen Demand between 50,000 and >100,000 mg/L. This composition load was 100 times more than municipal wastewater. Palm oil mill effluent was highly preferable for bioenergy production because of its enriched profile (Iwuagwu and Ugwuanyi 2014).

The profile of sewage sludge was more than spent digester effluents. Municipal wastewater was prominent with 99.9% water and 0.1% organic, and inorganic solids include detergents, food leftovers, fats and oils, grease, heavy metals, and varied biomolecules and their decomposed mixture of sand, grits, excrements, paper products and chemicals from process industries (Puyol et al. 2017; Qu et al. 2013). Pathogenic microbes are also used to coexist, and they cause foul odor and gases. Wastewater was enriched with carbohydrates, lipids, phosphorus, and nitrogen and collapsed the aquatic ecosystem and increased eutrophication.

Leachate was a major contaminant and disturbance in open landfill disposal. It was a dripped liquid from a sanitary landfill containing organic materials, ammonia, heavy metals, phenol, nitrogen, and phosphorus. Contaminants possessed by leachate are highly toxic to the environment (Aziz et al. 2015). Spent digester effluents are used to comprise the valuable byproducts themselves through anaerobic digestion. All these wastewaters are usually applied to the production of Acetone-butanol-ethanol, hydrogen, and biogas (Smoliński et al. 2019).

2.3 Integration of Wastewater and Bioenergy Production

In bioenergy production, different types of feedstocks are in use. They are classified into different generations. In first-generation biofuel production, large amounts of sugar, starch, and oil crops and animal fats are used. These are carried out using standard techniques such as fermentation, esterification, and distillation, etc. It exploited the food and feed for bioenergy production and competition with the food and feed industry. Common bioenergy production from first-generation feedstocks is vegetable oil, bio alcohols, biodiesel, biogas, and solid biofuels. First-generation biofuels need lots of land to cultivate, and it became costlier than petroleum (Perimenis et al. 2011).

Second-generation biofuels have emerged to cut off the limitations of first-generation biofuels. Non-consumable organic feedstocks like wood, agricultural residues, specific crops contained cellulose, hemicellulose, and lignin are employed in second-generation bioenergy production. Drawbacks are cost consuming, and many modifications are needed to make it on a large scale, so making commercials is difficult (Naik et al. 2010).

To minimize the disadvantages of first and second-generation biofuels, third-generation biofuels have arisen Algal biomass as the feedstock for this process. It yields 30 times more than plants and crops. They need many specifications and additional infrastructures for processing and cultivation. To overcome these drawbacks, algae was metabolically engineered in the fourth generation of biofuels. But investments are high in the preliminary stage even though the yield was better. Still, renewable energy has minimal limitations and technical issues. Research is still going on to eliminate the obstacles and improve biofuels' quality (Datta et al. 2019).

A new innovative approach was developed for the treatment of wastewater. This encouraged wastewater usage as a substrate for energy production instead of treating it as waste. It impacted the elimination of greenhouse gas emissions during the usual wastewater treatment and was used to attain energy sustainability, water, nutrient recovery, and decreased generation of residual solids and greenhouse gases. This approach includes combining the biological approach with physical and chemical processes and allows the organic carbon conversion into methane and nitrogen and phosphorous recovery (Sutton et al. 2011). The presence of organic content, nutrients, conserved thermal heat, kinetic energy in the wastewater was used to produce onsite energy production. This equals the energy required for wastewater treatment, which reduces the energy cost and removes the harmful contaminants in the wastewater (Ahmed et al. 2015).

While wastewater treatment, wastewater was dumped inside the anaerobic digester and broken down the complex mixture, which results in the production of biogas. It was the primary production of bioenergy from wastewater. Many technologies emerged consequently for using the substrate of low cost for bioenergy production. Wastewater got attention as the zero-cost substrate among all other different generations of feedstock as substrates due to organic matter and catalyzing microorganism. Biofuel production is a renewable, eco-friendly, and sustainable alternative platform for fossil fuels, and it has the annual growth increased by 30.28 billion liters in 2020. In substrate utilization, utilizing wastes with the mixture of components is featured in maintaining pH levels, reduced usage of nitrogen, and co-substrate management to yield biofuels to a large extent. Energy production from wastewater has multiple advantages, such as self-efficient energy production for industries, elimination of greenhouse gas emissions, waste management, and

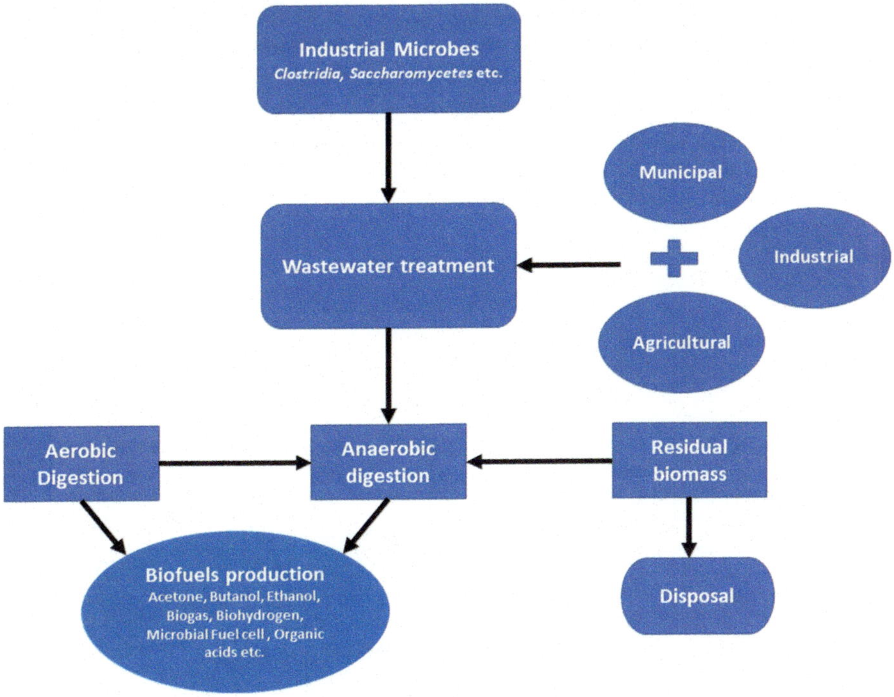

Fig. 2.2 Schematic approach of wastewater into bioenergy

financial profit. The schematic diagram for the conversion of wastewater into biofuels was presented in Fig. 2.2.

2.4 *Clostridia* Era for Bioenergy Production

The genus *Clostridium* was proposed by the type organism *Clostridium butyricum* in the year 1880 by Prazmowski. It was categorized as an anaerobic, spore-forming, and gram-positive bacterium. Currently, there are about 228 species and subspecies under the genus *Clostridium*. *Clostridium* was the well-known industrial microbial genus for its efficiency in producing organic acids and solvents such as ethanol, butanol, 1,3- propanediol, and acetone.

Fermentation and biochemical pathways are significant and specific in *Clostridium* species. Based on its end products, fermentation pathways were classified as Acetone- Butanol-Ethanol, Isopropanol-Butanol- Ethanol and Hexanol-Butanol- Ethanol. Each strain uses different metabolic pathways based on its substrate and genetic information. The Parnas pathway was used for hexose contained substrate, and for pentose contained substrate, the pentose phosphate pathway was used. To metabolize CO, CO_2, and H_2 in Hexanol- Butanol- Ethanol fermentation, the Wood Ljungdahl pathway was used (Liberato et al. 2019).

Clostridium strains are used in many bioenergy productions using different feedstocks, and it yields significant bioenergy for the energy crisis. The list of bioenergy production using *Clostridium* species with different substrates is followed in the given Table 2.1.

Table 2.1 List of bioenergy production using *Clostridium* species with different substrates

Clostridium sp.	Substrate used	Bioenergy	Yield	References
C. saccharoperbutylacetonium N1- 4	Pretreated switchgrass	Butanol	8.6 g/L	Wang et al. (2019)
C. beijerinckii ASU10	Food Processing solid waste	Acetone-butanol-ethanol	17.91 g/L	Abd-Alla et al. (2017)
C. actobutylicum B 527	Pineapple waste	Acetone-butanol-ethanol	5.23 g/L	Khedkar et al. (2017)
C. thermocellum	Sugarcane bagasse	Hydrogen	22.78 g/L	Cheng et al. (2014)
C. beijerinckii P260	Corn stover	Butanol	11.58 g/L	Qureshi et al. (2014)
C. ljungdahliii	Syngas	Bioethanol	166.1%	Kim et al. (2014)
C. acetobutylicum NCIMB 13357	Date fruit	Butanol	0.48 g/g	Khamaiseh et al. (2014)
C. acetobutylicum and C. aurantibutyricum	Ceylon moss, red algae	Acetate, Butyrate, Ethanol, Acetone, and Butanol	8.06 mol/m^3 (Acetate), 39.43 mol/m^3 (Butyrate), 1.60 mol/m^3 (Ethanol), 2.41 mol/m^3 (Acetone), 11.75 mol/m^3 (Butanol)	Kim et al. (2013)
C. acetobutylicum ATCC 824	Sago pith residues hydrolysate	Acetone-butanol-ethanol	8.84 g/L	Linggang et al. (2013)
C. acetobutylicum ATCC824	Maple hemicellulose hydrolysate	n- butanol	7 g/L	Sun and Liu (2012)

2.5 Bioenergy Production from Wastewater Resources Using *Clostridium* Species

Clostridium was significant in the assimilation of different types of substrates. They can perform fermentation with multiple substrates (Kaushal et al. 2019). The kinds of wastewater employed in the production of bioenergy using *Clostridium* species are followed below; it explains the challenges and drawbacks of the processing of different wastewater to make it suitable substrates for energy production and microbial growth.

2.5.1 *Biodiesel Industrial Wastewater*

Glycerol was an efficient substrate for bioenergy production and available in excess amounts in the biodiesel industries as a byproduct. It is used as the monomer to produce novel polyester, polytrimethylene terephthalate. It was the only low-cost substrate that could produce 1,3- propanediol and was also employed in cosmetics, pharmaceutical, and food industries. 1,3- propanediol acts as the precursor for the production of propionic acid, acetic acid, butyric acid, butanol, and hydrogen.

For the production of 1,3- propanediol from the crude glycerol, Gungormusler et al. (2010) compared the different aerobic and anaerobic microorganisms. Among them, *C. saccharobutylicum NRRL B- 643* showed the efficiency to consume glycerol completely. But 1,3- propanediol production yielded 0.54 mol/mol by using *C. beijerinckii* (B-593), which was more than *C. saccharobutylicum NRRL B- 643*. pH and temperature are controlled within 5–7.5 and 28–37 °C for high yield. Guo et al. (2017) characterized the *C. perfringens GYL* from soil sludge to produce 1,3-propanediol from glycerol. Optimum fermentation conditions like inoculation size between 5% and 10%, temperature, 4 g/L yeast extract, pH range from 6.5 to 7.0, and temperature between 30 and 40 °C were observed. *C. perfringens GYL* showed a significant glycerol tolerance feature of 200 g/L and yielded 40 g/L 1,3- propanediol production.

Hydrogen productivity was mainly based on the presence of hydrogenase enzyme in the microorganism. Trchounian et al. (2017) studied the *C. beijerinckii* DSM791 for its ability to utilize glycerol and the presence of hydrogenase enzymes. Optimum utilization of glycerol at a pH of 7.5 as compared with pH 5.5. In addition to that, the presence of trace metals such as 0.1 mM Fe^{2+} and 1 mM Ni^{2+} increased the enzyme activity up to ~50% at pH 7.5. It achieved a yield of 1.21 mol H_2/mol glycerol. This study implies the importance of pH and trace metals in enhanced enzyme activity in hydrogen production.

The Acetone-butanol-ethanol pathway in *Clostridium* is used to produce acetone, butanol, and ethanol in the ratio of 3:6:1 in its solventogenic phase. Acetone was less preferred among the market and disabled to be used as an automotive fuel. To prevent acetone production, Kaushal et al. (2019) used the non-acetone producing *clostridium* strain, *C. sporogenes NCIM 2918* in the fermentation process using rice straw hydrolysate and crude glycerol as the substrates. It achieved a high total alcohol titer of 23.7 g/L at the production rate of 0.52 g/L/h in batch fermentation was increased by continuous gas stripping and yields 44.4 g/L (butanol 21.5 g/L and ethanol 22.9 g/L) at the rate of 0.62 g/L/h.

Dams et al. (2016) studied *C. acetobutylicum 824* in pure and bioaugmentation with anaerobic sludge and goat ruminal liquid using glycerol as the substrate. In this study, *C. acetobutylicum* was grown in glycerol separately as well as in the mixture. Bioaugmentation of anaerobic sludge and ruminal liquid with *C. acetobutylicum* showed a high production of metabolites compared with the pure culture in glycerol. A combination of flocculated sludge, glycerol, and *C. acetobutylicum* showed the high production of acetic acid (1.2 g/L), 1,3- propanediol (1.6 g/L) and ethanol (1.6 g/L). Propionic acid (1.2 g/L) was increased in the combination of granular sludge, glycerol, and *C. acetobutylicum*. High production of butyric acid (0.9 g/L) and n-caproic acid (1.6 g/L) was reported using ruminal liquid, glycerol, and *C. acetobutylicum.*

All the above mentioned and most of the other studies of glycerol fermentation used only free cells. Even though it was beneficial, it had some disadvantages. Free cells had difficulty in downstream processes and no reuse. To tackle these disadvantages, immobilization can be used. Immobilization was economical and reusable. Khanna et al. (2013) employed the immobilization of cells using *C. pasteurianum,* which was the only known strain that can utilize glycerol to convert into n- butanol that also produces 1,3- propanediol, ethanol, butyrate, acetate, lactate, CO_2, and hydrogen. The cells of *C. pasteurianum* are immobilized in the mixed bed amberite. As per the concentration of glycerol in the medium, biofuels are produced. Increased yield of n- butanol was achieved in a medium containing 25 g/L glycerol, and 1,3-propanediol was maximum in the medium containing 10 g/L glycerol. Ethanol yield was increased in 5 g/L glycerol concentration. Glycerol levels of more than 25 g/L acts as the inhibition for the solvent production. This study explains the importance of immobilization by its glycerol tolerance and notable yield. Immobilized cells yielded the 1,3- propanediol and n- butanol at the rate of 0.07 g/g glycerol and 0.02 g/g glycerol. It was achieved at a high concentration of 150 g/L glycerol.

Krasňan et al. (2018) used the same *C. pasteurianum* for immobilization using polyvinyl alcohol in the glycerol fermentation for the production of n- butanol. It yields the butanol concentration at the rate of 3.08, 2.42 and 1.73 g L^{-1} h^{-1} by three different batches of glycerol such as pure glycerol, crude glycerol 01, and crude glycerol 02 of different impurities. Polyvinyl alcohol was less sensitive to contaminants and improved productivity.

2.5.2 Distillery Industrial Wastewater

Distillery industrial wastewater was different according to the raw materials used for ethanol production, and it produced wastewater in differing processing units. Vinasse is the byproduct of ethanol fermentation using sugar cane and thin stillage by using corn as the raw material. Sugar and starch content is highly advantageous for the production of bioenergy.

Thin stillage was the liquid waste stream generated by ethanol fermentation. It is important to treat the stillage for waste management. *C. pasteurianum* DSM 525 was used to ferment stillage to produce butanol. Glycerol was the main component in stillage. Stillage provides the perfect environment for the growth of *C. pasteurianum* DSM 525 because of its glycerol and other nutrients. The pH control was necessary for the cultivation of any *clostridium* species. In stillage, only the initial pH was adjusted to 5–7, because lactic acid acted as the buffering agent. Excess of acetic acid and lactic acid in the stillage showed the inhibition effects for butanol production. It suggested the importance of checking the acetic acid and lactic acid content in the stillage. Concentrations of 16 g/L and 10 g/L of lactic acid and acetic acid were applicable for butanol production and yield up to 6.2–7.2 g/L of butanol (Ahn et al. 2011).

Vinasse was the wastewater generated during ethanol production using sugar cane. Lazaro et al. (2014) used granular sludge from slaughterhouse wastewater digester effluent as inoculum to ferment vinasse for hydrogen production. They evaluated both temperature and substrate concentration. It yields hydrogen more in the thermophilic phase (2.31 mmol $H_2/g/COD_{influent}$) than the mesophilic phase (2.23 mmol $H_2/g/COD_{influent}$). Inoculum under thermophilic condition is mostly related to Thermoanaerobacterium and in the mesophilic state, the genus Clostridium dominates, which is involved in the production of hydrogen. A species like *C. carboxidivorans* acts as an inhibitor by consuming the gas. This represents the importance of the selection of microorganisms for biological production.

Anaerobic up-flow packed bed reactors are well known for biohydrogen production. In these kinds of reactors, supporting materials are important for the attachment of biomass. There are many different types of supporting materials used. To make it sustainable for hydrogen production, Ferraz Júnior et al. (2015) used four supporting materials for comparison while using vinasse as the substrate. They are expanded clay, Charcoal, Porous ceramic, and Low-Density Polyethylene. It concluded that low-density polyethylene was preferable supporting material with a volumetric hydrogen production rate of 509.5 ml/d/L with a maximum yield of 3.2 mol/mol $_{total\ sugar}$, under the conditions of 24 h hydraulic retention times, loading rate of 36.2 kg-COD.m^{-3}.d^{-1}, at 25 °C. Sequencing of samples from anaerobic up-flow packed bed reactors with low-density polyethylene found the presence of hydrogen-producing bacteria such as *Clostridium* and *Pectinatus* along with other lactic acid bacteria and non-fermenting bacteria. These microbial species might be involved in the production of hydrogen.

2.5.3 Beverage Industrial Wastewater

Beverage industries use large amounts of water for production and discharge large amounts of wastewater and effluents. It diversified into processing wastewater, left-over stocks, washing water, etc.

C.acetobutylicum DSM 792 was used in the fermentation of high sugar content beverages for the production of butanol. The study was significant by using high sugar content beverages as the substrate for butanol production. It also acts as an efficient remediation process for industrial beverage wastewater. To cultivate *C. acetobutylicum*, yeast and mineral salts are supplemented with high sugar content beverages. The assimilation of glucose and fructose was increased compared with sucrose and produced butanol at the maximum concentration of 10 g/L. It has two-fold advantages such as waste disposal and biofuel production (Raganati et al. 2015).

Liu et al. (2016) used beverage processed wastewater as the substrate for hydrogen production. Their inoculum includes *C. tyrobutyricum, Clostridium sp., Clostridium acetobutylicum, C. acetobutylicum,* and *Ethanoligenens harbinense.* They used four different hydraulic retention times and total sugar concentrations to compare their efficiency. It yields a maximum at 0.30 ± 0.06 mol H_2/mol hexose at hydraulic retention times 1 h and in the range of 11.39 ± 1.39 L/L/d at 5 g total sugar and 1 h hydraulic retention times. This concluded that using low concentrated beverage process wastewater was sustainable for hydrogen production and further produced electricity via Proton Exchange Membrane Fuel Cell. Sulfide-reducing bacteria exhibited inhibition in hydrogen production by simultaneous conversion of hydrogen into hydrogen sulfide.

Soft drink production was commercially increased in the market and is also used to produce effluents rich in carbohydrates. These carbohydrates are highly preferred for the production of biohydrogen. Peixoto et al. (2011) used a mixed culture of naturally occurring microbial communities in the soft drink wastewater as an inoculum to digest the same soft drink wastewater. It yielded 3.5 mol H_2mol^{-1} sucrose and was produced at the rate of 0.4 L/h/L. It was operated under pH 6.4 with 0.5 h hydraulic retention times in the substrate concentration of 1.94 g/L. Mixed inoculum mainly consists of *Clostridium sp., Klebsiella sp.,* and *Enterobacter sp.* The specific strain was not determined.

2.5.4 Food Industrial Wastewater

Food industrial wastewater was efficient in bioenergy production because of its composition. It includes starch-rich wastewater, maltose-rich wastewater, Palm oil mill effluent, etc. Enriched with fats, carbohydrates, and nitrogen content was found in these wastewaters.

C. acetobutylicum ATCC 824 was employed for the production of hydrogen from cassava wastewater. Cassava wastewater was starch-rich. pH was a must to maintain between 7 and 5 for hydrogen production by using cassava wastewater. The study showed that significant hydrogen production was attained with low Chemical Oxygen demand. It implies that factors such as pH and Chemical Oxygen demand are involved in the high production of hydrogen and yields 2.68 mol H_2/mol glucose with the Chemical Oxygen demand content of 5 g/L. High Chemical Oxygen demand yields low hydrogen production from cassava wastewater (Cappelletti et al. 2011). Cassava and molasses are relatively efficient feedstocks for the production of biofuels (Lépiz-Aguilar et al. 2011). Li et al. (2013b) investigated the co-culture of *C. beijerinckii* and *C. tyrobutyricum* using different substrates such as glucose, cassava, and molasses. They used both free cell and immobilized cell fermentation modes. It achieved the maximum yield of butanol (0.18 g/g) in the rate of 0.96 g/L/h with the cassava starch in both strains' continuous immobilization cell culture. It suggested the advantage of continuous co-culture. Maiti et al. (2016) screened the biobutanol production using different waste streams such as suspended brewery liquid waste, starch industry wastewater, and apple pomace ultra-filtration sludge as a substrate. *C. beijerinckii NRRL B- 466* was used in the fermentation process, and the butanol production of 11.04 g/L by starch industry wastewater, 9.3 g/L by apple pomace ultra-filtration sludge and 8.06 g/L by suspended brewery liquid waste was achieved.

Cassava wastewater was rich in carbohydrates but poor in nitrogen content. It has a high C/N and not preferable for hydrogen production. Co- digestion was the solution to this problem. Wadjeam et al. (2019) investigated hydrogen production by co-digestion of cassava wastewater with buffalo dung to enhance productivity. Polymerase chain reaction – Denaturing gradient gel electrophoresis profiles of the composition result in the domination of *Clostridium sp.*, batch mode and at the same time *Clostridium sp., Megasphaera sp.* and *Chloroflexi sp.*, are present in the continuous mode under optimal conditions. The maximum yield of 1787 ml H_2/L was achieved in batch mode. In continuous mode, it produced a high yield of 16.90 ml H_2/g-COD_{added} at the rate of 839 ml H_2/L.d., at 60 h hydraulic retention times. As same as cassava, tapioca wastewater was highly preferable for the production of bioenergy. *C. butyricum* and *C. acetobutylicum* are employed in the tapioca wastewater fermentation under three different pH of 4.5, 5.5, and 6.5. They evaluated both ethanol and butanol. It produced a high concentration of butanol and ethanol of 1.810 g/L and 0.144 g/L at pH 5.5 using *C. butyricum*. This showed the potential of *C. butyricum* to ferment tapioca wastewater at pH 5.5 than *C. acetobutylicum* (Ouephanit et al. 2011).

Corn was the most edible food and processed for the separation of proteins, vitamins, and starch. Corn processing industries are used to release the wastewater as corn steep liquor and processed wastewater. Medium for *Clostridium* includes yeast extract, vitamins, minerals, trace metals, reducing agents, and carbon sources. Yeast extract was costly and cost nearly \$1000 ton^{-1}. Instead of yeast extract, we can use corn steep liquor, which costs around \$200 ton^{-1}, which also acts as an alternative nitrogen source (Maddipati et al. 2011; Zhang and Jia 2018). Many studies used corn steep liquor instead of yeast extract. It also yields seven times more butanol than yeast extract and also produced such as ethanol, n- butanol, n-hexanol, acetic acid, butyric acid, and hexanoic acid in the concentrations of 2.78 g/L, 0.70 g/L, 0.52 g/L, 4.06 g/L, 0.13 g/L, and 0.42 g/L by *C. carboxidivorans and C. ragsdalei* (Ramachandriya et al. 2013). Zhang and Jia (2018) statistically studied the fermentation process to produce Acetone-butanol-ethanol using the cheapest feedstocks of Corncob and Corn steep liquor by *C. beijerinckii* SE- 2. It found the yield of 20.29 g/L of Acetone-butanol-ethanol and 11.92 g/L of butanol using statistical experimental designs.

Maltose was used as the sweetener extracted from grains, fruits, and vegetables. Maltose producing food industries used to discharge wastewater as same as starch wastewater. Wang et al. (2020) used the maltose fermentative biohydrogen production bacterial strain *C. butyricum* NH- 02 from humus-rich soil. It reached the maximum production of 1.80 L/L medium and yielded 1.90 mol H_2/mol reducing sugar using maltose rich wastewater.

Palm oil mill effluent was the industrial effluent released from food waste and oil mills. Palm oil mill effluent was suitable for hydrogen production because of its rich carbohydrate content. Preliminary pretreatment was required for the palm oil mill effluent to break into simple sugar from its lignocellulosic material. Acid heat pretreated palm oil mill effluent was used as the feedstock to produce hydrogen using *C. butyricum* in batch mode. It produced 4304 ml H_2/L palm oil mill effluent under the optimum conditions of pH 7 and temperature 37 °C (Kamal et al. 2012). Singh et al. (2013) employed the immobilized *C. butyricum* EB6 to produce biohydrogen using Palm oil mill effluent as a substrate. Polyethylene Glycol was used for cell immobilization for its characteristics of low toxicity, high porous structure, low cost, and simple process. It produced 510 ml H_2/L- palm oil mill effluent h at a load of 60,000 mg COD/L- palm oil mill effluent. Polyethylene Glycol immobilized cells are reused by washing before and after adding palm oil mill effluent, and the optimum bead size was 3.0 mm.

2.5.5 *Municipal Sewage Sludge*

Municipal sewage sludge was the solid, semi-solid, and muddy state waste from household and industrial leftovers. It was composed of organic and inorganic compounds and also toxic compounds. *Clostridium* can produce bioenergy production from them.

Microbial fuel cells were the new innovative technology for energy production and could convert organic matter into electricity. It also plays a major role in the treatment of wastewater. Organic matter generates bioenergy while removing carbon and nitrogen. Hisham et al. (2013) analyzed three different waste streams for bioenergy using Microbial fuel cells such as activated sludge, palm oil mill effluent, and leachate from food waste. As per the growth pattern in different wastewater, Microbial fuel cells produce a high voltage with leachate (0.455 V) and Palm oil mill effluent (0.444 V) and reduced with activated sludge (0.396 V). Consistency was recorded in activated sludge and inconsistent for leachate and Palm oil mill effluent. Activated sludge containing microbial communities are *β-Comamonas sp, γ-Enterobacter sp, Bacillus cereus sp.* and *Clostridium sp.*, for leachate and Palm oil mill effluent, it was unidentified. This may be the cause of the inconsistency among leachate and Palm oil mill effluent to optimize.

Fallen leaves and sewage sludge was co-fermented for the production of biohydrogen. It was investigated at different ratios, and mono fermentation was also carried out. The results proposed the optimum ratio of 20:80 of sludge and leaves. They showed the synergistic effect to produce the hydrogen yield of 37.8 mL/g- Volatile $solids_{added}$ compared with mono- fermentation of sludge (10.3 mL/g- Volatile $solids_{added}$) or the leaves (30.5 mL/g- Volatile $solids_{added}$). Inoculum molecular analysis showed the dominance of *Clostridium, Bacillus,* and *Rummeli bacillus* in the fermenter (Yang et al. 2019).

2.5.6 Anaerobic Digester Effluents

Wastewater treatment includes physical, chemical, and biological treatments and uses lots of energy while running the digesters. Treated digester effluents will not provide any recovery or reuse by its dilution or deficiency of resources even though it has limited efficiency to produce bioenergy production.

C. kluyveri was the well-studied microbe for the production of n- butyric acid and n- caproic acid from acetic acid and ethanol. Syngas fermentation effluent was the low-value product of ethanol production from syngas. Syngas fermentation effluent was a mixture of acetic acid and ethanol. It can be elongated into medium-chain carboxylic acids. It was a beneficial coupling process of syngas fermentation and chain elongation. *C. kluyveri* was a significant microbe for this production process. It yields 4.1 g/L/day of n- caproic acid from the effluent. Dilution and pH control are important factors to maintain the progressive production of MCCA (Gildemyn et al. 2017).

Most of the industrial effluents are rich in sulfide and nitrite. It was toxic to the environment, but genus *clostridium* can tolerate the toxicity of aromatic compounds such as sulfide and nitrite (Biswas et al. 2011). This makes it advantageous to treat effluents and produce energy by using *Clostridium*. Ho and Lee (2011) studied the efficiency of cellobiose containing effluent rich in sulfide and nitrite in the hydrogen production using *Clostridium sp.* R1 under mesophilic conditions. It showed the optimum production of >3.0 mol H_2 mol^{-1} in the presence of <75 mg/L sulfide and <5 mg/L nitrite, which suggested the strain lost the potential by the high nitrite content as compared with sulphide.

The list of wastewaters involved in the production of bioenergy using industrial microorganism *Clostridium* was mentioned in Table 2.2.

Table 2.2 Different types of wastewater in bioenergy production using *Clostridium* species

Clostridium sp.	Type of wastewater	Bioenergy	Production rate	References
C. butyricum NH- 02	Maltose rich wastewater	Hydrogen	1.90 mol-H_2/ mol- RS (reducing sugar)	Wang et al. (2020)
C. sporogenes NCIM 2918	Rice straw hydrolysate + crude glycerol	Butanol and ethanol	0.52 g L^{-1} h^{-1}	Kaushal et al. (2019)
Clostridium sp. in wastewater	Fallen leaves and sewage sludge	Hydrogen	37.8 ml/g-VS_{added} (VS – volatile solids)	Yang et al. (2019)
Clostridium sp. in wastewater	Cassava wastewater with buffalo dung	Hydrogen	1787 ml H_2/L	Wadjeam et al. (2019)
C. pasteurianum	Crude glycerol	n- butanol	3.08 g.L^{-1}.h^{-1}	Krasňan et al. (2018)
C. beijerinckii SE- 2	Corn cob and corn steep liquor	Acetone-butanol-ethanol	20.29 g L^{-1}	Zhang and Jia (2018)
C. perfingens GYL	Crude glycerol	1,3-propanediol	2.0 g L^{-1}.h^{-1}	Guo et al. (2017)
C. beijerinckii DSM791	Crude glycerol	Hydrogen	5 mmol/h/g cell mass	Trchounian et al. (2017)
C. kluyveri	Syngas fermentation effluent	n- caproic acid	4.64 g L^{-1}.day^{-1}	Gildemyn et al. (2017)
C. acetobutylicum 824	Anaerobic sludge + goat ruminal liquid + glycerol	Acetic acid, Propionic acid, Ethanol, 1,3-propanediol, Butyric acid and n-caproic acid	Acetic acid(1.2 g/L),1,3-propanediol (1.6 g/L), Ethanol (1.6 g/L), Propionic acid (1.2 g/L), Butyric acid (0.9 g/L) and n-caproic acid (1.6 g/L)	Dams et al. (2016)
C.tyrobutyricum, Clostridium sp., C. acetobutylicum and C. acetobutylicum	Beverage processed wastewater	Hydrogen	11.39 L/L/d	Liu et al. (2016)
C. beijerinckii NRRL B- 466	Brewery liquid waste, starch industrial wastewater and apple pomace ultra-filtration sludge	Butanol	1.8 g/L (Brewery liquid waste), 4.68 g/L (starch industrial wastewater), 1.4 g/L (apple pomace ultra- filtration sludge)	Maiti et al. (2016)

(continued)

Table 2.2 (continued)

Clostridium sp.	Type of wastewater	Bioenergy	Production rate	References
Clostridium sp. in wastewater	Vinasse	Hydrogen	509.5 ml/d/L	Ferraz Júnior et al. (2015)
C.acetobutylicum DSM 792	High Sugar Content Beverages	Butanol	12 g/L/h	Raganati et al. (2015)
Clostridium sp. in wastewater	Vinasse	Hydrogen	0.44 mmol H_2 g^{-1} $COD_{influent}$ (COD – Chemical Oxygen Demand)	Lazaro et al. (2014)
C. pasteurianum	Crude glycerol	n- butanol, 1,3-propanediol and ethanol	0.02 g/g glycerol	Khanna et al. (2013)
C. beijerinckii and C. tyrobutyricum	Cassava and molasses	Butanol	0.96 g/L/h	Li et al. (2013b)
C. carboxidivorans and C. ragsdalei	Corn steep liquor	Ethanol, n- butanol, n-hexanol, Acetic acid, Butyric acid, and Hexanoic acid	2.78 g/L (ethanol), 0.70 g/L(n- butanol), 0.52 g/L(n-hexanol), 4.06 g/ L(acetic acid), 0.13 g/L(butyric acid) and 0.42 g/L (hexanoic acid)	Ramachandriya et al. (2013)
C. butyricum EB6	Palm oil mill effluent	Hydrogen	510 mL H_2/L- Palm oil mill effluent /h	Singh et al. (2013)
Clostridium sp. in wastewater	Activated sludge, Palm oil mill effluent, and leachate	Microbial Fuel Cells	Leachate (0.455 V), Palm oil mill effluent (0.444 V) and activated sludge (0.396 V)	Hisham et al. (2013)
C. butyricum	Palm oil mill effluent	Hydrogen	2.18 mol H_2/mol total carbohydrate	Kamal et al. (2012)
C. pasteurianum DSM 525	Thin stillage	Butanol	6.2–7.2 g/L	Ahn et al. (2011)
Mixed inoculum with clostridium sp.	Soft drink wastewater	Hydrogen	3.5 mol H_2/mol of sucrose	Peixoto et al. (2011)
C. acetobutylicum ATCC 824	Cassava wastewater	Hydrogen	2.41 mol H_2/mol glucose	Cappelletti et al. (2011)
C. butyricum and C. acetobutylicum	Tapioca wastewater	Ethanol and butanol	1.810 g/L (butanol) and 0.144 g/L(ethanol)	Ouephanit et al. (2011)
Clostridium sp. R1	Cellobiose containing effluent	Hydrogen	3.0 mol H_2 mol^{-1}	Ho and Lee (2011)

(continued)

Table 2.2 (continued)

Clostridium sp.	Type of wastewater	Bioenergy	Production rate	References
C. saccharobutylicum NRRL B- 643 and C. beijerinckii B- 593	Crude glycerol	1,3-propanediol	0.54 mol/mol	Gungormusler et al. (2010)

2.6 Conclusion

Many optimization and processes are needed for the increased production of diversified wastewaters. Process parameters such as reactor type, operating mode, add-on substrates, pH, temperature, and hydraulic retention time are important for each type of wastewater according to its composition. Sugar, starch, lipids, and glycerol in wastewater are highly feasible for bioenergy production, as per reports. pH and temperature are important parameters for the growth of microbes. In *Clostridium*, the growth was mostly optimum at anaerobic conditions around 5.5 to 7 pH and mesophilic conditions. The versatility of *Clostridium* can assimilate the maximum substrates present in the wastewater to yield bioenergy. This coupled process was advantageous for the issues of the energy crisis and waste management. Integrating wastewater treatment with bioenergy production systems will develop bioenergy through its continuous supply of wastewater. This integrated approach can achieve increased energy generation and reduced process and substrate cost.

Toxic compounds and contamination of other microbes present in the wastewater may inhibit bioenergy production. Screening of the composition of wastewater is important to prevent these situations. Even though *Clostridium* assimilates a wide range of substrates, it is lagging in some places. To avoid those consequences, metabolic engineering and process engineering were encouraged for future aspects.

References

Abd-Alla MH, Zohri ANA, El-Enany AWE, Ali SM (2017) Conversion of food processing wastes to biofuel using clostridia. Anaerobe 48:135–143. https://doi.org/10.1016/j.anaerobe.2017.08.011

Ahmed M, Hasan CK, Rahman H, Ali Hossain M, Uddin SA (2015) Prospects of using wastewater as a resource-nutrient recovery and energy generation. Am J Environ Sci 11:99–114. https://doi.org/10.3844/ajessp.2015.99.114

Ahn JH, Sang BI, Um Y (2011) Butanol production from thin stillage using *Clostridium pasteurianum*. Bioresour Technol 102:4934–4937. https://doi.org/10.1016/j.biortech.2011.01.046

Amenorfenyo DK, Huang X, Zhang Y, Zeng Q, Zhang N, Ren J, Huang Q (2019) Microalgae brewery wastewater treatment: potentials, benefits and the challenges. Int J Environ Res Public Health 16. https://doi.org/10.3390/ijerph16111910

Amin M (2019) Conversion of food processing waste to bioenergy: Bangladesh perspective. Trends Renew Energy 5:1–11. https://doi.org/10.17737/tre.2019.5.1.0080

Aziz SQ, Aziz HA, Bashir MJK, Mojiri A (2015) Assessment of various tropical municipal landfill leachate characteristics and treatment opportunities. Glob NEST J 17:439–450

Baleta J, Mikulčić H, Klemeš JJ, Urbaniec K, Duić N (2019) Integration of energy, water and environmental systems for a sustainable development. J Clean Prod 215:1424–1436. https://doi.org/10.1016/j.jclepro.2019.01.035

Bertolino SM, Rodrigues ICB, Guerra-Sá R, Aquino SF, Leão VA (2012) Implications of volatile fatty acid profile on the metabolic pathway during continuous sulfate reduction. J Environ Manag 103:15–23. https://doi.org/10.1016/j.jenvman.2012.02.022

Biswas R, Bagchi S, Bihariya P, Das A, Nandy T (2011) Stability and microbial community structure of a partial nitrifying fixed-film bioreactor in long run. Bioresour Technol 102:2487–2494. https://doi.org/10.1016/j.biortech.2010.11.039

Blázquez E, Guisasola A, Gabriel D, Baeza JA (2019) Application of bioelectrochemical systems for the treatment of wastewaters with sulfur species, microbial electrochemical technology. Elsevier B.V. https://doi.org/10.1016/b978-0-444-64052-9.00026-1

Cappelletti BM, Reginatto V, Amante ER, Antônio RV (2011) Fermentative production of hydrogen from cassava processing wastewater by *Clostridium acetobutylicum*. Renew Energy 36:3367–3372. https://doi.org/10.1016/j.renene.2011.05.015

Cheng J, Yu Y, Zhu M (2014) Enhanced biodegradation of sugarcane bagasse by *Clostridium thermocellum* with surfactant addition. Green Chem 16:2689–2695. https://doi.org/10.1039/c3gc42494d

Dams RI, Guilherme AA, Vale MS, Nunes VF, Leitão RC, Santaella ST (2016) Fermentation of residual glycerol by *Clostridium acetobutylicum* ATCC 824 in pure and mixed cultures. Environ Technol (UK) 37:2984–2992. https://doi.org/10.1080/09593330.2016.1173114

Datta A, Hossain A, Roy S (2019) An overview on biofuels and their advantages and disadvantages. Asian J Chem 31:1851–1858. https://doi.org/10.14233/ajchem.2019.22098

Ferraz Júnior ADN, Etchebehere C, Zaiat M (2015) Mesophilic hydrogen production in acidogenic packed-bed reactors (APBR) using raw sugarcane vinasse as substrate: influence of support materials. Anaerobe 34:94–105. https://doi.org/10.1016/j.anaerobe.2015.04.008

Gildemyn S, Molitor B, Usack JG, Nguyen M, Rabaey K, Angenent LT (2017) Upgrading syngas fermentation effluent using *Clostridium kluyveri* in a continuous fermentation. Biotechnol Biofuels 10:1–15. https://doi.org/10.1186/s13068-017-0764-6

Gungormusler M, Gonen C, Ozdemir G, Azbar N (2010) 1,3-Propanediol production potential of *Clostridium saccharobutylicum* NRRL B-643. New Biotechnol 27:782–788. https://doi.org/10.1016/j.nbt.2010.07.010

Guo Y, Dai L, Xin B, Tao F, Tang H, Shen Y, Xu P (2017) 1,3-Propanediol production by a newly isolated strain, *Clostridium perfringens* GYL. Bioresour Technol 233:406–412. https://doi.org/10.1016/j.biortech.2017.02.116

Hisham NSN, Zain SM, Jusoh S, Anuar N, Suja F, Ismail A, Basri NEA (2013) Microbial fuel cells using different types of wastewater for electricity generation and simultaneously removed pollutant. J Eng Sci Technol 8:317–326

Ho KL, Lee DJ (2011) Harvesting biohydrogen from cellobiose from sulfide or nitrite-containing wastewaters using Clostridium sp. R1. Bioresour Technol 102:8547–8549. https://doi.org/10.1016/j.biortech.2011.04.031

Iwuagwu JO, Ugwuanyi JO (2014) Treatment and valorization of palm oil mill effluent through production of food grade yeast biomass. J Waste Manag 2014:1–9. https://doi.org/10.1155/2014/439071

Justo AJ, Junfeng L, Lili S, Haiman W, Lorivi MR, Mohammed MOA, Xiangtong Z, Yujie F (2016) Integrated expanded granular sludge bed and sequential batch reactor treating beet sugar industrial wastewater and recovering bioenergy. Environ Sci Pollut Res 23:21032–21040. https://doi.org/10.1007/s11356-016-7307-8

Kamal SA, Jahim JM, Anuar N, Hassan O, Daud WRW, Mansor MF, Rashid SS (2012) Pretreatment effect of Palm Oil Mill Effluent (POME) during hydrogen production by a local isolate *Clostridium butyricum*. Int J Adv Sci Eng Inf Technol 2:325. https://doi.org/10.18517/ijaseit.2.4.214

Kaushal M, Ahlawat S, Makut BB, Goswami G, Das D (2019) Dual substrate fermentation strategy utilizing rice straw hydrolysate and crude glycerol for liquid biofuel production by *Clostridium sporogenes* NCIM 2918. Biomass Bioenergy 127:105257. https://doi.org/10.1016/j.biombioe.2019.105257

Khamaiseh EI, Abdul Hamid A, Abdeshahian P, Wan Yusoff WM, Kalil MS (2014) Enhanced butanol production by *Clostridium acetobutylicum* NCIMB 13357 grown on date fruit as carbon source in P2 medium. Sci World J 2014. https://doi.org/10.1155/2014/395754

Khanna S, Goyal A, Moholkar VS (2013) Production of n-butanol from biodiesel derived crude glycerol using *Clostridium pasteurianum* immobilized on Amberlite. Fuel 112:557–561. https://doi.org/10.1016/j.fuel.2011.10.071

Khedkar MA, Nimbalkar PR, Gaikwad SG, Chavan PV, Bankar SB (2017) Sustainable biobutanol production from pineapple waste by using *Clostridium acetobutylicum* B 527: drying kinetics study. Bioresour Technol 225:359–366. https://doi.org/10.1016/j.biortech.2016.11.058

Kim Y, Mosier NS, Hendrickson R, Ezeji T, Blaschek H, Dien B, Cotta M, Dale B, Ladisch MR (2008) Composition of corn dry-grind ethanol byproducts: DDGS, wet cake, and thin stillage. Bioresour Technol 99:5165–5176. https://doi.org/10.1016/j.biortech.2007.09.028

Kim Y, Kim D, Kim T, Shin MK, Kim YJ, Yoon JJ, Chang IS (2013) Use of red algae, Ceylon moss (*Gelidium amansii*), hydrolyzatefor clostridial fermentation. Biomass Bioenergy 56:38–42. https://doi.org/10.1016/j.biombioe.2013.04.024

Kim YK, Park SE, Lee H, Yun JY (2014) Enhancement of bioethanol production in syngas fermentation with *Clostridium ljungdahlii* using nanoparticles. Bioresour Technol 159:446–450. https://doi.org/10.1016/j.biortech.2014.03.046

Krasňan V, Plž M, Marr AC, Markošová K, Rosenberg M, Rebroš M (2018) Intensified crude glycerol conversion to butanol by immobilized *Clostridium pasteurianum*. Biochem Eng J 134:114–119. https://doi.org/10.1016/j.bej.2018.03.005

Kushwaha JP (2015) A review on sugar industry wastewater: sources, treatment technologies, and reuse. Desalin Water Treat 53:309–318. https://doi.org/10.1080/19443994.2013.838526

Lazaro CZ, Perna V, Etchebehere C, Varesche MBA (2014) Sugarcane vinasse as substrate for fermentative hydrogen production: the effects of temperature and substrate concentration. Int J Hydrog Energy 39:6407–6418. https://doi.org/10.1016/j.ijhydene.2014.02.058

Lépiz-Aguilar L, Rodríguez-Rodríguez CE, Arias ML, Lutz G, Ulate W (2011) Butanol production by *Clostridium beijerinckii* BA101 using cassava flour as fermentation substrate: enzymatic versus chemical pretreatments. World J Microbiol Biotechnol 27:1933–1939. https://doi.org/10.1007/s11274-010-0630-1

Li HG, Luo W, Gu QY, Wang Q, Hu WJ, Yu XB (2013a) Acetone, butanol, and ethanol production from cane molasses using *Clostridium beijerinckii* mutant obtained by combined low-energy ion beam implantation and N-methyl-N-nitro-N-nitrosoguanidine induction. Bioresour Technol 137:254–260. https://doi.org/10.1016/j.biortech.2013.03.084

Li L, Ai H, Zhang S, Li S, Liang Z, Wu ZQ, Yang ST, Wang JF (2013b) Enhanced butanol production by coculture of *Clostridium beijerinckii* and *Clostridium tyrobutyricum*. Bioresour Technol 143:397–404. https://doi.org/10.1016/j.biortech.2013.06.023

Liberato V, Benevenuti C, Coelho F, Botelho A, Amaral P, Pereira N, Ferreira T (2019) Chemicals production in a biorefinery context. Catalysts 1:1–37

Linggang S, Phang LY, Wasoh H, Abd-Aziz S (2013) Acetone-butanol-ethanol production by *Clostridium acetobutylicum* ATCC 824 using sago pith residues hydrolysate. Bioenergy Res 6:321–328. https://doi.org/10.1007/s12155-012-9260-9

Liu CM, Zheng JL, Wu SY, Chu CY (2016) Fermentative hydrogen production potential from washing wastewater of beverage production process. Int J Hydrog Energy 41:4466–4473. https://doi.org/10.1016/j.ijhydene.2015.08.079

Loy DD, Lundy EL (2018) Nutritional properties and feeding value of corn and its coproducts. In: Corn: chemistry and technology, 3rd edn. Elsevier Inc. https://doi.org/10.1016/B978-0-12-811971-6.00023-1

Maddipati P, Atiyeh HK, Bellmer DD, Huhnke RL (2011) Ethanol production from syngas by Clostridium strain P11 using corn steep liquor as a nutrient replacement to yeast extract. Bioresour Technol 102:6494–6501. https://doi.org/10.1016/j.biortech.2011.03.047

Maiti S, Sarma SJ, Brar SK, Le Bihan Y, Drogui P, Buelna G, Verma M (2016) Agro-industrial wastes as feedstock for sustainable bio-production of butanol by *Clostridium beijerinckii*. Food Bioprod Process 98:217–226. https://doi.org/10.1016/j.fbp.2016.01.002

Mansouri SS, Udugama IA, Cignitti S, Mitic A, Flores-Alsina X, Gernaey KV (2017) Resource recovery from bio-based production processes: a future necessity? Curr Opin Chem Eng 18:1–9. https://doi.org/10.1016/j.coche.2017.06.002

Mazuchi M (2018) Life cycle assessment of sugarcane biorefinery: energy use, environmental assessment, and other sustainability indicators. In: Advances in sugarcane biorefinery: technologies, commercialization, policy issues and paradigm shift for bioethanol and by-products. Elsevier Inc. https://doi.org/10.1016/B978-0-12-804534-3.00011-2

Naik SN, Goud VV, Rout PK, Dalai AK (2010) Production of first and second generation biofuels: a comprehensive review. Renew Sust Energ Rev 14:578–597. https://doi.org/10.1016/j.rser.2009.10.003

Ouephanit C, Virunanon C, Burapatana V, Chulalaksananukul W (2011) Butanol and ethanol production from tapioca starch wastewater by Clostridium spp. Water Sci Technol 64:1774–1780. https://doi.org/10.2166/wst.2011.675

Peixoto G, Saavedra NK, Varesche MBA, Zaiat M (2011) Hydrogen production from soft-drink wastewater in an upflow anaerobic packed-bed reactor. Int J Hydrog Energy 36:8953–8966. https://doi.org/10.1016/j.ijhydene.2011.05.014

Perimenis A, Walimwipi H, Zinoviev S, Müller-Langer F, Miertus S (2011) Development of a decision support tool for the assessment of biofuels. Energy Policy 39:1782–1793. https://doi.org/10.1016/j.enpol.2011.01.011

Prakash J, Sharma R, Ray S, Koul S, Kalia VC (2018) Wastewater: a potential bioenergy resource. Indian J Microbiol 58:127–137. https://doi.org/10.1007/s12088-017-0703-z

Puyol D, Batstone DJ, Hülsen T, Astals S, Peces M, Krömer JO (2017) Resource recovery from wastewater by biological technologies: opportunities, challenges, and prospects. Front Microbiol 7:1–23. https://doi.org/10.3389/fmicb.2016.02106

Qu X, Brame J, Li Q, Alvarez PJJ (2013) Nanotechnology for a safe and sustainable water supply: enabling integrated water treatment and reuse. Acc Chem Res 46:834–843. https://doi.org/10.1021/ar300029v

Qureshi N, Singh V, Liu S, Ezeji TC, Saha BC, Cotta MA (2014) Process integration for simultaneous saccharification, fermentation, and recovery (SSFR): production of butanol from corn stover using *Clostridium beijerinckii* P260. Bioresour Technol 154:222–228. https://doi.org/10.1016/j.biortech.2013.11.080

Raganati F, Procentese A, Montagnaro F, Olivieri G, Marzocchella A (2015) Butanol production from leftover beverages and sport drinks. Bioenergy Res 8:369–379. https://doi.org/10.1007/s12155-014-9531-8

Ramachandriya KD, Kundiyana DK, Wilkins MR, Terrill JB, Atiyeh HK, Huhnke RL (2013) Carbon dioxide conversion to fuels and chemicals using a hybrid green process. Appl Energy 112:289–299. https://doi.org/10.1016/j.apenergy.2013.06.017

Setyawaty R, Katayama-Hirayama K, Kaneko H, Hirayama K (2011) Current tapioca starch wastewater management in Indonesia. World Appl Sci J 14:658–665

Singh L, Wahid ZA, Siddiqui MF, Ahmad A, Rahim MHA, Sakinah M (2013) Biohydrogen production from palm oil mill effluent using immobilized *Clostridium butyricum* EB6 in polyethylene glycol. Process Biochem 48:294–298. https://doi.org/10.1016/j.procbio.2012.12.007

Smoliński A, Karwot J, Bondaruk J, Bak A (2019) The bioconversion of sewage sludge to bio-fuel: the environmental and economic benefits. Materials (Basel) 12:1–9. https://doi.org/10.3390/ma12152417

Sun Z, Liu S (2012) Production of n-butanol from concentrated sugar maple hemicellulosic hydrolysate by *Clostridia acetobutylicum* ATCC824. Biomass Bioenergy 39:39–47. https://doi.org/10.1016/j.biombioe.2010.07.026

Sutton PM, Rittmann BE, Schraa OJ, Banaszak JE, Togna AP (2011) Wastewater as a resource: a unique approach to achieving energy sustainability. Water Sci Technol 63:2004–2009. https://doi.org/10.2166/wst.2011.462

Trchounian K, Müller N, Schink B, Trchounian A (2017) Glycerol and mixture of carbon sources conversion to hydrogen by *Clostridium beijerinckii* DSM791 and effects of various heavy metals on hydrogenase activity. Int J Hydrog Energy 42:7875–7882. https://doi.org/10.1016/j.ijhydene.2017.01.011

Wadjeam P, Reungsang A, Imai T, Plangklang P (2019) Co-digestion of cassava starch wastewater with buffalo dung for bio-hydrogen production. Int J Hydrog Energy 44:14694–14706. https://doi.org/10.1016/j.ijhydene.2019.04.138

Wang P, Chen YM, Wang Y, Lee YY, Zong W, Taylor S, McDonald T, Wang Y (2019) Towards comprehensive lignocellulosic biomass utilization for bioenergy production: efficient biobutanol production from acetic acid pretreated switchgrass with *Clostridium saccharoperbutylacetonicum* N1-4. Appl Energy 236:551–559. https://doi.org/10.1016/j.apenergy.2018.12.011

Wang Y, He L, Zhang Z, Zhao X, Qi N, Han T (2020) Efficiency enhancement of H2 production by a newly isolated maltose-preferring fermentative bio-hydrogen producer of *Clostridium butyricum* NH-02. J Energy Storage 30:101426. https://doi.org/10.1016/j.est.2020.101426

Yang F, Hanna MA, Sun R (2012) Value added use crude glycerol – a byproduct of biodiesel production. Biotechnol Biofuels 5(13). https://doi.org/10.1186/1754-6834-5-13

Yang G, Hu Y, Wang J (2019) Biohydrogen production from co-fermentation of fallen leaves and sewage sludge. Bioresour Technol 285:121342. https://doi.org/10.1016/j.biortech.2019.121342

Zhang J, Jia B (2018) Enhanced butanol production using *Clostridium beijerinckii* SE-2 from the waste of corn processing. Biomass Bioenergy 115:260–266. https://doi.org/10.1016/j.biombioe.2018.05.012

Chapter 3
Management of Phosphate in Domestic Wastewater Treatment Plants

Sumathi Malairajan and Vasudevan Namasivayam

Abstract Phosphorus, being an essential nutrient, accounts for about 0.8–1.1% of the total body weight and plays a major role in cellular metabolic activities. Phosphorus from phosphate rock is widely employed in the food and feed, detergent, fertilizer, and other chemical industries. The elimination of phosphorus from the environment to aquatic ecosystems occurs by point and non-point sources. The loading of phosphate from sewage causes phosphate enrichment in the aquatic ecosystem leading to eutrophication. In the aquatic ecosystem, many directives and regulations have been put forth worldwide targeting phosphate removal from wastewater to reduce eutrophication. Phosphate removal could be achieved by physical, chemical, and biological methods. The physical and chemical methods require high energy and chemical inputs to produce a huge volume of sludge, making it difficult for disposal. Biological phosphate removal is advantageous over the other phosphate removal methods due to its simplicity and economic viability and therefore adopted widely in wastewater treatment plants. Enhanced biological phosphate removal treatment plants remove phosphate by enriching phosphate accumulating organisms that store phosphate intra-cellularly by alternating between aerobic and anaerobic cycles. Phosphate, being a non-renewable resource, the present scenario at Municipal Wastewater treatment plants involves phosphate removal targeting its recovery in a usable form for application in various sectors.

Keywords Phosphorus · Phosphate rock · Eutrophication · Sewage · Phosphate accumulating bacteria · Enhanced biological phosphate removal

---

S. Malairajan · V. Namasivayam (✉)
Centre for Environmental Studies, Anna University, Chennai, Tamil Nadu, India

K. M. Gothandam et al. (eds.), *Environmental Biotechnology Volume 4*, Environmental Chemistry for a Sustainable World 68,
https://doi.org/10.1007/978-3-030-77795-1_3

3.1 Introduction

Phosphorus is one of the natural elements present in the rocks, soils, and organic material. Phosphate exists as apatite in rocks and reaches the soil by weathering, being available for plant's uptake, and from plants to animals. Phosphate from the soil and decayed matter enters the aquatic system through runoff and settles down in the sediment. Due to the uplifting of rock over time, phosphate exists in the rock again, and the cycle continues, termed as 'Phosphorus cycle' (Fig. 3.1). The phosphate rock production was 181 million metric tons (mmt) during 2010, and 220–256 mmt in 2014 and the phosphorus extracted from rock phosphate ranged from 36 mmt in 2010 to 42 mmt in 2014 (Jasinski 2015).

Phosphorus is a required nutrient for all living beings. The different forms of organic phosphorus in biological sources are orthophosphates (Phosphate (PO_4^{3-}), hydrogen phosphate (HPO_4^{2-}), Dihydrogen phosphate ($H_2PO_4^-$)), organic phosphate (nucleic acids, phospholipids, inositol phosphates, phosphoamides, phosphoproteins, sugar phosphates, amino phosphoric acids, organic condensed phosphates) and inorganic phosphates (polyphosphates) (Majed et al. 2012). They are essential for cellular functions (Kornberg et al. 1999) and adaptation to environmental stress (Alcantara et al. 2014; Rao et al. 2009). Apart from its biological role, phosphate acts as a major ingredient in the food and feed, detergent, fertilizer and other chemical industries (Scholz et al. 2013).

The domestic and industrial wastewater, agricultural runoff, and stormwater drain into the aquatic system, causing nutrient enrichment resulting in eutrophication. Most of the lakes have become deteriorated worldwide (Khan and Ansari 2005) due to eutrophication, which creates hypoxia or anoxia leading to the limitation of

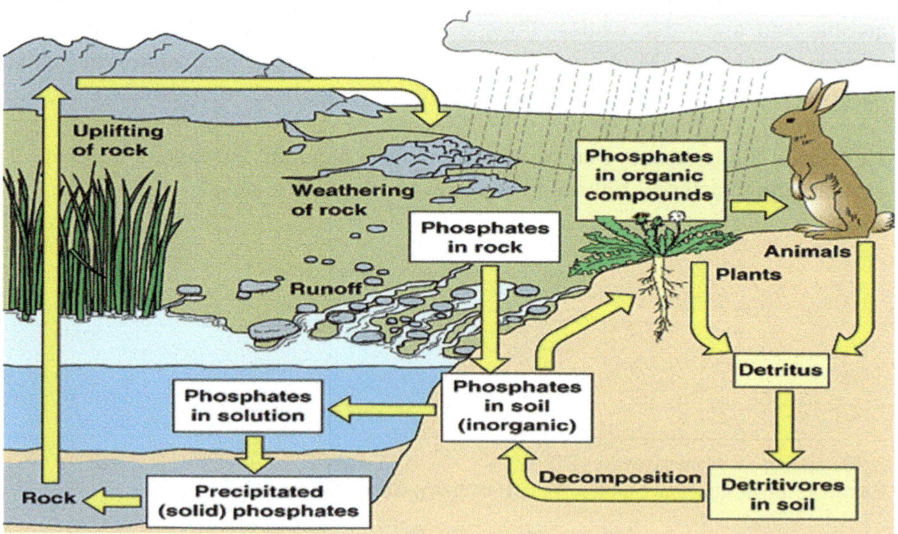

Fig. 3.1 Phosphorus Cycle: phosphorus cycling in the atmosphere (Henry 2013)

primary productivity in the aquatic ecosystem (Correll 1998), causing loss of spawning habitats, excessive growth of planktivorous and carnivorous fish populations resulting in the decrease of economically important fish populations (Roy et al. 2013) and death of aquatic lives (Leng 2009). Eutrophication can be controlled by restricting any limiting nutrient, either nitrate or phosphate, into the aquatic ecosystem. The maximum permissible concentration of phosphate for discharge into inland water bodies recommended by the World Health Organization (WHO) is 1 mg/L, United States Environmental Protection Agency (USEPA) is 2 mg/L, and Central Pollution Control Board (CPCB) is 5 mg/L. To maintain the stringent discharge limits, treatment of wastewater becomes mandatory.

3.2 Global Phosphate Rock Resources and Consumption

The search for global phosphate rock deposits began in the twentieth century due to the 'Peak phosphorus' event in 2008. Phosphate rock exists in sedimentary (70–80%) and igneous rock deposits (10–20%). Northern Africa, China, The Middle East, and United States have the largest sedimentary rock deposits, whereas, in Brazil, Canada, Finland, Russia, and South Africa, a significant amount of igneous rock occurs. According to United States Geological Survey (USGS) data, phosphate rock resources were more than 300 billion tonnes during 2017 (Jasinski 2017) (Table 3.1). World phosphate rock production is expected to increase by 2% per year through 2020. Major producers of global phosphate rock (75%) are China (81 mmt), United States of America (USA) (28.1 mmt), Morocco (28 mmt), and Russia (11.2 mmt) (International Fertilizer Industry Association (IFA 2011)). World consumption of phosphorus pentoxide (P_2O_5) in all sectors is expected to increase from 44.5 million tonnes in 2016 to 48.9 million tonnes in 2020 (Jasinski 2017). The majority of the mined phosphate rock (90%) is used for food production as fertilizer (approx. 82%) and feed additives (approx. 7%), and only the residual fraction (9–10%) is used for non-food-related industrial purposes (Schroder et al. 2010). An increase in phosphate consumption from 40 to 42 mmt was recorded during 2010–2014 (Fact book 2015). China and India are the largest consumers of phosphorus fertilizers, demanding 34% and 19% of global consumption respectively (Fig. 3.2).

Table 3.1 Phosphate rock reserves, resources and mine production

Country	Mine production (Million Metric Tons (mmt)) 2015	2016[a]	USGS 2017[b] reserves (mmt)	IFDC reserves 2010 (mmt) Product	Resources[c]
United States	27,400	27,800	1100,000	1800	49,000
Algeria	1400	1500	2,200,000	–	–
Australia	2500	2500	1100,000	82	3500
Brazil	6100	6500	320,000	400	2800
China[d]	120,000	138,000	3,100,000	3700	16,800
Egypt	5500	5500	1200,000	51	3400
India	1500	1500	65,000	–	–
Israel	3540	3500	130,000	220	1600
Jordan	8340	8300	1200,000	900	1800
Kazakhstan	1840	1800	260,000	–	–
Mexico	1680	1700	30,000	–	–
Morocco & Western Sahara	29,000	30,000	50,000,000	51,000	170,0000
Peru	3880	4000	820,000	–	–
Russia	11,600	11,600	1,300,000	500	4300
Saudi Arabia	4000	4000	680,000	–	–
Senegal	1240	1250	50,000	50	250
South Africa	1980	1700	1500,000	230	7700
Syria	750	–	1800,000	250	2000
Togo	1100	900	30,000	34	1000
Tunisia	2800	3500	100,000	85	1200
Vietnam	2500	2800	30,000	–	–
Other countries	2470	2410	810,000	600[e]	22,000[f]
World total (rounded)	241,000	261,000	68,000,000	60,000	290,000

USGS United States Geological Survey
IFDC – International Fertilizer Development Centre
[a]Estimated
[b]Reserves – usable or marketable product
[c]Resources-unprocessed phosphate rocks of varying grades or concentrate
[d]Production data for large mines only, as reported by the National Bureau of Statistics of China
[e]Includes data from Algeria, Finland, Peru and Saudi Arabia
[f]Includes data from Algeria, Angola, Finland, Kazakhstan, Peru and Saudi Arabia

3.3 Sources of Phosphate Pollution

3.3.1 *Point Source*

The term point source means any discernible, confined and discrete conveyance, including but not limited to any pipe, ditch, channel, tunnel, conduit, well, discrete fissure, container, rolling stock, concentrated animal feeding operation, or vessel or other floating craft, from which pollutants are or may be discharged. This term does

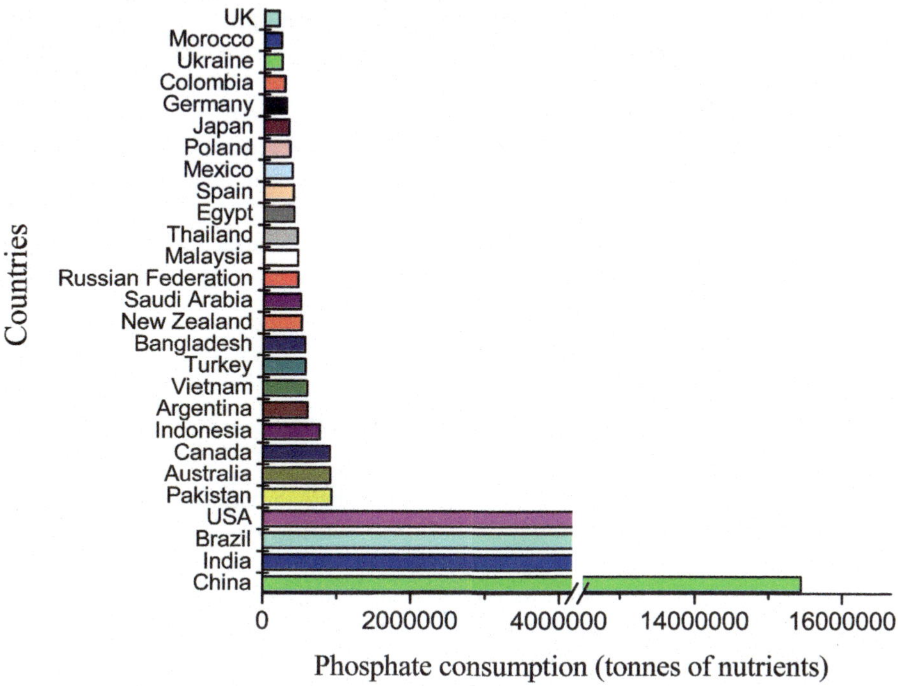

Fig. 3.2 Global Phosphate Consumption (Food and Agricultural Organisation (FAO)-AQUASTAT 2014)

not include agricultural stormwater discharges and return-flows from irrigated agriculture (Section 502(14) of the Clean Water Act). The sources of point chemical inputs recognized by United States statutes include wastewater effluent (municipal and industrial), runoff and leachate from waste disposal sites, runoff and infiltration from animal feedlots, runoff from mines, oil fields and unsewered industrial sites, storm sewer outfalls from cities with populations more than 1,00,000 overflows of combined storm and sanitary sewers, and runoff from construction sites with an area more than 2 hectare (Smith et al. 1999). The major sources of phosphate contamination through point sources are effluent discharges from sewage, fertilizer, and other chemical industrial facilities.

3.3.2 Non point Source

Non-point source pollution results from land runoff, precipitation, atmospheric deposition, drainage, seepage or hydrologic modification. The term 'non-point source' refers to any source of water pollution that does not meet the legal definition of a point source in section 502(14) of the Clean Water Act. Agricultural and urban activities such as excessive fertilization and manure production, volatilization of phosphorus into the atmosphere and redeposition are the primary sources of non-point nutrient pollution (Carpenter et al. 1998). The non-point sources of chemical inputs recognized by United States (US) statutes include runoff from agriculture (including return flows from irrigated agriculture), runoff from pastures and rangelands, urban runoff from unsewered areas and sewered areas with populations more than 1,00,000 septic tank leachate and runoff from failed septic systems, runoff from construction sites with an area more than 2 hectare, runoff from abandoned mines, atmospheric deposition over a water surface, activities on land that generate contaminants, such as logging, wetland conversion, construction and development of land or waterways (Smith et al. 1999).

3.4 Phosphate Leaching in the Atmosphere

Earth's crust contains about 1200 mg/kg of phosphate, a majority present in rocks and the remaining in soil ranging from 200 to 800 mg/kg (Tiessen et al. 2008). Natural weathering releases phosphate in the range of 0.05–1.0 kg/ha/year by direct means and 5.0 kg/ha/year by less direct means (Newman 1995). An estimate of 20 Metric ton (Mt) of phosphate/year (Ruttenberg 2003) and 13 Mt. of phosphate/year (Carpenter and Bennett 2011) is weathered from rocks to the soil. In soil, phosphorus occurs both in organic and inorganic forms. The phosphorus source in the soil is apatite minerals and weathered by reaction with dissolved carbon dioxide as follows (Filippelli and Rosso 2002).

$$Ca_5\left(PO_4\right)_3 OH + 4CO_2 + 3H_2O \rightarrow 5Ca^{2+} + 3HPO_4^{2-} + 4HCO_3^{4-}$$

The limited availability of phosphorus in soil necessitates its application in the form of fertilizer. The application of phosphorus to cropland is accounted for both as harvest and as losses by erosion and runoff (Zhang et al. 2003). The amount of nutrient loss from agricultural land depends on soil type, fertilizer added, management practices, crop type, pasture and grazing operations, animal feedlots, and manure storage facilities (Pitois et al. 2001). A net loss of phosphorus from the world's cropland is estimated at 10.5 mmt P/year, equivalent to one-half of phosphorus extracted each year (Liu et al. 2008). Atmospheric deposition of phosphorus on Earth's surface was 460 × 10^{10} g/year, of which 320 × 10^{10} g/year falls on the continent and 140 × 10^{10} g/year falls on the Ocean (Graham and Duce 1979). The total annual transfer of phosphorus from the atmosphere is estimated to be 3.7 Teragram (Tg)/year, of which about 85% gets deposited on land and freshwater (Tipping et al. 2014). Another estimate by atmospheric transport model simulation showed 3.5 Tg/year to be the total global emissions of atmospheric phosphorus of which 2.7 Tg/year deposits on land and 0.8 Tg/year deposits on the Ocean, the predominant source being combustion-related emissions (1.8 Tg/year) (Wang et al. 2010).

3.5 Phosphate in Sewage

Food and Agricultural Organisation (FAO)-AQUASTAT (database) data predicts the global freshwater withdrawal of 3928 km^3 per year, of which 44% (1716 km^3/year) is used for agriculture. Of the used water generated per annum, 56% (2212 km^3/year) of it emanates from domestic wastewater, industrial effluent and agricultural runoff, which drains back into the environment (Mateo et al. 2015). Wastewater is roughly composed of 99% water and 1% suspended, colloidal and dissolved solids (United Nations World Water Development Report (UNWWDR) 2015). The wastewater from residences and institutions, carrying bodily wastes (primarily feces and urine), washing water, food preparation wastes, laundry wastes, and other waste products of normal living, are classified as domestic or sanitary sewage. Sewage derives nutrients, including phosphorus from several sources (Table 3.2).

Domestic wastewater has higher nutrient input (75%) than the industrial sources (17%) due to the population density and anthropogenic activities in the municipal wastewater treatment plant. The phosphorus loading in municipal wastewaters is mainly due to human excreta (30–50%), detergents (50–70%) and industrial effluents (2–20%) (Rybicki 1998). The domestic contribution of phosphate to sewage from excretory waste is estimated to be 0.9 g/person/day in urine and 0.4 g/person/day in faeces (Jonsson et al. 2005). Apart from the excretory waste, the detergents and dishwashers contribute 1.89 g of total phosphate/person/day (Gilmour et al. 2008) to the sewage. US sewered population of 137 million people discharged 376 million pounds of phosphorus that corresponds to 2.74 pounds/capita/year (National Emission Reduction Plan (NERP) 1970). From the total amount of phosphate discharged (3.5 pounds/person/year), 1.2 pounds comes from the excretory source and 2.3 pounds from other sources (Barth 1971).

The per capita phosphorus loading to wastewater was estimated to be 2.2 g/day in USA (Hetling and Carcich 1973), 2.5–4.1 g/day in England (Devey and Harkness 1973), 3.8 g/day in Sweden (Olsson et al. 1968), 1.8 g/day in Northern Ireland (Alexander and Stevens 1976) and 1.96 ± 0.11 g/d in Tehran (Mesdaghinia et al. 2015). Phosphorus loading in natural waters due to detergent was 1.6 pounds/capita/year during 1950, 1.9 pounds/capita/year during 1955, and 2.1 pounds during 1958 (Engelbrecht and Morgan 1959; Anon 1967). The annual consumption of

Table 3.2 Domestic load of phosphorus in UK Sewage Treatment Works (Comber et al. 2013)

Source of phosphorus	Phosphorus loading (%)	Phosphorus loading (tonnes/day)
Natural diet	40	17,600
Food additives	29	12,760
Domestic laundry	14	6160
Dishwashing detergents	9	3960
P dosing to reduce lead in tap water	6	2640
Food waste disposal	1	440
Personal care products	1	440
%Total	100	44,000

phosphate-containing laundry detergents in India is 2.88 million tones, and the total outflow of phosphate is estimated to be 146 thousand tonnes per year (Kundu et al. 2015). It is estimated that 0.6 million tonnes of nitrogen and 0.1 million tonnes of phosphorus reach the country's East and West coastal waters through the channels (Jayatilake 2014).

Sewage with a nitrogen/phosphorus weight ratio of about 2–3 adds phosphorus at a proportionately much higher rate than nitrogen in the aquatic ecosystem (Dean 1971). The total phosphorus loading in Lake Washington during 1957 was between 100,000 and 108,000 kg, of which 42% (42,000 kg) was from sewage effluent, 23,200 m^3/day during 1959, and 150,000 kg by 1962 (USEPA 1978). Higher phosphorus concentration in the aquatic system causes undesirable changes such as excessive production of aquatic plants, depletion of dissolved oxygen, fish kills, and appearance of nuisance algal species collectively termed as 'eutrophication' (Eutrophication Research Newsletter (ERN) 1970).

The direction under the Indian Water (Prevention and Control of Pollution) Act, 1974 clearly defines the treatment and utilization of sewage for controlling pollution in the receiving water bodies. Even after secondary treatment, municipal wastewater stimulates algal growth when added to natural waters (Shapiro and Ribeiro 1965). Some of the examples for lake eutrophication due to sewage inputs include Lake Minnetonka in Minnesota (Megard 1970), lakes at Madison, Wisconsin, Zurichsee in Switzerland, Lake Washington at Seattle, Lake Erie, and the Potomac Estuary (Aalto et al. 1970), Cam Brook (Howell 2010), River Kennet, United Kingdom (UK) (Neal et al. 2000; Neal et al. 2010) and many more. Globally 80% of the wastewater generated is released into the environment without adequate treatment (World Water Assessment Programme (WWAP) 2012). In Europe, most of the generated municipal and industrial wastewater gets treated (71%), whereas in Latin American countries, only 20% of the generated wastewater is treated, and in the Middle East and North Africa, 51% of the generated wastewater is treated (United Nations Educational, Scientific and Cultural Organisation (UNESCO) 2017). In Arab states, only 69% of generated wastewater was collected, of which only 46% of wastewater underwent secondary treatment, and from which only 23% of secondary treated wastewater underwent tertiary treatment (United Nations Economic and Social Commission for West Asia (UNESCWA) 2016). In Asia and the Pacific region, 80–90% of the wastewater generated is being released untreated (United Nations Economic and Social Commission for Asia and Pacific (UNESCAP) 2015). In India, an assessment during 2010–2011 by Central Pollution Control Board (CPCB 2016) revealed that out of 38,254 million litres per day of sewage generated by class I cities and class II towns; only 11,787 million litres per day has been treated. A reassessment in 2015 revealed that the treatment facility is existing only for 23,277 million litres per day in 816 STP's for a total sewage generation of 62,000 million litres per day all over the country (CPCB 2016).

To control eutrophication in inland water bodies and the Great Lakes Basin, USEPA imposed a ban for the use of phosphate in detergents from 1971 onwards (USEPA 1977) later on with an emphasis on the establishment of nutrient removal technologies at Wastewater treatment plants (USEPA 2009). Similarly, European

Urban Wastewater Treatment Directive necessitates the removal of nutrients from large Sewage Treatment Plants (population more than 10,000) (Department of Environment, Food and Rural Affairs (DEFRA) 2002) and the European Water Framework Directive 2000/60/European Commission (EC) directs control of nutrient inputs to the river to maintain the good ecological status of freshwater (Arnscheidt et al. 2007). The capital, operation and maintenance cost for nutrient removal is based on several factors, including the type of effluent and treatment technologies. Most treatment schemes for phosphorus removal generated effluent concentrations of 1 mg/L or less (USEPA 2015). The upgraded system of nutrient removal technologies at treatment plants decreased eutrophication potential impact by 40% with additional environmental benefits such as wastewater reuse, fertilizer production as biosolids, biogas, and electricity generation reduction in global warming, thereby saving energy demand and generating revenue (USEPA 2017).

3.6 Treatment Technologies for Phosphate Removal

The different forms of phosphorus found in Wastewater treatment plants are particulate phosphorus (5–13 mg/L), soluble phosphorus (3–7 mg/L), total phosphorus (10–16 mg/L) and total phosphate (3–7 mg/L), of which only particulate phosphate is removed completely in the treatment process of the available phosphate forms (Duenas et al. 2003). Hence the additional treatment for nutrient removal is essential to eliminate phosphate from the effluent to meet the requirement of several directives. The treatment option for phosphate removal includes (1) chemical (precipitation, adsorption), (2) physical (filtration for particulate phosphorus, membrane technologies) and (3) biological, which includes assimilation by plants (rooted or floating plants), algae (attached or planktonic), cyanobacteria, and microbe (enhanced biological phosphorus removal), biological nutrient removal methods (Strom 2006).

3.6.1 Physical Treatment

The physical treatment methods for phosphate removal include filtration, magnetite separation, and membrane technologies. The physical separation process, such as clarification and filtration is carried out after chemical precipitation to remove the precipitate (Jiang et al. 2004).

3.6.1.1 Filtration

The filtration process includes sand filters, trickling filters, moving bed filters, activated granular carbon filters, and mechanical filters. The mechanism is by adsorptive phosphate removal. The dosing of chemicals such as alum and ferric salts can be added to the filtration unit if necessary. Filtration is carried out at the tertiary level after secondary treatment. It is also termed as the effluent polishing treatment where suspended solids are also removed and phosphate before the effluent is let into the water bodies (Spellman 2008).

3.6.1.2 Magnetic Separation

De Latour (1975) found a technology to remove phosphorus using a magnetic field after alum or iron salt addition. Magnetic carriers such as magnetite (Fe_3O_4) are added to the insoluble phosphorous to strengthen the magnetic properties. A lower phosphate concentration of 0.1–0.5 g P/m^3 in the effluent, high elimination performance, compact process and low power input are the advantages of this process.

3.6.1.3 Membrane Technologies

Membrane technologies gained much importance in the last decade due to minimal space requirement and high process reliability such that the treated effluent characteristics meet the stringent discharge limits (Smith et al. 2015). A hybrid MicroFiltration-Forward Osmosis Membrane Bioreactor was operated in parallel for 98 days to remove and recover phosphate from municipal wastewater. The nutrient-rich reject from the Forward Osmosis membrane acts as permeate for Microfiltration. Phosphate removal of 97.9% and recovery of 71.7% as calcium phosphate was achieved by adding seawater brine (Qiu et al. 2015). An integrated anoxic/aerobic membrane bioreactor containing two hollow fiber microfiltration membrane module made of polyvinylidenefluoride was employed for the treatment of hypersaline municipal wastewater at Tianjin, China, where NH_3-N removal of 95%, total nitrogen of 50–70%, and total phosphate removal of 60–80% were achieved (Sun et al. 2011). The main obstacle remains in the membrane lifespan, high energy requirements, and membrane fouling. The chemical treatment for membrane cleaning using sodium hypochlorite and citric acid is chemically cost expensive (Bashar et al. 2018).

3.6.2 Chemical Treatment

The chemical treatment of phosphate removal includes the addition of metal salts and minerals. The mechanism is based upon the solubility of phosphorus-metallic compounds to precipitate down. The point of chemical addition (primary, secondary and tertiary) is dependent on the unit process involved (Table 3.3) (Jenkins et al. 1991). Many factors such as influent phosphorus level, effluent discharge standard, wastewater characteristics (alkalinity etc.), plant size, chemical and transportation costs, sludge handling facilities, ultimate disposal alternatives are involved in the choice of chemical to be added and the point of addition for phosphate removal. The cost involved in the chemical operation and maintenance, excessive sludge handling difficulties, and pH maintenance is the limiting factors.

Table 3.3 Phosphate removal process based upon the point of chemical dosing

Process	Point of chemical dosing	Advantages	Disadvantages
Pre-precipitation	Primary settlement tank	Removal of colloids and solids resulting in reduced organic load on secondary treatment. Reduced aeration costs and secondary sludge production.	Organically-bound phosphate not precipitated. Expensive due to chemical usage and primary sludge production. Impairs secondary treatment efficiency. High Operation and maintenance cost. An Effluent generation with total phosphate >1 mg/L.
Co-precipitation	Aeration tank of activated sludge plant.	Phosphate removal occurs by continuous recirculation, coagulation-flocculation and adsorption. Reduced chemical consumption, Costs reduction and easy implementation.	Separation of biological and chemical sludge is difficult. Excess sludge production and sludge handling difficulties. Effluent with the phosphate concentration of about 1 mg/L. (85% Phosphorus removal).
Post-precipitation	After secondary treatment.	High efficiency phosphate removal (95%) with effluent phosphate concentration lower than 0.5 mg/L. Removal of suspended solids along with phosphate.	High Operation and maintenance costs (big ponds, mixing devices, filtration units). Higher metallic ion dosage (1.5–2.5 ions for every phosphorus ion). Risk of having iron and residual colouration in effluent due to ferric salts.

3.6.2.1 Chemical Dosing with Iron and Aluminium

Chemical precipitation of phosphate is carried out by adding metal ions such as aluminium and iron in the form of sodium aluminate, ferric chloride, and ferric sulfate (Jenkins et al. 1991). Phosphate removal is carried out either by (i) precipitation as an insoluble phosphate salt of the metallic ion of the coagulant (ii) adsorption on the colloidal floc formed by the coagulant in water (Lea et al. 1954). The chemistry of phosphate (PO_4^{-3}) precipitation with aluminium [Al^{+3}] and ferric iron [Fe^{+3}] are similar with the formation of Aluminium phosphate ($AlPO_4$) and ferric phosphate ($FePO_4$) which is represented as follows are represented as follows.

$$Al^{+3} + PO_4^{-3} \Rightarrow AlPO_4 \downarrow$$

$$Fe^{+3} + PO_4^{-3} \Rightarrow FePO_4 \downarrow$$

The stoichiometric weight ratio in both the reactions is 0.87:1 for $Al:PO_4^{-3}$ and 1.8:1 for $Fe:PO_4^{-3}$. The coagulant at a dose of 1.25–1.75 mole of metal ion per mole of phosphorus is required for precipitation. Phosphorus removal is efficient at a pH range of 5–7 for alum and 6.5–7.5 for ferric salts (Convery 1970). Alum ($Al_2(SO_4)_3.14H_2O$ dosage of 100 mg/L, 200 mg/L, and 300 mg/L were employed for phosphate removal from Secondary treated municipal wastewater containing 4.6 mg/L of soluble phosphorus at Nine Springs Sewage treatment plant at Madison. Phosphate removal of 66%, 96%, and 99% was achieved at a pH range of 7.1–7.7, but alum concentration of 185 mg/L was the optimum (Lea et al. 1954).

Recently ferrate (VI) was substituted for ferric sulphate due to lower dosage requirements and a lesser volume of sludge production. Pilot-scale treatment with ferrate (VI) at a dosage of 0.03 mg/L removed 56% PO_4-P from the sewage effluent at Hailsham North Wastewater treatment plant of Southern Water Ltd., United Kingdom, similarly equal to the removal of 59% PO_4-P by 37 mg/L ferric sulphate. Full-scale treatment with 0.16 mg/L Ferrate (VI) showed 44% PO_4-P removal equal to phosphate removal by 25 mg/L ferric sulphate (Jiang et al. 2012). The chemical requirement is dependent on the influent phosphorus concentration, solubility of metals, stoichiometric mass balance etc. The precipitation with ferric salts is pH-dependent and is based upon the water alkalinity.

The addition of base (either calcium hydroxide or sodium hydroxide) or anionic polymers is necessary to optimize flocculation while adding ferrous salts for precipitating phosphate. Increased sludge production is observed when the chemicals are added during the primary treatment (40%) and secondary treatment (26%) in activated sludge plants (Convery 1970). The removal of non-reactive lime, restoration of water alkalinity and pH before discharge, water coloration due to ferric salts, excess sludge production and handling, high operation and maintenance costs are the main disadvantages.

3.6.2.2 Addition of Minerals

The minerals such as calcium, magnesium, ferric iron, aluminium (sodium aluminate) are employed for phosphorus removal. The minerals are generally added to the secondary clarifier to precipitate phosphorus. The molar concentration of 1.4–1.6 moles of aluminium/mole of phosphorus resulted in effluent containing 0.5–1.0 mg/L of phosphate precipitate (Jenkins et al. 1991). The main drawback of this method is the generation of turbid effluent due to the change in pH. The addition of polyelectrolyte could control the turbidity.

3.6.2.3 Lime Clarification

When slaked lime ($Ca(OH)_2$) is added to the wastewater, it reacts with bicarbonate alkalinity precipitating as calcium carbonate ($CaCO_3$). Calcium ions react with orthophosphate to precipitate as hydroxyapatite ($Ca_5\,OH(PO_4)_3$) at the pH range of 8.5–10 (Nassef 2012).

$$5Ca^{++} + 7OH^{-} + 3H_2PO_4^{-} \rightarrow Ca_5OH(PO_4)_3 + 6H_2O$$

Approximately 3–4 g/L of lime is added for precipitation and is dependent on the alkalinity of wastewater. The amount of chemical required (1.5–2.5 times) is determined on the basis of jar tests or alkalinity of wastewater. The lime requirement could be minimized by air blowing, such as CO_2 stripping. It is mainly employed for the treatment of digester supernatant. The costs involved in the lime addition, CO_2 stripping, operation and maintenance costs, and difficulty in sludge handling are the main drawbacks of this method. The production of sludge in excess and neutralization of pH in wastewater before discharge are tedious to implement.

3.6.2.4 Adsorptive Removal of Phosphate

The adsorptive removal of phosphate employs an iron reactive filtration system (Moller 2006), gas concrete (Oguz et al. 2003), magnetite (Woodard 2006), and iron-oxide tailings (Zeng et al. 2004). The biosorbents such as wood particles (Eberhardt and Min 2008), microalgae (e.g., *Scenedesmus*) (Fierro et al. 2007; Martinez et al. 2000), seaweed (Eg. Carrageenan) (Lau et al. 1998), and orange waste gel loaded with zirconium (Biswas et al. 2008) was also employed for the adsorptive removal of phosphate. The waste materials or industrial by-products such as coir pith carbon (Namasivayam and Sangeetha 2004), fly ash (Cheung and Venkitachalam 2000), blast furnace slags (Johansson and Gustafsson 2000; Kostura et al. 2005), aluminium based water treatment residual (Barth 1971) have been employed in the effective removal of phosphate from aqueous solutions. The layered double hydroxides with Zinc (zn) and Aluminium (Al) in the molar ratio of 2, calcined at 300 °C for 4 h (Zn-Al-2-300) adsorbed ~50 mg/g of phosphate from sewage sludge with the regeneration of ~60% after six cycles of adsorption and desorption (Cheng et al. 2009). Magnesium-Aluminium (Mg-Al) layered double hydroxides synthesized by co-precipitation adsorbed 2.25 mmol/g of phosphate at pH 5 (Luengo et al. 2017).

3.6.2.5 Crystallization

The crystallization process involves pH adjustment to >8.0 by stripping (Aeration or agitation) followed by the addition of magnesium (Mg) and simultaneous pH adjustment by Sodium hydroxide (NaOH), Magnesium hydroxide ($Mg(OH)_2$) or an Magnesium Oxide (MgO) slurry (Le et al. 2009). The soluble phosphorus from wastewater can be recovered by precipitation from the soluble phase of the digested sludge, or concentrated chemical leach stream from ash, and digester supernatant in the form of struvite. Many pilot and full-scale crystallization reactors are fluidized beds that operate at a hydraulic retention time of less than 1 h.

3.6.3 Biological Phosphate Removal

Biological phosphate removal was first observed in wastewater treatment plants by Srinath et al. (1959). In biological phosphate removal systems, phosphate removal of 80–90% with intracellular phosphate storage of 5–30% on the dry weight basis were observed when compared to conventional activated sludge process, where the mixed liquor suspended solids (MLSS) contain 1.5–2.0% phosphorus with a phosphate removal efficiency of 20%. Later, Levin and Shapiro (1965) postulated the hypothesis that phosphate removal was a biologically mediated phenomenon where phosphate release occurred under non-aerated conditions and phosphate uptake under aerobic conditions (Levin and Shapiro 1965). Shapiro et al. (1967) reported that phosphate release during anaerobic conditions was due to the redox potential (Shapiro et al. 1967). Randall et al. (1970) reported that phosphate release occurred due to the conditions that adversely affect cellular metabolism. The research was focused on the process design and later moved towards microbiological and biochemical aspects of organisms involved in phosphate release and uptake. Harold (1966) called the phenomena of phosphate-uptake as 'overplus' or 'luxury' phosphate uptake and was later termed as 'Enhanced biological phosphate removal (EBPR)' or anaerobic/aerobic process (AAO).

3.6.3.1 Enhanced Biological Phosphate Removal

Enhanced Biological Phosphate Removal (EBPR) (Fig. 3.3) occurs by the enrichment of phosphate accumulating organisms (PAOs) in activated sludge (UNESCAP 2015; Sedlak 1991), which is dependent on the influent volatile fatty

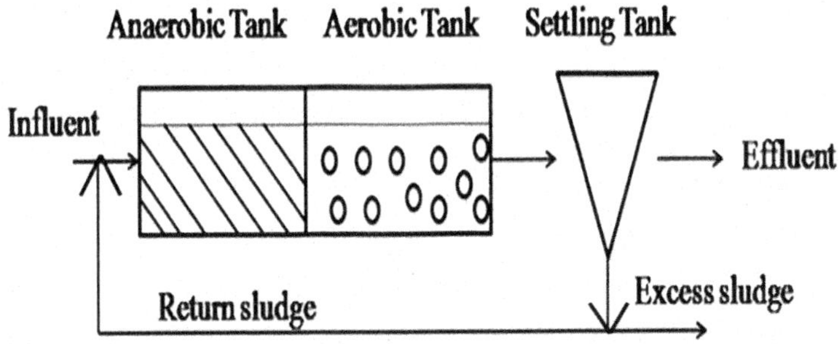

Fig. 3.3 Enhanced Biological Phosphate Removal (EBPR) process flow diagram: activated sludge alternates between anaerobic-aerobic cycles to enrich phosphate accumulating organisms for phosphate removal

acids, total biochemical oxygen demand (TBOD): chemical oxygen demand (COD) ratio, dissolved oxygen (DO) concentration and the carbon source provided. The operational and environmental factors such as carbon source, Carbon: Phosphate (CP) ratio, temperature, pH, cations, DO, solid retention time (SRT), and secondary phosphorus release influences phosphate removal in EBPR (Gurtekin 2014). The total phosphorus concentration of 0.7 mg/L in the effluent can be achieved by maintaining the influent biochemical oxygen demand (BOD): Phosphorus(P) ratio of 30:1 to 40:1 and DO concentration of 0.5–1.0 mg/L (Minnesota Pollution Control Agency 2006).

Phosphate accumulating organisms substitute between carbon-rich anaerobic and carbon-depleted aerobic cycles resulting in the removal of phosphate (Kortstee et al. 2000; Blackall et al. 2002) as polyphosphates with the generation of phosphate-rich sludge (Fig. 3.4). With the advent of EBPR, volatile fatty acids (VFAs) such as acetate and propionate were considered to be the favourable substrates for efficient EBPR (Fuhs and Chen 1975; Potgieter and Evans 1983; Malnou et al. 1984; Engelbrecht and Morgan 1959; Arvin and Kristensen 1985; Comeau et al. 1987). The influent VFA concentration corresponding to 25 mg COD/L accomplished EBPR significantly (Reddy 1998). Later, the mixed proportions of acetate and propionate (Pijuan et al. 2004; Chen et al. 2004; Oehmen et al. 2007; Li et al. 2008; Wang et al. 2010) were employed to study EBPR performance. However, other VFAs such as butyrate, lactate, valerate, and isovalerate were also used as carbon sources (Hood and Randall 2001; Carucci et al. 1999) and still showed good EBPR performance. The readily assimilable carbon source such as glucose was initially considered to deteriorate EBPR, but the experimental evidence to predict the fate of glucose in EBPR systems (Carucci et al. 1999; Sudiana et al. 1997; Jeon and Park 2000; Jeon et al. 2001) revealed efficient EBPR without polyphosphate degradation during anaerobic metabolism of glucose by Embden-Meyerhof-Parnas pathway (EMP pathway) (Wang et al. 2002) and also the reduction in EBPR process economics. Another prediction for glucose consumption in EBPR by two kinds of bacterial populations, lactic acid-producing organism (LAO) and PAO was also proposed (Jeon and Park 2000).

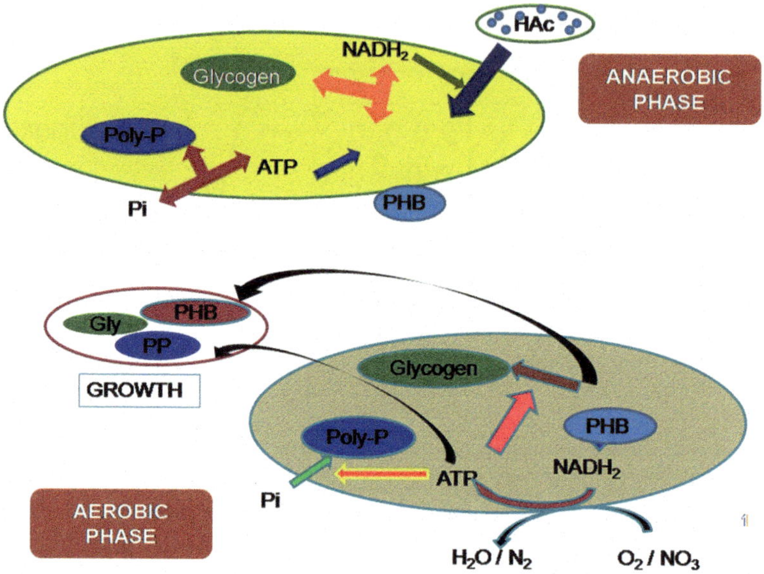

Fig. 3.4 Metabolic cycle of Phosphate accumulating organisms (PAO) under aerobic and anaerobic conditions: fate of carbon and reducing sources and storage polymers

The biochemical pathways for PAOs involve the tricarboxylic acid cycle (TCA), glyoxylate shunt and split tricarboxylic acid cycle (Fig. 3.5). The two metabolic models exist for EBPR (i) Comeau model and (ii) Mino model, depending upon the reducing power utilized for EBPR. Comeau model states that under anaerobic condition, PAOs take up organic substrate (preferably volatile fatty acids) and store them as polyhydroxyalkanoates (PHA), for which the reducing power is produced by TCAcycle (Comeau et al. 1986, 1987), whereas Mino model proposes the reducing equivalent supplied by glycolysis of internally stored glycogen (Mino et al. 1987). The main competitor for the organic substrate in EBPR for PAO is the glycogen accumulating organisms (GAO) whose enrichment deteriorates EBPR. GAOs compete for a substrate for glycogen accumulation rather than the polymer without phosphate removal. Hence the energy for GAOs is provided by glycogen degradation and not the polyphosphates. GAO population can be controlled in EBPR by altering the carbon source to propionate (Oehmen et al. 2005) and maintaining higher pH and temperature (Filipe et al. 2001).

Polyphosphates are linear chains of a few to hundred phosphate molecules linked by phosphoanhydride bonds (Kornberg 1995). Polyphosphates are the storage polymer that acts as energy reserves and are involved in regulating enzyme activity, regulation of gene expression and stress adaptation (Kulakovskaya et al. 2014). The maintenance of stability in phosphate removal efficiency over a period of time is difficult. The handling and disposal of sludge generated are also problematic.

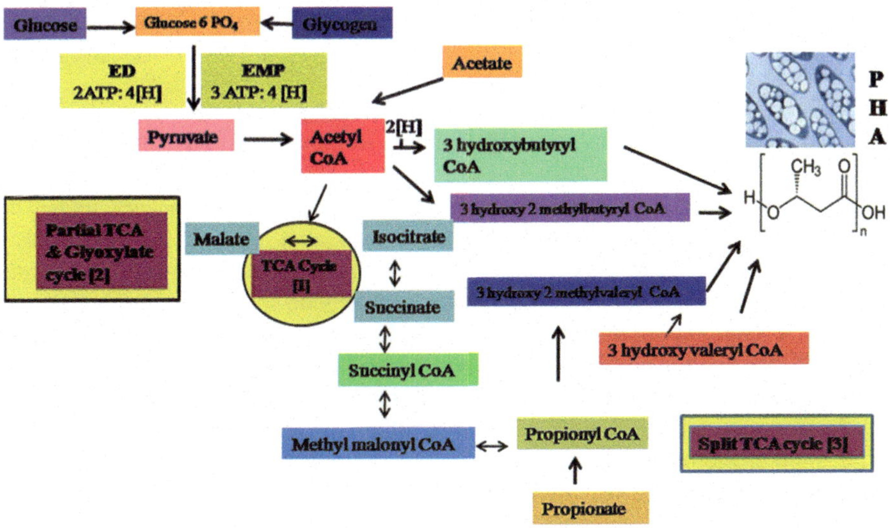

Fig. 3.5 Biochemical pathway of Phosphate Accumulating Organisms: three main cycles for polymer production, (1) Tricarboxylic acid cycle (TCA cycle) (Conversion of Acetate to storage polymer [polyhydroxyalkanoate (PHA)]), (2) Partial TCA and Glyoxylate cycle (conversion of Pyruvate to PHA), and (3) Split TCA cycle (Conversion of Propionate to PHA)

Fuhs and Chen (1975) conducted many isolation tests and reported '*Acinetobacter*' as the bacteria responsible for phosphate removal. However, they behave differently when isolated as pure cultures. Hence, it was understood that a community of organisms mediated phosphate uptake and removal. Later in the 2000's the molecular ecology techniques revealed that bacteria belonging to *Rhodocyclus* related to *Betaproteobacteria* and *Gammaproteobacteria* were the principal organisms for phosphate removal. The research on the process design and bacterial community involved in phosphate removal, biochemistry and metabolism, isolation of phosphate accumulating organisms in pure form, stabilization of phosphate removal process and many other aspects are studied widely to understand the process of phosphate removal.

Denitrifying Phosphate Removal

The establishment of an anoxic zone between the anaerobic and aerobic zones favours both nitrate and phosphate removal. Instead of oxygen, denitrifying phosphate accumulating organisms (DNPAOs) use nitrate as the electron acceptor. In the anaerobic zone, DNPAOs convert the readily degradable substrate to polyhydroxyalkanoate using stored polyphosphate and glycogen as an energy source. The nitrifiers convert ammonia to nitrate in the aerobic zone, and the recirculation of activated sludge from aerobic to anoxic zone aids in the conversion of nitrate to nitrogen by the denitrifiers and DNPAOs. Phosphate removal occurs in the aerobic zone by DNPAOs using nitrate as an electron source (Spagni et al. 2001; Oehmen et al. 2007; Garcia et al. 2010).

Alteration in Enhanced Biological Phosphorus Removal (EBPR)

Unlike the alternating anaerobic-aerobic phase, EBPR also occurs under the completely aerobic condition with phosphate removal of less than 0.1 mg/L. This is achieved by temporarily separating the simultaneous addition of acetate and phosphate, where acetate is added at the start of the feed stage and phosphate during the famine stage. The community analysis by denaturing gradient gel electrophoresis (DGGE) revealed *Candidatus, Accumulibacter phosphatis*, beta-proteobacterial *Dechloromonas* and alpha-proteobacterial tetrad-forming organisms related to *Defluviicoccus,* which predominates in aerobic-EBPR system (Ahn et al. 2007). The single-stage oxic process involves removing phosphate to 0.22–1.79 mg/L by maintaining DO at 3 ± 0.2 mg/L and pH 7 during the aerobic phase without any anaerobic phase. PHA accumulation was minimal, whereas polyphosphate storage was not affected by this process (Wang et al. 2007).

EBPR generated sewage sludge containing 3–5% of dry weight as phosphorus. Heating the phosphorus rich sludge to 70 °C for about 1 h released heat-labile phosphorus, and the addition of calcium salts aided in the recovery of phosphate, termed as heat phos process (Kuroda et al. 2002; Takiguchi et al. 2019). A pilot-scale plant employing the heat phos process was established at Kobe City, Japan, to check its feasibility. Approximately, waste sludge of 10–20 kg MLSS/day was subjected to heat at 70–90 °C for 1 h in a 1 m^3 capacity heating tank. Phosphate rich liquid separated from sludge was precipitated with Calcium chloride ($CaCl_2$) removing 90–95% of influent total phosphate (TP) and capturing about 50–60% of stored phosphate in biomass (Hirota et al. 2010). The excess removal of phosphate than required for cellular growth by *Rhodocyclus*-related organisms was observed in the aerated-anoxic Orbal process. However, the anaerobic zone in EBPR is replaced by an anoxic zone in Orbal (Zilles et al. 2001). Sulfate reduction, autotrophic denitrification and nitrification integrated process was coupled with EBPR in an alternating anaerobic/limited oxygen aerobic cycle in a sequencing batch reactor (SBR).

3.6.3.2 Phosphate Removal by Microalgae

The use of microalgae for nutrient removal was proposedby Oswald et al.(1957). Microalgae offer an efficient tertiary treatment system to remove inorganic nitrate and phosphate by utilizing these nutrients from the secondary treated domestic wastewater for their growth. The feasible options such as low operation costs, recycling of assimilated nutrients in algal biomass as fertilizer, thereby avoiding sludge handling problems, discharge of oxygenated effluent into the water body, unnecessary requirement of carbon for nutrient removal makes microalgal technology an attractive tool for nutrient removal (Aslan and Kapdan 2006). The microalgae such as *Euglena, Oscillatoria, Chlamydomonas, Scenedesmus, Chlorella, Nitzchia, Navicula* and *Stigeoclonium* are employed for phosphate removal. Batch cultivation of *Chlorella vulgaris* in synthetic wastewater removed 21.2 mg/L ammonia-nitrogen and 7.7 mg/L phosphate (78%) (Aslan and Kapdan 2006). Three freshwater microalgae (*Chlorella sp., Monoraphidium minutum sp.,* and *Scenedesmus sp.*) and three marine microalgae (*Nannochloropis sp., N. limnetica sp.,* and *Tetraselmis suecica sp.*) were screened to find the best phosphate removing microalgae. The freshwater microalga *Monoraphidium minutum sp.* removed 67.1% and 79.3% of phosphate on day 8 and 16, respectively, whereas the marine microalgae, *Tetraselmis suecica sp.* removed 79.4% and 83% of phosphate on day 8 & 16 (Patel et al. 2012). The enriched microbial biomass inoculum containing both bacteria and microalgae was employed to treat municipal wastewater in high rate algal ponds. The total nitrogen and phosphate removal of 72–83% and 100% were achieved at a temperature of 15 °C & 25 °C. Nitrate uptake of 62%, ammonia uptake of 42.2% and phosphate uptake of 64.7% were reported in immobilized cells of *Dunaliella salina* (Thakur and Kumar 1999).

Immobilized cells of *Chlorella sorokiniana* GXNN 01 removed 87.49%, 88.65% and 84.84% of phosphate under autotrophic, heterotrophic and micro-aerobic conditions (Liu et al. 2012). Immobilization of *Microcystis aeruginosa* in PVA-alginate complex removed 76.7–80.3% of phosphate (Li et al. 2015). Continuous operation of *Phormidium laminosum* in a hollow fiber photo bioreactor removed nitrogen of 3.36 mg/d and phosphate of 3.30 mg/d (Sawayama et al. 1998). The removal of phosphate due to osmotic-stress and storage of assimilated phosphate in the form of polyphosphate in acidic vacuoles have been visualized in *Dunaliella salina* (Pick and Weiss 1991) and *Phaeodactylum tricornutum* (Leitao et al. 1995). *Scenedesmus, Phormidium laminosum* and cyanobacteria have also been reported to have high nutrient uptake. They are advantageous due to cost-effectiveness, recovery of algal biomass for pigment and biofuel production (Abdel et al. 2012). The major drawback is the requirement of sunlight (poor in polar countries) and land.

3.6.3.3 Phosphate Removal by Macrophyte

Floating Macrophyte-Based Treatment Systems offers an effective system to remove organics, suspended solids and nutrients such as nitrate and phosphate. Water hyacinth (*Eichhorniacrassipes*), pennywort (*Hydrocotyle umbellate*) and duckweed (*Lemnasp*) are widely employed as floating macrophytes. The mechanism of nutrient removal includes plant uptake, ammonia-N volatilization, phosphate precipitation or adsorption, and denitrification. Water hyacinth offers an excellent phosphate removal rate with effluent phosphate concentration ranging from 0.3 to 5.7 mg/L for an influent phosphate concentration ranging from 0.7 to 11 mg/L (Busk and Reddy 1991). Phosphate removal from eutrophicated Lake Donghu, China using five different submerged macrophytes showed the highest phosphate removal by *C. demersum* (91.75%), *E. canadensis* (84.71%), and *P. crispu* (76.16%)

in spring and *C. demersum* (92.44%), *V. spiralis* (84.63%), *M. spicatum* (68.29%) in the autumn (Gao et al. 2009). The phosphate removal efficiency of 23% was achieved by employing the free-floating hydrophytes and emergent macrophytes in floating treatment wetland (Prajapati et al. 2017). The hydraulic retention time, leaf surface area, wastewater constituents and seasonal variation are highly influential in nutrient removal. They are advantageous in cost-effectiveness, and the main drawback includes land requirements and frequent harvest of macrophytes after decay.

3.6.3.4 Phosphate Removal in Constructed Wetland

The use of wetland plants to treat municipal wastewater started back in the 1950s by Dr. Kathe Seidel of Germany. The full-scale wetland treatment system was established in the 1960s and is followed in many countries until now. The horizontal and vertical flow systems with different vegetation types, such as free-floating, floating-leaved, submerged, and emergent, are followed worldwide. The common species employed in Free water surface constructed wetland are *Phragmites australis* (common reed), *Scirpus lacustris* (Schoenoplectus), *Typha sp.* (Cattail), *Scirpus spp.* (Bulrush), *Sagittaria latifolia* (Arrowhead), *Phragmites australis*, *Bolboschoenus fluviatilis* (Marsh clubrush), *Eleocharis sphacelata* (Tall spikerush), *Scirpus tubernaemontani* (Table 3.4).

Constructed wetland with free-floating macrophytes consists of one or more shallow ponds where macrophytes such as *Eichhornia crassipes* (water hyacinth), *Pistia stratiotes* (water lettuce), duckweeds (*Lemna spp., Spirodela polyrhiza, Wolffia spp.*) float on the surface. Phosphorus removal occurs by microbial assimilation, precipitation with divalent and trivalent cations, adsorption onto clays or organic matter, plant uptake, and media with high phosphate sorption capacity (Vymazal et al. 2008). The total nitrogen removal of 48%, 38%, and 43% and total phosphate removal of 40%, 50% and 56% were observed in free water surface systems, horizontal subsurface flow, and vertical subsurface flowsystems (Vasudevan et al. 2011). The land requirement, removal of decayed plants, and the cost involved in constructing constructed wetland are the main drawbacks in the constructed wetland system. The phosphate storage capacity varies depending on the plant type, plant uptake rate, and storage capacity (Shilton et al. 2012) (Table 3.5).

Table 3.4 Phosphate storage capacity of the plants (Shilton et al. 2012)

S. no.	Plant	Phosphorus per dry weight (%)	Biomass yield (tonnes/ha/year)
1.	Switchgrass	0.3	8.7–12.9
2.	Maize	0.3	9–30
3.	Macrophytes	1–3	35–106
4.	Algae	1–3	69–91

Table 3.5 Efficiency of phosphate removal in different treatment systems

Treatment	Chemical Oxygen Demand removal (%)	Phosphate removal (%)	Ortho phosphate as Phosphorus (mg/L)		Total phosphate as Phosphorus (mg/L)		Advantages	Disadvantages
			Influent	effluent	Influent	effluent		
Chemical treatment with lime clarification	77.5	96–97	44	1	96	3.6	Higher phosphate removal efficiency.	Increased sludge production, higher cost investment in purchase of chemicals, O&M and sludge disposal. Maintaining higher pH.
Chemical treatment with iron salts.	29	81				1.1		
Chemical treatment with aluminium		74–94						
Physical-chemical treatment. a. Primary		72–83			15	0.3		Increased sludge production, costs involved is higher, higher pH requirement.
b. Activated sludge		87–90			7			Increased sludge production, difficulty in sludge handling.
Mineral addition in secondary treatment (Activated Sludge Process)					50–100	10	Higher settling characteristics of sludge.	Turbid effluent due to change in pH, costs involved in polyelectrolyte addition.
Filtration a. Trickling filter					5–15	0.1–0.5	Low power and cost requirement. Suitable for small scale community. High organic removal rates	Frequent backwashing required due to clogging. Low loading rate. Additional treatment required to maintain effluent discharge quality.

(continued)

Table 3.5 (continued)

Treatment	Chemical Oxygen Demand removal (%)	Phosphate removal (%)	Ortho phosphate as Phosphorus (mg/L)		Total phosphate as Phosphorus (mg/L)		Advantages	Disadvantages
			Influent	effluent	Influent	effluent		
b. Moving bed filtration		90	13.2	0.6	21.5	2.2	Less land requirement and improved settling characteristics. No periodic backwashing	High-cost investment.
a. Commercial filters (hydrotech disc filters)						0.1	Lesser size requirement, operation and maintenance at reduced costs.	Chemical pretreatment with metallic coagulant and polymer required.
Biological phosphate removal a. Activated Sludge Process		20–30					Enrichment of bacterial population. Higher organics removal rates. Use of sludge as fertilizer.	Low phosphate removal efficiency. More sludge production. Necessity to return sludge for activation. Higher area requirement.
b. Luxury P uptake-Enhanced Biological Phosphorus Removal					10–20	0.5	Higher organics and nutrient removal rates.	Higher operating cost Maintenance of stable phosphate removal efficiency is difficult.
c. Macrophytes					0.7–11	0.3–5.7	Feasible option for small-low community. Higher organic and nutrient removal rates depending on wastewater characteristics.	Frequent harvest of decayed plants. Disposal of harvested plants. Costs involved in the design of constructed wetland.
d. Constructed wetland		40–56						

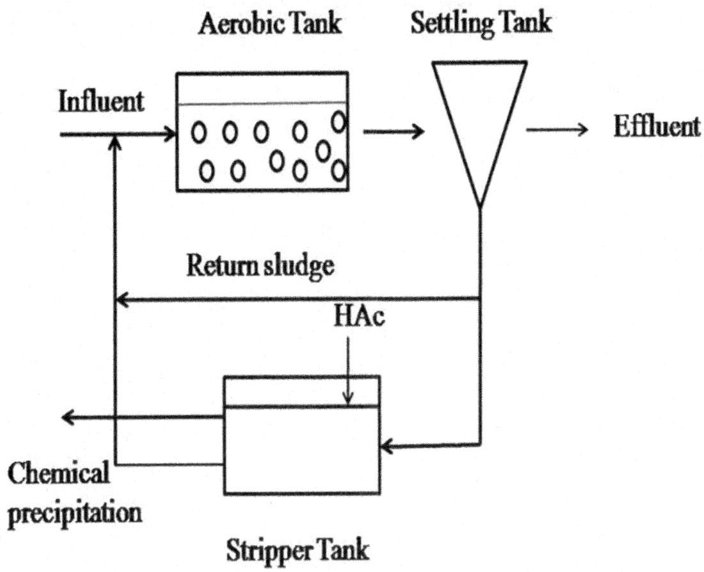

Fig. 3.6 Phostrip flow diagram: a combined chemical (precipitation of nutrients as struvite by addition of magnesium) and biological process (phosphate removal by PAOs)

3.6.4 Combined Biological/Chemical Method

The combined biological/chemical method for phosphorus removal is phostrip (Levin 1966), where the phosphate-rich wastewater from the anaerobic zone is treated with magnesium/potassium salt for phosphate and ammonia removal as struvite (Fig. 3.6). The phostrip process is also known as the sidestream process. The phostrip process is not affected by the influent total biochemical oxygen demand and can achieve less than 1 mg/L effluent total phosphate concentration. It requires pH change and the generation of increased waste activated sludge. The performance efficiency of different treatment technologies is provided in Table 3.5.

3.7 Conclusion

Although phosphate is an essential nutrient and is widely employed in many industrial, domestic and commercial sectors, it acts as a potent contaminant in freshwater bodies resulting in eutrophication. The detrimental effects of phosphate contamination in aquatic ecosystems necessitate its removal from the receiving waters. The standard and stringent discharge limits for phosphate put forth by the legislative, globally urges phosphate removal from wastewater. The phosphate contamination by point sources can be easily treated and eliminated rather than non-point sources. Direct sewage inputs and partially treated sewage without nutrient removal are the major contributors to phosphate inputs in aquatic bodies. Hence, it is high time for installing phosphate removal plants and retrofitting the existing conventional treatment plants with phosphate removal technologies either by physical, chemical, or biological means. The physical and chemical methods involve high-cost investment for its maintenance and operation. The biological means of phosphate removal employs bacteria, cyanobacteria, algae, microalgae, macrophytes and plants. Of these, bacterial phosphate removal by EBPR process benefits the energy and cost-saving aspect with efficient phosphate removal. The advancement of EBPR has paved the way for removing phosphate and removing nitrate and sulphate with the production of biogas and biopolymer. Both the biological and chemical phosphate removal techniques help remove phosphate and convert it into struvite, a slow-release fertilizer. Since phosphate is a non-renewable source, the treatment process for phosphate removal should also focus on its recovery in a usable form.

References

Aalto JA, Jaworski NA, Lear DW (1970) Current water quality conditions and investigations in the upper Potomac River tidal system. Chesapeake Tech. Support Lab, Middle Atlantic Region, Florida Water Quality Association (FWQA)

Abdel-Rauf N, Al-Homaiden AA, Ibraheem IBM (2012) Microalgae and wastewater treatment. Saudi J Biol Sci 19:257–275. https://doi.org/10.1016/j.sjbs.2012.04.005

Ahn J, Schroeder S, Beer M, McIlroy S, Bayly RC, May JW, Vasiliadis G, Seviour RJ (2007) Ecology of the microbial community removing phosphate from wastewater under continuously aerobic conditions in a sequencing batch reactor. Appl Environ Microbiol 73:2257–2270. https://doi.org/10.1128/AEM.02080-06

Alcantara C, Blasco A, Zuniga M, Monedero V (2014) Accumulation of polyphosphate in *Lactobacillus spp.* and its involvement in stress resistance. Appl Environ Microbiol 80:1650–1659. https://doi.org/10.1128/AEM.03997-13

Alexander GC, Stevens RJ (1976) Per capita phosphorus loading from domestic sewage. Water Res 10:757–764. https://doi.org/10.1016/0043-1354(76)90093-2

Anon (1967) Sources of nitrogen and phosphorus in water supplies. American Works Association Task group Report 2610P. J Am Water Works Assoc 59:344–366. https://ec.europa.eu/jrc/sites/jrcsh/files/lb-na-26822-en-n.pdf

Arnscheidt J, Jordan PLS, McCormick S (2007) Defining the sources of low-flow phosphorus transfers in complex catchments. Sci Total Environ 382:1–13. https://doi.org/10.1016/j.scitotenv.2007.03.036

Arvin E, Kristensen GH (1985) Exchange of organics, phosphate and cations between sludge and water in biological phosphorus and nitrogen removal processes. Water Sci Technol 17:147–162. https://doi.org/10.2166/wst.1985.0229

Aslan S, Kapdan IK (2006) Batch kinetics of nitrogen and phosphorus removal from synthetic wastewater by algae. Ecol Eng 28:64–70. https://doi.org/10.1016/j.ecoleng.2006.04.003

Barth EJ (1971) Environment Protection Agency, Robert A. Taft Water Research Centre, Advanced Waste Treatment Lab., Cincinnati, Ohio. Personal Communication

Bashar R, Gungor K, Karthikeyan KG, Barak P (2018) Cost effectiveness of phosphorus removal processes in municipal wastewater treatment. Chemosphere 197:280–290. https://doi.org/10.1016/j.Chemosphere.2017.12.169

Biswas BK, Inoue K, Ghimire KN, Harada H, Ohto K, Kawaita H (2008) Removal and recovery of phosphorus from water by means of adsorption onto orange waste gel loaded with zirconium. Bioresour Technol 99:8685–8690. https://doi.org/10.1016/j.biortech.2008.04.015

Blackall LL, Crocetti GR, Saunders AM, Bond PL (2002) A review and update of the microbiology of enhanced biological phosphorus removal in wastewater treatment plants. Antonie Van Leeuwenhoek 81:681–691. https://doi.org/10.1023/A:1020538429009

Carpenter RS, Bennett EM (2011) Reconsideration of the planetary boundary for phosphorus. Environ Res Lett 6:1–12. https://doi.org/10.1088/1748-9326/6/1/014009

Carpenter SR, Caraco NF, Correll DL, Howarth RW, Sharpley AN, Smith VH (1998) Nonpoint pollution of surface waters with phosphorus and nitrogen. Ecol Appl 8:559–568. https://doi.org/10.1890/1051-0761(1998)008[0559:NPOSWW]2.0.CO;2

Carucci A, Lindrea K, Majone M, Ramadori R (1999) Different mechanisms for the anaerobic storage of organic substrates and their effect on EBPR. Water Sci Technol 39:21–28. https://doi.org/10.1016/S0273-1223(99)00119-5

Central Pollution Control Board (CPCB) Bulletin (2016). From: http://www.cpcb.nic.in/upload/Latest/Latest_123_SUMMARY_BOOK_FS.pdf.1, pp 1–26

Chen Y, Randall AA, McCue T (2004) The efficiency of enhanced biological phosphorus removal from real wastewater affected by different ratios of acetic to propionic acid. Water Res 38:27–36. https://doi.org/10.1016/j.watres.2003.08.025

Cheng X, Huang X, Wang X, Zhao B, Chen A, Sun D (2009) Phosphate adsorption from sewage sludge filtrate using zinc-aluminium layered double hydroxides. J Hazard Mater 169:958–964. https://doi.org/10.1016/j.jhazmat.2009.04.052

Cheung KC, Venkitachalam TH (2000) Improving phosphate removal of sand infiltration system using alkaline fly ash. Chemosphere 41:243–249. https://doi.org/10.1016/S0045-6535(99)00417-8

Comber S, Gardne M, Georges K, Blackwood D, Gilmour D (2013) Domestic source of phosphorus to sewage treatment work. Environ Technol 34:1349–1358. https://doi.org/10.1080/09593330.2012.747003

Comeau Y, Hall KJ, Hancock REW, Oldham WK (1986) Biochemical model for enhanced biological phosphorus removal. Water Res 20:1511–1521. https://doi.org/10.1016/0043-1354(86)90115-6

Comeau Y, Oldham WK, Hall KJ, Ramadori (eds) (1987) Dynamics of carbon reserves in biological dephosphatation of wastewater. In Advances in water pollution control. Proceedings of the International Association on Water Pollution Research and Control (IAWPRC) Conference on Biological Phosphate Removal from Wastewater, Rome, Italy. Pergamon Press, Oxford, pp 39–56

Convery JJ (1970) Treatment techniques for removing phosphorus from municipal wastewaters. Environment Protection Agency (EPA) Program 17010:1–44

Correll DL (1998) The role of phosphorus in the Eutrophication of receiving waters: a review. J Environ Qual 27:261–266. https://doi.org/10.2134/jeq1998.00472425002700020004x

De Busk TA, Reddy KR (1991) Wastewater treatment by floating macrophytes, wastewater treatment and biomass production by floating aquatic macrophyte. In: Isaacson R (ed) Methane from community wastes. Elsevier Barking, London, pp 21–36

De Latour C (1975) Process for removing dissolved phosphorus from water magnetically. Patent no. US 3983033A, 1–7

Dean RD (1971) Environmental Protection Agency (EPA), Robert A. Taft, Water Research Center, Advanced Waste Treatment Lab., Cincinnati, Ohio. Personal Communication

Department of Environment, Food and Rural Affairs (DEFRA) (2002) UK Implementation of the European Commission (EC) Urban Wastewater treatment directive. http://www.defra.gov.uk/environment/quality/water/waterquality/sewage/uwwtd/index.htm

Devey DG, Harkness N (1973) The significance of man-made sources of phosphorus: detergents and sewage. Water Res 7:35–54. https://doi.org/10.1016/0043-1354(73)90151-6

Duenas JF, Alonso JR, Rey AF, Ferrer AS (2003) Characterization of phosphorous forms in wastewater treatment plants. J Hazard Mater 97:193–205. https://doi.org/10.1016/S0304-3894(02)00260-1

Eberhardt TL, Min SH (2008) Biosorbents prepared from wood particles treated with anionic polymer and iron salt: effect of particle size on phosphate adsorption. Bioresour Technol 99:626–630. https://doi.org/10.1016/j.biortech.2006.12.037

Engelbrecht RS, Morgan JJ (1959) Studies on the occurrence and degradation of condensed phosphate in surface waters. Sewage Ind Waste 31:458–478

Eutrophication Research Newsletter (ERN) (1970) National Eutrophication research program: annual newsletter, pp 1–16

Fact book (2015) Agrium fact book 2014–15, pp 1–121. Available from: www.agrium.com/system/files/2014-15factbook.pdf

FAO-AQUASTAT (2014) [Online] http://www.fao.org/nr/water/aquastat/main/index.stm

Fierro S, Saavedra MS, Del P, Copalcua C (2007) Nitrate and phosphate removal by chitosan immobilized *Scenedesmus*. Bioresour Technol 99:1274–1279. https://doi.org/10.1016/j.biortech.2007.02.043

Filippelli GM, Rosso JJ (2002) Chapter 10: the global phosphorus cycle. In Kohn MJ, John R, Hughes JM (eds) Phosphates – geochemical, geobiological, and materials importance. Mineralogical Society of America (Paul H. Ribbe, Series editor), Geochemical Society. Rev Mineral Geochem 48:391–425

Filipe CDM, Daigger GT, Grady CPL Jr (2001) Effects of pH on the rates of aerobic metabolism of phosphate-accumulating and glycogen-accumulating organisms. Water Environ Res 73(2):213–222

Fuhs GW, Chen M (1975) Microbiological basis of phosphate removal in the activated sludge process for treatment of wastewater. Microb Ecol 2:119–138. https://doi.org/10.1007/BF02010434

Gao J, Xiong Z, Zhang J, Zhang W, Mba FO (2009) Phosphorus removal from water of eutrophic Lake Donghu by five submerged macrophytes. Desal 242:193–204. https://doi.org/10.1016/j.desal.2008.04.006

Garcia-Usach F, Ribes J, Ferrer J, Seco A (2010) Calibration of denitrifying activity of polyphosphate accumulating organisms in an extended ASM2d model. Water Res 44:5284–5297. https://doi.org/10.1016/j.watres.2010.06.061

Gilmour D, Blackwood D, Comber S, Thornell A (2008) Identifying human waste contribution of phosphorus loads to domestic wastewater. 11th international conference on urban drainage, Edinburgh, Scotland

Graham WF, Duce RA (1979) Atmospheric pathways of the phosphorus cycle. GCA 43:1195–1208. https://doi.org/10.1016/0016-7037(79)90112-1

Gurtekin E (2014) Factors affecting enhanced biological phosphate removal (A3-ISEM2014ID83. pdf-water pollution and control). Presented at 2014-2nd International Symposium on Environment and morality (ISEM), Adiyaman University, pp 491–499

Harold FM (1966) Inorganic polyphosphates in biology: structure mechanisms and function. Bacteriol Rev 30:772–794. https://doi.org/10.1128/mmbr.30.4.772-794.1966

Henry HAL (2013) Soil extracellular enzyme dynamics in a changing climate. Soil Biol Biochem 56:53–59

Hetling LJ, Carcich IG (1973) Phosphorus in wastewater. Water Sewage Works 120:59–62

Hirota R, Kuroda A, Kato J, Ohtake H (2010) Bacterial phosphate metabolism and its application to phosphorus recovery and industrial bioprocesses. J Biosci Bioeng 109:423–432. https://doi.org/10.1016/j.jbiosc.2009.10.018

Hood CR, Randall AA (2001) A biochemical hypothesis explaining the response of enhanced biological phosphorus removal biomass to organic substrates. Water Res 35:2758–2766. https://doi.org/10.1016/S0043-1354(00)00573-X

Howell JA (2010) The distribution of phosphorus in sediment and water downstream from a sewage treatment works. Biosci Horiz: Int J Student Res 3:113–123. https://doi.org/10.1093/biohorizons/hzq015

International Fertilizer Industry Association (IFA) (2011) Feeding the Earth – global phosphate rock Production trends from 1961 to 2010: reasons for the temporary set-back in 1988–1994. Paris

Jasinski SM (2015) Phosphate rock. US Geological survey, mineral commodity summaries, pp 1–2. https://www.usgs.gov/centers/nmic/phosphate-rock-statistics-and-information
Jasinski SM (2017) Phosphate rock. US Geological Survey, mineral commodity summaries, pp 124–125. https://minerals.usgs.gov/minerals/pubs/commodity/phosphate_rock/mcs-2017-phosp.pdf
Jayatilake SMDPA (2014) Nutrient loading and eutrophication of coastal waters of the South Asian seas – a scoping study, pp 1–56. https://www.sacep.org/pdf/ScopingstudyonNutrientloadinginSASRegion.pdf
Jenkins D, Hermanowicz SW, Sedlak R (1991) Chapter 4: principles of chemical phosphate removal. In: Phosphorus and nitrogen removal from municipal wastewater principles and practice, 2nd edn. Lewis Publishers, CRC Press, Routledge, pp 91–108
Jeon CO, Park JM (2000) Enhanced biological phosphorus removal in a sequencing batch reactor supplied with glucose as a sole carbon source. Water Res 34:2160–2170. https://doi.org/10.1016/S0043-1354(99)00383-8
Jeon CO, Lee DS, Park JM (2001) Enhanced biological phosphorus removal in an anaerobic-aerobic sequencing batch reactor: characteristics of carbon metabolism. Water Environ Res 73:295–300. https://doi.org/10.2175/106143001X139308
Jiang F, Beck MB, Cummings RG, Rowles K, Russell D (2004) Estimation of costs of phosphorus removal in wastewater treatment facilities: construction De Novo. Water Policy Working paper 2004–2010. https://www.issuelab.org/resources/4705/4705.pdf
Jiang JQ, Stanford C, Mollazeninal A (2012) The application of ferrate for sewage treatment: pilot-to full-scale trials. Global NEST J 14:93–99. https://doi.org/10.30955/gnj.000821
Johansson L, Gustafsson JP (2000) Phosphate removal using blast furnace slags and opoka-mechanisms. Water Res 34:259–265. https://doi.org/10.1016/S0043-1354(99)00135-9
Jonsson H, Baky A, Jeppsson U, Hellstrom D, Karrman E (2005) Composition of urine, faeces, greywater and biowaste for utilization in the URWARE model, Urban water report 2005–2006. https://vdocuments.mx/composition-of-urine-faeces.html
Khan FA, Ansari AA (2005) Eutrophication: an ecological vision. Bot Rev 71:449–482. https://www.jstor.org/stable/4354503
Kornberg A (1995) Inorganic polyphosphate: toward making a forgotten polymer unforgettable. J Bacteriol 177:491–496. https://doi.org/10.1128/jb.177.3.491-496.1995
Kornberg A, Rao NN, Ault-Riche D (1999) Inorganic polyphosphate: a molecule of many functions. Annu Rev Biochem 68:89–125. https://doi.org/10.1146/annurev.biochem.68.1.89
Kortstee GJJ, Appeldoorn KJ, Bonting CFC, van Niel EWJ, van Veen HW (2000) Recent developments in the biochemistry and ecology of enhanced biological phosphorus removal. Biochem Mosc 65:332–341
Kostura B, Kulveitova H, Lesko J (2005) Blast furnace slags as sorbents of phosphate from water solutions. Water Res 39:1795–1802. https://doi.org/10.1016/j.watres.2005.03.010
Kulakovskaya TV, Lichko LP, Ryazanova LP (2014) Diversity of phosphorus reserves in microorganisms. Biochem Mosc 79:1602–1614. https://doi.org/10.1134/s0006297914130100
Kundu S, Vassanda C, Rajendiran M, Ajay S, Subba Rao A (2015) Phosphates from detergents and eutrophication of surface water ecosystem in India. Curr Sci 108:1320–1325. https://doi.org/10.18520/CS%2FV108%2FI7%2F1320-1325
Kuroda A, Takiguchi N, Gotanda T, Nomura K, Kato J, Ikeda T, Ohtake H (2002) A simple method to release polyphosphate from activated sludge for phosphorus reuse and recycling. Biotechnol Bioeng 78:333–338. https://doi.org/10.1002/bit.10205
Lau PS, Tam NFY, Wong YS (1998) Effect of carrageenan immobilization on the physiological activities of *Chorella vulgaris*. Bioresour Technol 63:115–121. https://doi.org/10.1016/j.biortech.2007.02.043
Le PD, Aarnink AJA, Jongbloed AW (2009) Odour and ammonia emission from pig manure as affected by dietary crude protein level. Livestock Sci 121:267–274
Lea WL, Rohlich GA, Katz WJ (1954) Removal of phosphate from treated sewage. Sewage Ind Waste 26:261–275. https://doi.org/10.3354/meps121279

Leitao JM, Lorenz B, Bachinski N, Wilhelm C, Muller WEG, Schroder HC (1995) Osmotic-stress-induced synthesis and degradation of inorganic polyphosphates in the alga *Phaeodactylum tricornutum*. Mar Ecol Prog Ser 121:279–288

Leng R (2009) The impact of cultural eutrophication on lakes: a review of damages and nutrient control measures. Writing 20: freshwater systems and society, Professor Sandra Cooke

Levin GV (1966) Sewage treatment process, US patent 3236766

Levin GV, Shapiro J (1965) Metabolic uptake of phosphorus by wastewater organisms. J Water Pollut Control Fed 37:800–821. https://www.jstor.org/stable/25035307

Li HJ, Chen YG, Gu GW (2008) The effect of propionic to acetic acid ratio on anerobic-aerobic (low dissolved oxygen) biological phosphorus and nitrogen removal. Bioresour Technol 99:4400–4407. https://doi.org/10.1016/j.biortech.2007.08.032

Li F, Mao WJ, Li X, Wang XY, Xiao ZH, Zhou YY, Zeng GM (2015) Characterization of *Microcystis aeruginosa* immobilized in complex of PVA and sodium alginate and its application on phosphorus removal in wastewater. J Cent South Univ 22:95–102. https://doi.org/10.1007/s11771-015-2499-5

Liu Y, Villalba G, Ayres RU, Schroder H (2008) Global phosphorus flows and environmental impacts from a consumption perspective. J Ind Ecol 12:229–247. https://doi.org/10.1111/j.1530-9290.2008.00025.x

Liu K, Li J, Qiao H, Lin A, Wang G (2012) Immobilization of *Chlorella soroiniana* GXNN01 in alginate for removal of N and P from synthetic wastewater. Bioresour Technol 114:26–32. https://doi.org/10.1016/j.biortech.2012.02.003

Luengo C, Volpe MA, Avena MJ (2017) High sorption of phosphate on Mg-Al Layered double hydroxides: kinetics and equilibrium. J Environ Chem Eng 5:4656–4662. https://doi.org/10.1016/j.jece.2017.08.051

Majed N, Li Y, Gu AZ (2012) Advances in techniques for phosphorus analysis in biological sources. Curr Opin Biotechnol 23:1–8. https://doi.org/10.1016/j.copbio.2012.06.002

Malnou D, Meganck M, Faup GM, Rostu M (1984) Biological phosphorus removal: study of the main parameters. Water Sci Technol 16:173–185. https://doi.org/10.2166/wst.1984.0222

Martinez ME, Sanchez S, Jimenez M, El Yousfi F, Munoz L (2000) Nitrogen and phosphorus removal from urban wastewater by the microalgae *Scenedesmus obliquus*. Bioresour Technol 73:263–272. https://doi.org/10.1016/S0960-8524(99)00121-2

Mateo-Sagasta J, Raschid-Sally L, Thebo A (2015) Chapter 10 – global wastewater and sludge production, treatment and use. In: Drechsel P, Qadir M, Wichelns D (eds) Wastewater. Springer, Dordrecht, pp 15–38

Megard RO (1970) Lake Minnetonka: nutrients, nutrient abatement and the photosynthetic system of the phytoplankton. Interim report no. 7. Limnological Research Center, University of Minnesota, pp 210–215

Mesdaghinia A, Nasseri S, Mahvi AM, Tashauoei HR, Hadi M (2015) The estimation of per capita loadings of domestic wastewater in Tehran. J Environ Health Sci Eng 13:1–9. https://doi.org/10.1186/s40201-015-0174-2

Minnesota Pollution Control Agency (2006) Phosphorus treatment and removal technologies. https://www.pca.state.mn.us/sites/default/files/wq-wwtp9-02.pdf

Mino T, Arun V, Tsuzuki Y, Matsuo T (1987) Effect of phosphorus accumulation on acetate metabolism in the biological phosphorus removal. In: Ramadori R (ed) Biological phosphate removal from wastewaters, Advances in water pollution control. Pergamon Press, Oxford, pp 27–38

Molle RG (2006) Nutrient removal – Water Environment Research Foundation Workshop, Washington DC. Water Environment Research Foundation, Alexandria

Namasivayam C, Sangeetha D (2004) Equilibrium and kinetic studies of adsorption of phosphate onto ZnCl2 activated coir pith carbon. J Colloid Interface Sci 280:359–365. https://doi.org/10.1016/j.jcis.2004.08.015

Nassef E (2012) Removal of phosphates from Industrial wastewater by chemical precipitation. IRACST Eng Sci Technol: Int J (ESTIJ) 2:409–413. https://www.academia.edu/4368723

National Eutrophication Research Program (NERP) (1970) Original data or information from the NERP-Environmental Protection Agency. Pacific Northwest Water Laboratory, Corvallis

Neal C, Jarvie HP, Howarth SM, Whitehead PG, Williams RJ, Neal M, Harrow M, Wickham H (2000) The water quality of the River Kennet: initial observations on a lowland chalk stream impacted by sewage inputs and phosphorus remediation. Sci Total Environ 251–252:477–495. https://doi.org/10.1016/S0048-9697(00)00400-9

Neal C, Martin E, Neal M, Hallett J, Wickham HD, Harman SA, Armstrong LK, Bowes MJ, Wade AJ, Keay D (2010) Sewage effluent clean-up reduces phosphorus but not phytoplankton in lowland chalk stream (River Kennet, UK) impacted by water mixing from adjacent canal. Sci Total Environ 108:5306–5316. https://doi.org/10.1016/j.scitotenv.2010.08.010

Newman EI (1995) Phosphorus inputs to terrestrial ecosystems. J Ecol 83:713–726. https://doi.org/10.2307/2261638

Oehmen A, Vives MT, Lu HB, Yuan ZG, Keller J (2005) The effect of pH on the competition between polyphosphate accumulating organisms and glycogen accumulating organisms. Water Res 39:3727–3737. https://doi.org/10.1016/j.watres.2005.06.031

Oehmen A, Zeng RJ, Keller J, Yuan Z (2007) Modeling the aerobic metabolism of polyphosphate-accumulating organisms enriched with propionate as a carbon source. Water Environ Res 79:2477–2486. https://doi.org/10.1002/j.1554-7531.2007.tb00347.x

Oguz E, Gurses A, Canpolat N (2003) Removal of phosphate from wastewaters. Cem Concr Res 33:1109–1113. https://doi.org/10.1016/S0008-8846(03)00016-4

Olsson E, Karlgren L, Tullander V (1968) Household wastewater. Report 24. The National Swedish Institute for Building Research, Stockholm

Oswald W, Gotaas H, Golueke C, Kelle W, Gloyna E, Hermann E (1957) Algae in waste treatment. Sewage Ind Waste 29:437–457

Patel A, Barrington S, Llefsrud M (2012) Microalgae for phosphorus removal and biomass production: A six species screen for dual-purpose organisms. GCB Bioenergy 4(5):484–495

Pick U, Weiss M (1991) Polyphosphate hydrolysis within acidic vacuoles in response to amine-induced alkaline stress in the halotolerant alga *Dunaliella salina*. Plant Physiol 97:1234–1240. https://doi.org/10.1104/pp.97.3.1234

Pijuan M, Saunders AM, Guisasola A, Baeza JA, Casas C, Blackall LL (2004) Enhanced biological phosphorus removal in a sequencing batch reactor using propionate as the sole carbon source. Biotechnol Bioeng 85:56–67. https://doi.org/10.1002/bit.10813

Pitois S, Jackson MH, Wood BJB (2001) Sources of the Eutrophication problems associated with toxic algae: An overview. J Environ Health 64:25–32

Potgiete DJJ, Evans BW (1983) Biochemical changes associated with luxury phosphate uptake in a modified Phoredox activated sludge system. Water Sci Technol 15:105–116. https://doi.org/10.2166/wst.1983.0110

Prajapati M, van Bruggen JJA, Dalu T, Malla R (2017) Assessing the effectiveness of pollutant removal by macrophytes in a floating wetland for wastewater treatment. Appl Water Sci 7:4801–4809. https://doi.org/10.1007/s13201-017-0625-2

Qiu G, Law YM, Das S, Ting YP (2015) Direct and complete phosphorus recovery from municipal wastewater using a hybrid microfiltration-forward osmosis membrane bioreactor process with seawater brine as draw solution. Environ Sci Technol 49:6156–6163. https://doi.org/10.1021/es504554f

Randall CW, Marshall DW, King PH (1970) Phosphate release in activated sludge process. J Sanitary Eng Div: Proc Am Soc Civ Eng 96:395–408. https://cedb.asce.org/CEDBsearch/record.jsp?dockey=0017231

Rao NN, Gomez-Garcia MR, Kornberg A (2009) Inorganic polyphosphate: essential for growth and survival. Annu Rev Biochem 78:605–647. https://doi.org/10.1146/annurev.biochem.77.083007.093039

Reddy M (1998) Biological and chemical systems for nutrient removal. Water Environ Fed, special publication, 1–398

Roy K, Chari MS, Gaur SR (2013) Eutrophication in Lentic systems and its impact on fisheries. Int J Res Fish Aquac 3:170–175

Ruttenberg KC (2003) The Global Phosphorus cycle. Treatise Geochem 8:585–643

Rybicki SM (1998) New technologies of phosphorus removal from wastewater. http://citeseerx.ist.psu.edu/viewdoc/download?doi=10.1.1.470.2711&rep=rep1&type=pdf

Sawayama S, Rao KK, Hall DO (1998) Nitrate and phosphate ion removal from water by *Phormidium laminosum* immobilized on hollow fibres in a photobioreactor. Appl Microbiol Biotechnol 49:463–468. https://doi.org/10.1007/s002530051199

Scholz RW, Ulrich AE, Eilitta M, Roy A (2013) Sustainable use of phosphorus: a finite resource. Sci Total Environ 461–462:799–803. https://doi.org/10.1016/j.scitotenv.2013.05.043

Schroder JJ, Cordell D, Smit AL, Rosemarin A (2010) Sustainable use of phosphorus. Report 357. Plant Research International BV, Wageningen. Available from: http://ec.europa.eu/environment/natres/pdf/sustainableusephosphorus.pdf; https://ec.europa.eu/environment/natres/pdf/sustainable_use_phosphorus.pdf

Sedlak RL (1991) Phosphorus and nitrogen removal from municipal wastewater, 2nd edn. Lewis Publishers, New York

Shapiro J, Ribeiro R (1965) Algal growth and sewage effluent in the Potomac Estuary. J Water Pollut Control Fed 37:1034–1043. https://www.jstor.org/stable/25035329

Shapiro J, Levin GV, Zea HG (1967) Anoxically induced release of phosphate in -wastewater treatment. J Water Pollut Control Fed 39:1810–1818. https://www.jstor.org/stable/25035858

Shilton AN, Powell N, Guieysse B (2012) Plant based phosphorus recovery from wastewater via algae and macrophytes. Curr Opin Biotechnol 2:884–889. https://doi.org/10.1016/j.copbio.2012.07.002

Smith VH, Tilman GD, Nekolaazm JC (1999) Eutrophication: impacts of excess nutrient inputs on freshwater, marine, and terrestrial ecosystems. Environ Pollut 100:179–196. https://doi.org/10.1016/S0269-7491

Smith AL, Stadker LB, Love NG, Skerlos SJ, Raskin L (2015) Perspectives on anaerobic membrane bioreactor treatment of domestic wastewater: a critical review. Bioresour Technol 122:149–159. https://doi.org/10.1016/j.biortech.2012.04.055

Spagni A, Stante L, Bortone G (2001) Selection of denitrifying phosphorus accumulating organisms in activated sludge. Environ Technol 22:1429–1437. https://doi.org/10.1080/09593330.2001.11090877

Spellman FR (2008) Chapter 18: water treatment operations. In: Handbook of water and wastewater treatment plant operations, 2nd edn. CRC Press, Taylor & Francis Group, pp 617–620. https://doi.org/10.1201/9781003038351

Srinath EG, Sastry CA, Pillai SC (1959) Rapid removal of phosphorus from sewage by activated sludge. Experientia 15:339–340. https://doi.org/10.1007/BF02159818

Strom PF (2006) Technologies to remove phosphorus from wastewater. Invited presentation for water quality trading, 91st annual meeting, New Jersey Water Environment Association (NJWEA), Atlantic City, New Jersey

Sudiana M, Mino T, Satoh H, Nakamura K, Matsuo T (1997) Metabolism of enhanced biological phosphorus removal and non enhanced biological phosphorus removal sludge with acetate and glucose as carbon source. Water Sci Technol 1999 39(6):29–35

Sun J, Wang X, Li R, Zhu W, Li Y (2011) Hypersaline municipal wastewater treatment of a processing zone through pilot-scale A/O MBR, Part II: nitrogen and phosphorus removal. Procedia Environ Sci 8:781–788. https://core.ac.uk/download/pdf/82166735.pdf

Takiguchi N, Kuroda A, Ohtake H, Tsuneda S (2019) Heatphos process for recovering phosphorus from bio-P sludge before anaerobic digestion. In: Phosphorus recovery and recycling. https://doi.org/10.1007/978-981-10-8031-9_35

Thakur A, Kumar HD (1999) Nitrate, ammonium and phosphate uptake by the immobilized cells of *Dunaliella salina*. Bull Environ Contam Toxicol 62:70–78. https://link.springer.com/article/10.1007/s001289900843

Tiessen H, White PJ, Hammond JP (2008) Chapter 1 – phosphorus in the global environment. In: The ecophysiology of plant-phosphorus interactions. Springer. https://www.springer.com/gp/book/9781402084348

Tipping E, Benham S, Boyle JF, Crow P, Davies J, Fischer U, Guyatt H, Helliwell R, Blake JL, Lawlor AJ, Monteith DT, Rowe EC, Toberman H (2014) Atmospheric deposition of phosphorus to land and freshwater. Environ Sci Process Impacts 16:1608–1617

United Nations Economic and Social Commission for Asia and the Pacific (UNESCAP) (2015) United Nations Human Settlements Programme (UN-Habitat) and Asian Institute of Technology (AIT). Policy guidance manual on wastewater management with a special emphasis on decentralized wastewater treatment systems. Bangkok, Thailand. ISBN: 978-974-8257-87-7, pp 1–144

United Nations Economic and Social Commission for West Asia (UNESCWA) (2016) Annual report. E/ESCWA/OES/2017/1, pp 1–45

United Nations Educational, Scientific and Cultural Organisation (UNESCO) (2017) The United Nations world water development report. Wastewater the untapped resource: facts and figures, pp 1–12. http://unesdoc.unesco.org/images/0024/002475/247553e.pdf

United Nations World Water Development Report (UNWWDR) (2015) Water for a sustainable world. http://unesdoc.unesco.org/images/0023/002318/231823E.pdf

United States Environmental Protection Agency (USEPA) (1977) Detergent phosphate ban: U.S. Environmental Protection Agency Report 905/2-77-003

United States Environmental Protection Agency (USEPA) (1978) Impact of inorganic phosphates in the environment: final report, pp 1–94

USEPA (2009) Nutrient control design manual: state of technology review report. EPA/600/R-09/012

USEPA (2015) A compilation of cost data associated with the impacts and control of nutrient pollution, pp 1–110. www.epa.gov/sites/production/files/2015-04/documents/nutrient-economics-report-2015.pdf

USEPA (2017) Life Cycle Assessment and cost analysis of water and wastewater treatment options for sustainability: upgrade of bath, New York Wastewater Treatment Plant, pp 1–120

Vasudevan P, Griffin P, Warren A, Thapliyal A, Tandon M (2011) Localized domestic wastewater treatment: part I – constructed wetlands (an overview). JISR 70:583–594. http://nopr.niscair.res.in/handle/123456789/12485

Vymazal J, Sengupta M, Dalwani R (eds) (2008) Constructed wetland for wastewater treatment: a review. Proceedings of Taal 2007: the 12th word lake conference, pp 965–980. http://fos.darvodgeo.ru/CW_Vymazal_review_1.pdf

Wang N, Peng J, Hill G (2002) Biochemical model of glucose induced enhanced biological phosphorus removal under anaerobic condition. Water Res 36:49–58. https://doi.org/10.1016/s0043-1354(01)00236-6

Wang D, Li X, Yang Q, Zeng G, Liao D, Zhang J (2007) Biological phosphorus removal in sequencing batch reactor with single-stageoxic process. Bioresour Technol 99:5466–5473. https://doi.org/10.1016/j.biortech.2007.11.007

Wang Y, Jiang F, Zhang Z, Xing M, Lu Z, Wu M, Yang J, Peng Y (2010) The long-term effect of carbon source on the competition between polyphosphate accumulating organisms and glycogen accumulating organisms in a continuous plug-flow anaerobic/aerobic (A/O) process. Bioresour Technol 101:98–104. https://doi.org/10.1016/j.biortech.2009.07.085

Woodard S (2006) Magnetically enhanced coagulation for phosphorus removal. Session B2 in Water Environment Research Foundation (WERF).

World Water Assessment Programme (WWAP). (2012) Managing water under uncertainty and risk. The UN World Water development report 4, vol 1. http://unesdoc.unesco.org/images/0021/002156/215644e.pdf

Zeng L, Li X, Liu J (2004) Adsorptive removal of phosphate from aqueous solutions using iron oxide tailings. Water Res 38:1318–1326. https://doi.org/10.1016/j.watres.2003.12.009

Zhang HC, Cao ZH, Shen QR, Wong MH (2003) Effect of phosphate fertilizer application on phosphorus losses from paddy soils in Taihu Lake region I: effect of phosphate fertilizer rate on phosphorus losses from paddy soil. Chemosphere 50:695–701. https://doi.org/10.1016/S0045-6535(02)00207-2

Zilles JL, Peccia J, Hung CH, Noguera DR (2001) Molecular microbial ecology of enhanced biological phosphorus removal in aerated-anoxic orbal processes. Proceedings of the Water Environment Federation (WEF), WEF's Annual Technical Exhibition Conference, session 15, pp 718–731. https://doi.org/10.2175/193864701790902725

Chapter 4
Agricultural Waste: A Potential Solution to Combat Heavy Metal Toxicity

Rachana Singh, Kavya Bisaria, Parul Chugh, Lashika Batra, and Surbhi Sinha

Abstract Heavy metal contaminated soil and water have been a matter of concern for decades now. The numerous commercial, agricultural, domestic, technical and medical applications have caused their broad environmental distribution. This has raised concerns and questions regarding their possible influence on the environment and human health. Due to a high degree of perniciousness, these systemic toxicants such as arsenic, chromium, Mercury and cadmium rank among the priority metals of public health importance. They cause multiple organ damage, albeit at low exposure levels. Despite several methods of detoxifying the water, they are still limited from their widespread use due to certain disadvantages. Bioremediation, such as the use of biosorbents, however, is a promising substitute to the current means. Along with the available options, recent studies have reported the potential use of agricultural wastes like sawdust, orange peels, groundnut shells, banana peels, coconut shells, potato peels, etc., in mitigating the effect of heavy metals on the environment, and the same has been discussed here. In this chapter, we have committed our contributions to a thorough and detailed delineation of potential agricultural wastes applications in heavy metals treatment, their modified and unmodified forms, and factors such as pH, temperature, initial concentration of metal ions, biosorbent dosage, ionic strength on the removal process and so on.

Keywords Heavy-metals · Agricultural wastes · Bioremediation · Toxicity · Environment

---

R. Singh (✉) · K. Bisaria (✉) · P. Chugh · L. Batra · S. Sinha
Amity Institute of Biotechnology, Amity University Uttar Pradesh, Noida, Uttar Pradesh, India
e-mail: rsingh2@amity.edu

K. M. Gothandam et al. (eds.), *Environmental Biotechnology Volume 4*, Environmental Chemistry for a Sustainable World 68,
https://doi.org/10.1007/978-3-030-77795-1_4

4.1 Introduction

Heavy metals are characterized as metals having a density exceeding 5 g per cubic cm (Fergusson 1990). In view of the inter-relationship between toxicity and heaviness, heavy metals also contain metalloids like arsenic, responsible for causing toxicity at low exposure rates. With the increase in anthropogenic activities, especially after the industrial revolution, there has been a simultaneous hike in heavy metal contamination in the environment (Gough 2017; Christopher et al. 2012). It has caused considerable worldwide concern for environmental and public health safety. Both natural phenomena and human activities cause the release of toxic heavy metals in nature. Anthropogenic activities entail coal burning, metal processing, textiles, gasoline combustion, microelectronics, paper processing plants, wood preservation plants, nuclear and high-voltage power plants, chemicals (Ojuederie and Babalola 2017). Natural phenomena like volcanic eruptions and weathering also add considerably to heavy metal pollution (RoyChowdhury et al. 2018). Heavy metals immediately threaten freshwater since they act as an immediate sink for these pollutants from human and natural activities. Further, untreated wastewater used for irrigation purposes has led to soil contamination by accumulating heavy metals in the soil, which enter the food chain and further lead to biomagnification. Due to the latest environmental approaches used to tackle environmental concerns, there has been increased awareness of the importance of providing impacts. This has moved the research community towards developing economically viable, reliable and environmentally friendly methods that can extract toxins from water, safeguarding the health of the communities affected.

Many heavy metal removal procedures include chemical precipitation, adsorption, electrochemical, ion exchange, treatments, and coagulation-flocculation. Due to its easy operation, chemical precipitation is one of the most frequently used industry methods to eliminate heavy metals from inorganic effluent (Gunatilake 2015). Disadvantages of chemical precipitation, however, include the process's conventionally active nature. Chemicals must be procured, energy inputs and manual supervision are necessary, and a waste stream is produced. This may amount to relatively high treatment costs (Chen et al. 2018).

Likewise, the coagulation-flocculation method has drawbacks such as sludge production, chemical application, and toxic compounds' shifting to the solid phase (Sinha et al. 2020). Also, a very low concentrated metal solution can be used in electrolysis, and this process is extremely pH-sensitive. Another possible substitute for decontamination of heavy metals from wastewater is adsorption, in which substances are bound to a solid surface by physical and chemical interactions (Tan et al. 2017). Conventional adsorbents include commercially used activated carbon, coal fly ash, inorganic materials such as activated alumina, zeolites, silica gel and ion exchange resins. Despite the high quality of activated carbon, they have certain disadvantages such as non-selective nature, rapid saturation, poor regeneration and disposal. The hydrophilicity of silica gel, bauxite's ability to accumulate bacteria, use of strong acid or base for the regeneration of zeolites, and other such disadvantages limit their widespread use.

Biosorption is an effective substitute for traditional metal ion removal and/or recovery technologies from both water and soil (Singh and Prasad 2000; Singh and Sinha 2013). This method's significant advantages over traditional treatment methods include high performance, reduced biological or chemical sludge, low cost, biosorbent regeneration, and metal recovery potential (Prasad et al. 2006; Singh 2012). Generally, biosorption biomass includes algae, bacteria, fungus, sea material, kitchen waste, agricultural waste, etc. (Sinha et al. 2019a; Mehrotra et al. 2019). However, microbiological biomass has to be mass cultured, which requires additional costs due to instrumentation, media requirements, etc. (Sinha et al. 2019b). Therefore, adsorption by biomass, including agriculture or kitchen waste, is a better option for pollution control. It would reduce the negative environmental burden and enhance waste management and ensure sustainable resource use. Its process mechanisms include chemisorption, surface adsorption, and ion exchange due to the functional groups present in agricultural waste biomass like hydroxyl, alcoholic, carbonyl, amino, amide sulfhydryl, phenolic etc., by forming metal complexes or chelates. Basic components present in agricultural wastes such as lignin, hemicelluloses, lipids, simple sugars, proteins, water, starch, and hydrocarbons have various functional groups with potential sorption properties. Amongst many materials, cellulosic waste is a useful source of metal biosorption because it is available in abundance, non-toxic, cheap and has good adsorption properties (Acharya et al. 2018).

This chapter critically outlines the potential types of agricultural waste biosorbents in heavy metal adsorption based on their adsorption capability, operating factors, pretreatment methods and progress over the years. They can be treated to improve efficiency and recyclability to enhance their industrial applicability.

4.2 Heavy Metals and Their Effect on Human Health

Heavy metals are toxic, non-biodegradable elements that accumulate in the environment and toxify it. Industries release untreated effluent loaded with heavy metals into the surrounding water body, thereby destructing its flora and fauna. This ultimately reaches us, consequently damaging our health. A few heavy metals and their effects on humans, animals, and plants are discussed here:

Arsenic is found in lower concentrations in all the environmental matrices. Its sources include natural phenomena like soil erosion, volcanic eruptions and industrial activities, including manufacturing of glass, ceramics, pesticide, refining of metallic ores etc. It is also found in various products like pesticides, wood preservatives, dyes, veterinary medicines and other medicinal drugs. Toxicity due to arsenic causes various diseases in humans such as heart disease, neurological, neurobehavioral disorders, hearing loss, diabetes, peripheral vascular disease and hematological disorders. It also interferes with metabolic processes and inhibits plant growth and development by inducing phytotoxicity (Shrivastava et al. 2015).

Cadmium sources comprise pollutants from industrial activities, namely, smelting, battery production, mining, pigments, paint and fertilizer production, stabilizers and alloys. It causes various diseases like changes in pulmonary function, emphysema, osteoporosis, etc. and a severe gastrointestinal irritant. Cadmium-contaminated soil damage the plants include oxidative stress, photosynthesis inhibition, and inhibition of root metabolism.

Chromium is a metal that occurs naturally in the earth's crust, with oxidation states varying from Cr (II) to Cr (VI) and Cr (VI) being its most toxic form. Industries that release chromium into water bodies involve tannery plants, metal refining, chromate manufacturing, pigment manufacturing from ferrochrome and chrome and stainless steel welding. Naturally occurring Cr (VI) is found in groundwater and surface water at values above the World Health Organisation drinking water cap of 50μg Cr (VI) per liter (WHO 1996). Subjection to Cr (VI) containing compounds affects the respiratory tract, cardiovascular functioning, gastrointestinal tract, skin, liver and kidney (Achmad et al. 2017).

Lead is a bluish-gray, naturally occurring metal found in the earth's crust in limited quantities. It is often used in paints, automobile batteries, ammunitions, leaded glass, crystal, fossil fuels, caulking, pipe solder and pesticides. Accumulation of Lead in the soil causes oxidative stress to plants (Sidhu et al. 2016). Lead poisoning is also associated with various cancers and kidney diseases (Ibrahem et al. 2020).

Mercury, another innate heavy metal, enters the water bodies through water discharged from coal-burning power plants, oil refineries, and pharmaceutical products like dentistry, medical waste disposal facilities and cremation grounds. Combustion of jet fuel, heating oils, and diesel add Mercury to air, which then gets deposited in water and land and then enters into the food chain. It exists in 3 forms, elemental, inorganic and organic, and all forms are toxic and have harmful effects on humans, such as gastrointestinal toxicity, neurotoxicity, and nephrotoxicity. Mercury was reported to induce DNA methylation and post-translation modification of renal cells causing problematic kidney function (Khan et al. 2019).

4.3 Agricultural Waste as Biosorbents for Heavy Metal Removal

Agricultural waste-based biosorbents can be derived from various plant parts, like stem, leaves, flower, husk, root, skin, bran and fruit. They consist primarily of lignin, hemicellulose and cellulose that have strong hydroxyl group content. Thus, they have excellent properties to attach with heavy metals (Qadeer et al. 2018). Additionally, they include different functional groups like sulfhydryl, acetamide, ester, phenolic, amino, carboxyl, alcohols and amide. Such functional groups either substitute the metal ions in solution with hydrogen ions or create complexes by electron sharing with the metal ions. Because of the presence of a large number of binding groups, agricultural waste biomass can be a potential candidate as adsorbent materials in wastewater decontamination (De Gisi et al. 2016).

4.3.1 Benefits of Using Agricultural Waste as Biosorbents

Many studies are based on utilizing agricultural waste to substitute the existing traditional methods in sequestering heavy metals due to their considerable merits. The main benefit of agricultural wastes biosorbent over other adsorbent materials is their intense attraction to heavy metals and high selectivity due to the abundance of binding groups on their surface (Singh et al. 2018). Moreover, they can be conveniently handled, utilized and recovered without harmful environmental effects (Naoufal et al. 2018). Such attributes of agricultural waste biosorbents in industrial applications are likely to play critical roles, making them preferable over traditional adsorbents. Also, the recycling of agricultural waste and by-products is assumed to minimize waste in an eco-friendly way for heavy metal treatment purposes.

4.3.2 Selection of Agricultural Waste as Biosorbents

The selection of good agricultural waste as biosorbent for sequestration of heavy metals among various waste and by-products is not a simple task. As per Ngo et al. (2016), when choosing agricultural waste for practical application, availability and cheapness are critical factors. On the other hand, based on his observations, Nguyen and Pignatello (2013) suggest that adsorption capacity can be used as a deciding element. Also, this observation is inconsistent with the results of Shakoor et al. (2016). The two research groups proposed that only certain bio-sorbents with reasonably high binding potential and heavy metal selectivity would be appropriate for use in a large-scale biosorption process. Most researchers agree that several requirements need to be met by a good biosorbent, including high regeneration capability, abundant availability, easy desorption and high cost-effectiveness, in aqueous solutions.

4.3.3 Mechanisms Involved in Heavy Metal Removal

Research into the adsorption mechanisms is important to understand the cycle, which will help to develop the method (Vardhan et al. 2019). Many variables influence the removal of heavy metal through agricultural waste biosorbents, including forms of waste, metal solution composition, and environmental factors. Consequently, the exact mechanism of metal biosorption is unknown, although several mechanisms, including ion exchange, diffusion, surface adsorption, precipitation and chelation, have been suggested. For instance, Zhao et al. (2017) reported that the size of the heavy metal and surface morphology of biochar determined the ion-exchange mechanism's effectiveness. Different mechanisms may function to varying degrees at the same time. Although there is a dynamic existence of biosorption tools, the application of conventional approaches like Transmission Electron Microscopy, Scanning Electron Microscopy, Fourier Transform Infra-Red, etc., and traditional methods such as titration are essential for the discovery of a dominant mechanism for specific agricultural waste biosorbent and heavy metal interactions. By producing complexes with functional groups (Kołodyńska and Bąk 2018), in an FTIR performed before and after Cu sorption on biochar, carboxyl and hydroxyl groups were found to play a dominant role in Cu sorption onto biochar. In FTIR spectra of Brassica adsorbed with Ni^{2+}, Cr^{6+}, and Pb^{2+}, slightly shifted peaks were seen at 1365, 1362, 1352 cm^{-1}, which attributed to Ni-O, Cr-O and Pb-O bonds, respectively (Baby Shaikh et al. 2018).

4.4 Modification Methods for Agricultural Waste Biosorbents

In several studies, untreated agricultural waste is used to obliterate heavy metals and other pollutants from effluent (Tatah et al. 2017; Prabha 2018; Singh et al. 2018). Its limitations include the high release of soluble organic compounds into the solution, and low adsorption performance may have significant drawbacks. Hence, it is suitable for agricultural waste to be modified before it is used for biosorption (Sharma et al. 2018). Modification techniques include physical as well as chemical modifications and many different approaches, as illustrated in Fig. 4.1.

Physical modifications are commonly regarded as simple and inexpensive but are not frequently used due to their poor performance. Thus, chemical modifications are more favored because of their flexibility and reliability. The influencing factors may be categorized as bases, acids and salts, organic compounds, oxidizing factors etc. The literature tells that chemically-activated agricultural waste biosorbents demonstrated higher adsorption ability than their unmodified counterparts, which could be due to improved ion-exchange efficiency, more binding sites, and new functional development groups that support metal absorption. Chemical treatments may increase the capacity of the biosorbent to adsorb (Melia et al. 2018). New materials such as coconut shell, hazelnut shell, jackfruit, rice husk, corn straw, pecan shells and maize cob etc., can be used as an adsorbent for heavy metal absorption after chemical alteration or as activated carbon or biochar formed after heating (Khatoon and Rai 2016). Activated carbon is derived from agro products, and waste materials generally contain low ash content and a high percentage of volatile matter, making them favorable precursors for activated carbon production. Ash content is the standard measure of minerals present in the carbon as impurities. They are the remains that linger during pyrolysis after the carbonaceous part is burned off. Products with

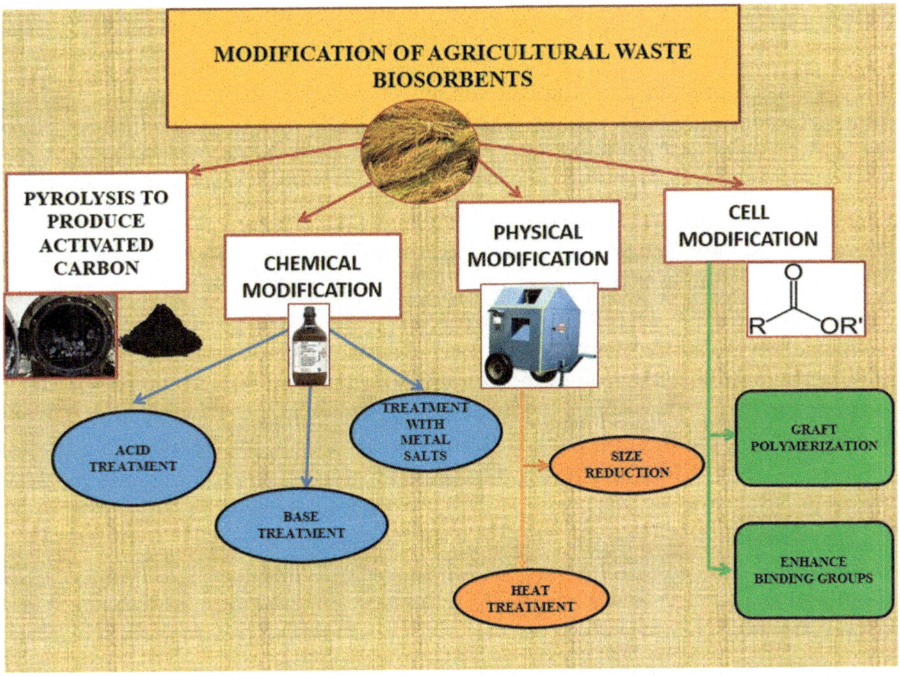

Fig. 4.1 Modification methods for producing effective Agricultural Waste Biosorbents

the lowest precursor ash content are chosen for providing the most active areas on the activated carbon surface.

4.5 Factors Affecting the Biosorptive Removal of Heavy Metals

Many chemical and physical factors, such as temperature, ionic strength, initial heavy metals concentration, pH, biosorbent size, biosorbent dose, co-ions, etc., could impact the biosorption process by affecting adsorption rate, selectivity, adsorption rate, and a number of heavy metals removed. Significant work into the impacts of these operating parameters has been conducted. This section briefly discusses the effects of such process parameters.

4.5.1 pH

pH appears to play an essential role among process factors in regulating biosorption. pH values may influence the degree of speciation and heavy metals ionization, the surface charge of the biosorbent, and the competition of metal ions in solution with coexisting ions (Ittrat et al. 2014). The intervention of functional groups in metal chemistry and metal uptake causes this pH dependency (Kumar et al. 2016). Biosorptive removal of cationic metals rises with the increase in pH of the solution, while that of anionic metals decreases. The overall surface charge of agricultural waste biosorbents at a lower pH is positive. H^+ ions interact efficiently with metal cations allowing sorption ability to reduce. As pH values rise, electrostatic interaction facilitates the absorption of metal ions, causing the surface of the biosorbent to become progressively negatively charged. Biosorption stops at a very high pH, and the hydroxide precipitation begins (Beni and Esmaeil 2020). In a study regarding adsorption of Cr, Mn, Cu, Co, Pb, Ni, Zn and Cd on the unmodified and modified tangerine peel, it was found that pH range 3–5 for sorption of Cr, Mn, Cu, Co, Pb, Ni, (5 mg/L) and Zn and Cd (2 mg/L) had a significant impact. For unmodified sorbent, very low adsorption was observed for mostly all the tested metals, which increased to pH 5. This was attributed to high H^+ concentration at low pH, which occupies active adsorbent sites, and possible metal hydroxide formation and phenolic polymerization take place above pH 5, decreasing the % removal efficiency. It was also cited that the removal should only be a function of adhesion to the sorbent and not by metal precipitation (Abdić et al. 2018). The oxidized biochar of Eichhornia crassipes showed increased adsorption of Pb^{2+}, Zn^{2+}, Cd^{2+} and Cu^{2+} at high initial pH due to stronger electronegativity under less acidic conditions (Lin et al. 2020).

4.5.2 Temperature

Many investigations have been conducted to examine the influence of temperature on metal adsorption by sorbents. The rise in the solution's temperature influences the metal ions diffusion rate and metal ion solubility (Mohubedu et al. 2019). The temperature has a characterized effect on the adsorption efficiency, which depends on the surface functional groups of the given agricultural waste biosorbent. Besides that, several reports generally assume that the temperature effect is restricted to a degree and within a specific temperature range. Depending on the endothermic or exothermic nature of the process, the temperature can affect the biosorption process in several ways (Iftekhar et al. 2018).

4.5.3 Initial Concentration of Metal Ions

Initial metal concentration can cause metal ions to transfer from the solution to the surface of the biosorbent. It has been perceived in multiple studies that as the initial concentration decreases, the overall adsorption potential of particular agricultural waste biomass rises.

4.5.4 Biosorbent Dosage

Several studies found that with rising dosages of the agricultural waste biosorbents, the percentages of metal ions removal from the medium increase. Pyrzynska (2019) reported that Cd (II) elimination rose steadily with the rise in cashew nut shell dosage. At 3 g L^{-1} cashew nutshell concentration, the elimination of Cd (II) ion was 75.35%. They linked this activity to the increased count of adsorption sites available in a large amount of biosorbents (De Gisi et al. 2016). A similar result was shown in adsorption of Pb^{2+}, Ni^{2+} and Cr^{6+} by *Brassica campestris* stems (Baby Shaikh et al. 2018).

4.5.5 Ionic Strength

Actual industrial effluent includes ions of different metals like K^+, Na^+, Mg^{2+}, Ca^{2+}, and heavy metal ions. Therefore, KCl, NaCl, $CaCl_2$ $MgCl_2$ generally added to heavy metal solutions to study the impact of ionic strength on heavy metal ion biosorption. He et al. (2018) observed that the increase in ionic strength caused a diminished capacity of adsorbents to absorb metal. This could be attributed to the elevated concentration of competing cations in the medium diminish the metal ion activity.

Along with these parameters, particle size and co-ions also influence the biosorption process. The particle size of the biosorbents may affect their adsorption ability due to the alteration of the total surface area required for metal adsorption. However, the wastewaters typically contain several heavy metals rather than single metals. Heavy metal can impede the uptake of another heavy metal. The study of the effects of co-ions and particle size is crucial in improving the industrial-scale implementation of biosorbents.

4.6 Different Agriculture Wastes Used as Sorbents

Various agricultural waste biomass such as waste tea, durian shells, sawdust, rice bran, coffee husks, herb residues, spent grains, cotton stalks, giant reeds, apricot shell, almond shell, cashew nutshell, barley straw, pongam seed shell, pine sawdust, groundnut shell, plum kernel, olive stone potato peel, pomegranate peel, rice shell, pomelo peel, sugarcane bagasse, rice straw, walnut shell, cane pith, banana peel, orange peel, coir pith, sunflower stalk, yellow passion fruit, soy meal hull, rice husk, white ash, sawdust carbon, etc. have previously been utilized as inexpensive adsorbents for heavy metal decontamination. A few of these wastes chosen on the basis amount in which they are produced have been discussed.

4.6.1 Potato Peels

Potato is an important, carbohydrate-rich crop that can be grown in any climate and condition. In 2017, the cumulative world potato production was approximated to be at 388,191,000 tons. China is the largest producer globally, and almost one-third of all potatoes are harvested in India and China. In view of the statistics of potato production, an enormous amount of waste is produced from it as most of the potato peel is thrown, and only a low amount of it is used as a fertilizer, supplementary animal feed or is composted. Hence, potato peels can be considered an excellent choice for inexpensive, green adsorbents. Much research has been done to determine potato peels' adsorption capacity in varied forms (raw, burnt and treated) for heavy metals.

El-Maghraby and Lamiaa (2019) have reported that dried potato peels' removal capacity for Cu^{2+} and Fe^{3+} from aqueous solution was 83% and 77%, respectively. Other trends observed from his study included that as the adsorbent dosage increases, the adsorption capacity of Cu^{2+} increased from 45% to 85%. This increase in the extent of adsorption is due to more active sites along with higher surface area. The adsorption capacity for Cu^{2+} and Fe^{3+} also rose with pH elevation from 2–8, and maximum adsorption for both Cu^{2+} and Fe^{3+} was obtained at pH 4.5–5. At pH above 5, giving lower removal percentage is observed due to hydrolysis of copper ion, which forms the colloidal complex making it challenging to be adsorbed on the sorbent. One gram of potato peel was found enough to remove 83% and 77% of the copper (II) and ferric (III) from a solution. The adsorption efficiency of Burnt Potato Peels was found comparatively more than Raw Potato Peels for Cu (II), Co (II), Fe (II), Ni (II), La (III) and Pb (II) ions (El-Azazy et al. 2019). The potential of potato peels as an effective sorbent can also be increased by means of chemical or biological methods. A few studies have been summarized in Table 4.1.

Table 4.1 Different Agricultural Wastes used for heavy metals removal

Agricultural Waste	Heavy Metal	Modifications	Source of wastewater	Initial concentration of the metal	pH	Dosage of adsorbent	Temperature	Adsorption Capacity (mg g^{-1} or %)	References
Dried persimmon leaves	Cu(II) Pb(II) Cd(II)	None	Synthetic wastewater	0.5-20 mg/L	5	3 g/l	40 °C	19.42 mg/g 22.59 mg/g 18.26 mg/g	Lee and Cho (2018)
Dried tangerine Peel	Cr Cu Mn Co Ni Pb Cd Zn	Treatment with 0.25 M nitric acid	Synthetic wastewater	5 ppm 5 ppm 5 ppm 5 ppm 5 ppm 5 ppm 2 ppm 2 ppm	5	200 mg	25 °C	88.92% 97.04% 92.48% 94.70% 93.50% 93.00% 97.90% 96.80%	Abdić et al. (2018)
Loofah biochar	Cr^{6+} Cu^{2+}	Chitosan combined with magnetic Loofah biochar	Synthetic wastewater	40 mg/l 40 mg/l	3 7	40 wt%	30 °C	30.14 mg/g 54.68 mg/g	Xiao et al. (2019)
Pistachio hull powder	Ni^{2+}	None	Synthetic wastewater	100 mg/l	4–6	10 g/l	25 ± 3 °C	14 mg/g	Beidokhti et al. (2019)
Lotus seedpod	Pb^{2+}	Functionalized with carboxyl group (CLSP)	Synthetic wastewater	100 mg/l	5	5 mg	298.15 k	111.1 mg/g	Liu et al. (2019)
Ripe Banana peels Unripe Banana peels	Cd^{2+} Pb^{2+} Cd^{2+} Pb^{2+}	None	Synthetic wastewater	50μg /ml	3 5	30 g/l 40 g/l 30 g/l 40 g/l	25 °C	2.6185 mg/g 2.8810 mg/g 1.9051 mg/g 1.630 mg/g	Sirilert and Maikrang (2018)
Oil palm frond	Cr^{6+} Pb^{2+}	None	Synthetic wastewater	100 ppm/50 ml	3	0.5 g	20 °C	2.044 mg/g 5.322 mg/g	Ehishan and Sapawe (2018)

(continued)

Table 4.1 (continued)

Agricultural Waste	Heavy Metal	Modifications	Source of wastewater	Initial concentration of the metal	pH	Dosage of adsorbent	Temperature	Adsorption Capacity (mg g^{-1} or %)	References
Dried soya bean seeds	Pb^{2+} As	None	Industrial wastewater	1.24 mg/ml 0.24 mg/ml	4.0 ± 0.26 2	3 g/100 ml	37 °C	80% 40%	Gaur et al. (2018)
Activated Carbon of Camellia oleifera seed shell (COSs-AC)	Cr^{6+}	KOH activation	Synthetic wastewater	30 mg/l	2–3	250 mg/l	25 °C	307.26 mg/g	Guo et al. (2018)
Chickpea husk activated carbon	Pb^{2+} Cr^{6+} Cu^{2+}	Activation by K_2CO_3 and KOH	Synthetic wastewater	2 g/l 3 g/l 3 g/l	8	400 mg/l	40 °C	135.8 mg/g 59.6 mg/g 56.2 mg/g	Özsin et al. (2019)
Soy waste biomass	Pb(II) Cu(II) Ni(II)	Functionalized with industrial Sulphur based chelating agent	Synthetic wastewater	0.40 mmol/l	5.5	5 g/l	20 °C	196% 131% 128%	Bulgariu and Bulgariu (2018)
Activated Carbon from Rice husk	Cr(VI)	None	Synthetic wastewater	50 mg/l	2	0.2 g	25 °C	91.23% (34.85 mg/g)	Mullick et al. (2017)
Dried apple peels	$AsO_2^{\ -}$ $AsO_4^{\ 3-}$ PO_4^{3-} $Cr_2O_7^{2-}$	Immobilization of zirconium (Zr^{4+}) on apple peels	Synthetic wastewater	100 mg/l	10 2 2 2	0.1 g	30 °C	15.64 mg/g 15.68 mg/g 20.35 mg/g 25.28 mg/g	Mallampati and Valiyaveettil (2013)
Activated carbon prepared from apple peels	Cr(VI)	None	Synthetic wastewater	50 mg/l	2	0.05 g/50 ml	28 °C	36.01 mg/g	Enniya et al. (2018)
Cucumis sativus peel	Cd^{2+}	HCl treatment	Synthetic wastewater	20 mg/l	5	2 g/l	298 K	84.85%	Pandey et al. (2014)

Agricultural Waste	Heavy Metal	Modifications	Source of wastewater	Initial concentration of the metal	pH	Dosage of adsorbent	Temperature	Adsorption Capacity (mg g^{-1} or %)	References
Dried cucumber peels	Cd^{2+} Cr^{6+} Cu^{2+} Hg^{2+} Pb^{2+} Ni^{2+}	Enclosed in sodium alginate beads	Drinking water	0.1 mg/l	7	25 ml	25 °C	90.8% 16.6% NA 89.9% 82.1% 66.2%	Singh et al. (2019)
Burnt potato peels	Pb^{2+}	In the presence of nanobubbles In the absence of nanobubbles	Synthetic wastewater	–	6	0.05 g/50 ml	25 °C	171 mg/g 167 mg/g	Kyzas et al. (2019)
Raw Banana peels	Pb^{2+}	–	–	–	–	–		90.99%	Taralgatti (2019)
Dried coconut Shell	Pb^{2+} Ni^{2+}	None	Synthetic wastewater	40 mg/l	4.5 5	0.01 g 0.03 g	20 ± 1 °C	36.9 mg/g 4.48 mg/g	Correiaa et al. (2018)
Activated coconut Shell	Pb^{2+}	Activated using 15% $ZnCl_2$		49.96µg/l				97.06%	Rohmah et al. (2018)
Banana peel activated carbon by H3PO4	Ni^{2+}	Activation by hydrothermal treatment (BAC-H samples) Activation by pyrolysis (BAC-P samples)	Synthetic wastewater	100 mg/l	6	0.02 g/20 ml	303 K	76.08 mg/g 84.64 mg/g	Bibaj et al. (2018)
Dried Banana peels	Cr^{6+}	Treated with NaOH (10%) and HCl(10%)	Synthetic wastewater	5µg/ml	2	1 g/50 ml	37 °C	90%	Ahmed and Misganaw (2019)

(continued)

Table 4.1 (continued)

Agricultural Waste	Heavy Metal	Modifications	Source of wastewater	Initial concentration of the metal	pH	Dosage of adsorbent	Temperature	Adsorption Capacity (mg g^{-1} or %)	References
Dried groundnut shells	Pb^{2+}		Synthetic wastewater		6	2 g/l	25 °C	42.64 mg/g	Bayuo et al. (2018)
Groundnut shell powder	Cu^{2+} Pb^{2+}	None	Synthetic wastewater	10-1000 mg/l	5 ± 0.01 5 ± 0.02	50 g/l 30 g/l	30 °C	68.2% 77.8%	Shruthi and Pavithra (2018)
Orange peel biochar Sugarcane biochar	Pb^{2+}	None	Synthetic wastewater	57.36 mg/g	5	1 g/l	25 °C	27.86 mg/g 86.96 mg/g	Abdelhafez and Li (2016)
Orange peels powder	Cr^{6+}	–	Synthetic wastewater	5 mg/l	7	2 g/40 ml	37 °C	4.69 mg/g	Malook and Ul-Haquea (2019)
Sawdust carbon	Cd^{2+}	Loaded with magnetic iron oxide nanoparticles (Fe3O4/SC) EDTA modified Fe3O4/SC (EDTA@ Fe3O4/SC) nanocomposites	Synthetic wastewater	30 mg/l	6.5	0.4 g/l	27 °C	51 mg/g 63.3 mg/g	Kataria and Garg (2018)
Jackfruit wood sawdust	Pb^{2+}	Treated with phosphoric acid	Synthetic wastewater	25 ppm/100 ml		5 g/100 ml		1.796 mg/g	Mutiara et al. (2018)
Pyrolyzed sawdust	Pb^{2+}	Coating with molybdenum *disulfide* (MoS2)	Synthetic wastewater	360 mg/l	6	150 mg/l	25 °C	189 mg/g	Zhu et al. (2018)

4.6.2 Coconut Shell

Coconuts (*Cocos nucifera*) are fibrous one-seeded drupes of coconut palms that grow in tropical climates. In 2018, approximately 6.1 M tonnes of coconuts were produced all over the world. In 2018, a combined 72% share of global coconut consumption was accounted to Indonesia (19 M tonnes), the Philippines (14 M tonnes) and India (12 M tonnes). Due to this massive amount of coconut production and consumption, the waste produced, i.e., the coconut shell is in abundance, and its disposal is a big problem for environmentalists; hence using them as adsorbents can help in better waste management. A significant amount of research has been done to harness coconut shells' adsorption property as raw or modified forms to eliminate heavy metals. Coconut shells can also be converted to activated carbon as they contain high carbon content, high strength, high lignin and low ash content. Waste tea, coconut shell and coconut husk and activated carbon were used to treat different effluent samples contaminated with Ni^{2+} and Cr^{6+}. Coconut shells had maximum adsorption capacity for Ni^{2+} but Cr^{6+}ions. Coconut husk exhibited the best results in this investigation. Results also revealed that the adsorption was because of acidic functional groups (C=O) on the sorbent, which was involved in electrostatic interaction between the negative charges on the adsorbent and the positive charges of metal ions.

Another biosorbent prepared from Coconut shell was converting it into activated carbon and further chemically activated with zinc chloride (ACS). This was utilized to eliminate Fe (II), Cu (II), Zn (II) and Pb (II) ions from electroplating industrial wastewater. At optimum reaction parameters, maximum recoveries obtained were; 93.37% removal of Fe (II), 92.22% of Cu (II), 60.52% of Zn (II) and 100% removal of Pb (II). Furthermore, in the research conducted by Babel and Kurniawan (2004) to ascertain the influence of chemical modification on the adsorption capacity of coconut shell charcoal and commercially activated carbon in metal decontamination from synthetic electroplating wastewater, the surfaces of coconut shell charcoal and commercially activated carbon were chemically activated with strong oxidizing agents (sulphuric acid and nitric acid) and coconut shell charcoal coated with chitosan. Their adsorption capacity was measured at different parameters, and it was observed that given the scale of Cr concentration (5 to 25 mg l^{-1}), the adsorption capacity of coconut shell charcoal activated with nitric acid was highest (10.88 mg of Cr (VI) g^{-1}) to comparison to all other types of modifications. Thus, considering the technical feasibility, engineering applicability and low-cost, it is said that coconut shell charcoal is better than commercially activated carbon for Cr (VI) elimination. More studies of various forms of coconut are cited in Table 4.1.

4.6.3 Banana Peels

Banana is the second most important fruit crop in India. It is available throughout and has a variable range with nutritional and medicinal values and affordable. With a yearly production of around 14.2 million tonnes, India is one of the world's biggest producers. Disposal of banana waste in rivers or ponds causes the formation of methane and other gases due to its slow degrading nature, leading to the formation of putrescible smell affecting the ecosystem (Ahmad and Danish 2018). Hence, banana peels have been a popular subject in adsorbent studies.

In an investigation, charred banana peels modified with 1:2 (w/v) concentrated sulfuric acid were found to be a suitable bio-sorbent for Fe^{2+} ions removal from industrial wastewater (Shrestha 2018). It was found that at optimum pH 3, the adsorption of Fe^{2+} ions on charred banana peels was approximately 98%, with the highest adsorption capacity of 33.79 mg g^{-1}. Also, in a different study, Amin and co-workers worked upon the use of biochar prepared from the banana peels to prepare biochar for the mitigation of Cu^{2+} and Pb^{2+} from wastewater (Amin et al. 2017). The optimum pH for removal of Cu^{2+} and Pb^{2+} were found 5.5 and 9 respectively, with 1.4 g adsorbent after 30 min of equilibration time. Functional groups like N-H, O-H, Si-O, aliphatic groups and aromatics rings were found to be responsible for the adsorption.

In recent research, the adsorbent made from the mixture of banana peels and chitosan was utilized as a sorbent for Cr ions removal from the liquid waste of Batik. Maximum 69.8% elimination of Cr ions was reported after 80 min of equilibrium time. Rengganis and co-workers found that the adsorption mixture could adsorb the Cr ions due to amino and carboxylic groups' presence on its surface as they are reactive to the metal ions (Rengganis et al. 2018). More studies are mentioned in Table 4.1.

4.6.4 Groundnut Shells

Groundnut, a leguminous crop plant, has high-oil edible seeds. It is the third most important vegetable protein source and the fourth most important source of edible oil. With 166.24 lakh tonnes, China is the largest producer as well as consumer of groundnut in the world, followed by India (68.57 lakh tonnes), Nigeria (30.28 lakh tonnes) and the United States (25.78 lakh tonnes). Groundnut shells can be considered an inexpensive, locally available resource to be utilized as an adsorbent for heavy metal ions from aqueous solutions (Joseph et al. 2019). It has been used as an adsorbent in powder or activated carbon to eliminate heavy metals. The characterization of groundnut shell powder revealed several functional groups like alcohols, carboxylic acids and amines (Bayuo et al. 2019) that hold heavy metals through electrostatic interactions. Activated carbon obtained from peanut shells was found mesoporous, chiefly amorphous with surface hydroxyl, carboxyl and carbonyl groups, which helped in the adsorption of Cd^{2+} ions from industrial wastewater. It exhibited 55.43 ± 6.82 mgg^{-1} adsorption capacity for Cd^{2+} ions, and equilibrium was attained within 180 min (Villar da Gama et al. 2018).

In a study, the effect of slow (G350) and fast pyrolysis (G700) of groundnut shell biochars produced at pyrolysis temperature 350 ± 5 °C and 700 ± 5 °C, respectively, was determined to understand its effect on the adsorption capacity of biosorbents in the removal of Hg^{2+}, Cd^{2+}, Pb^{2+}. At 0.50, 0.20, 0.50 mgL-1 concentrations of Hg^{2+}, Cd^{2+}, Pb^{2+}, respectively, G350 was able to remove 100% of each metal whereas G700 removed 100% lead and mercury, 99.93% cadmium (Cobbina et al. 2018).

4.6.5 *Orange Peels*

Orange is one of the top citrus fruits grown in most of the countries after banana and apple. The global annual production of orange in 2019 was 47.5 million tones, with Brazil being the largest producer globally, followed by the US and China. Various studies have been done to utilize the discarded orange peels in adsorbents for heavy metals. They contain lignin, pectin, cellulose and hemicellulose, where these constituents have various surface functional groups like O-H, C-H, C=O, C-O, etc., that attract the heavy metal ions.

Orange peels treated with 0.2 M NaOH and 0.2 M $CaCl_2$ were used to uptake heavy metal ions Cd^{2+} and Zn^{2+} from the solution, and the adsorption was observed to be 30.7 mg g^{-1} and 46.9 mg g^{-1}, respectively (Flores et al. 2018). Similarly, activated carbon from orange peels treated with H_3PO_4 acid solution and Orange peels coated with chitosan were used for the adsorption of Cr^{6+}. The batch experiment was conducted for the given study. The pseudo-second-order kinetics model seems to fit well with the adsorption data for activated carbon from orange peels and Orange peel coated with chitosan beads. The adsorption for C^{6+} ions was calculated to be 17.27 mg g^{-1} for activated carbon from orange peels and 17.24 mg g^{-1} for Orange peel coated with chitosan, respectively (Tejada-Tovar et al. 2018).

Furthermore, the uptake of Cd^{2+}, $Pb^{2+}$and Ni^{2+} ions was tested by grafted copolymerization-modified orange peel. It was observed that the maximum uptake capacities of copolymerization-modified orange peel for Pb^{2+}, Cd^{2+} and Ni^{2+} ions were 476.1, 293.3 and 162.6 mg g^{-1}, respectively, using the Langmuir equation. As compared with its untreated counterpart, the treated biomass's adsorption capacity increased 4.2, 4.6 and 16.5 folds for Pb^{2+}, Cd^{2+} and Ni^{2+}, respectively.

4.6.6 *Sawdust*

Sawdust is considered an appealing material to be used as an adsorbent to remove salts, dyes, and heavy metals from wastewater, among other agricultural wastes. Sawdust contains lignin, cellulose, and hemicellulose, and the surface study of the wooden sawdust revealed the existence of phenolic and alcohol hydroxyl functional groups that bind with the heavy metals. The adsorption capacity of sawdust and rice husk in natural form and modified form, i.e., coated with silver nanoparticles (AgNPs), was compared in adsorption of Pb^{2+} Cd^{2+} from the water. Comparable studies showed the adsorption of metals in order of the used adsorbents as AgNPs loaded rice-husk>rice-husk>AgNPs loaded sawdust>sawdust (Abdou et al. 2017).

In addition, modified raw sawdust was prepared by grafting its surface with phosphorus oxychloride for the elimination of Cd (II), Cr (III) and Pb (II) metal ions. According to monolayer Langmuir adsorption, the adsorption capacity for Cr (III), Cd (II) and Pb (II) metal ions were learnt to be 325, 244.3 and 217 mg g^{-1}, respectively, at 298 K. The adsorbent could easily be regenerated using 0.1 M HCl solution (Alhumaimess et al. 2018). In a similar study, the activated carbon of pinewood sawdust, modified with H_3PO_4 in a different mass ratio of H_3PO_4 to sawdust at various activation temperature as 1:1, 500 °C (IR1–500) and 4:1, 800 °C (IR4–800) in a spouted bed were used for biosorption of Cu (II) ions. It was well described using the pseudo-second-order reaction kinetics, indicating chemisorption dominated in the process. The uptake capacity of Cu (II) on activated carbon IR1–500 was much higher than that of IR4–800 as the adsorption area of IR1–500 was much larger with a larger number of carbonyl and carboxyl groups along with P-containing groups (Phosphates, metaphosphates and pyrophosphates), resulting in increased adsorption capacity for Cu(II) ions (Gao et al. 2018).

Apart from the above-mentioned agricultural wastes, many other natural agricultural wastes have also been studied for eliminating heavy metals. A few of these are mentioned in Table 4.1.

4.7 Conclusion and Future Aspects

The current chapter discusses various agricultural waste biosorbents and different factors affecting their adsorption efficiency for heavy metals removal from wastewater. Based on the literature, it can be concluded that agricultural waste biosorbents are a low-cost alternative to traditional methods. Since most of the literature is lab-scale studies, it should be thoroughly investigated for potential industrial use in multiple metals systems in real wastewater. Even though agricultural waste biosorbents cannot fully replace the currently used methods, they can do an extent to reduce the use of chemicals as much as possible and make wastewater treatment more economically and ecologically feasible. Sorbent recovery, regeneration and end life are other important aspects that need to be addressed to scale the current studies.

References

Abdelhafez AA, Li J (2016) Removal of Pb(II) from aqueous solution by using biochars derived from sugar cane bagasse and orange peel. J Taiwan Inst Chem Eng 61:367–375. https://doi.org/10.1016/j.jtice.2016.01.005

Abdić Š, Memić M, Šabanović E, Sulejmanović J, Begić S (2018) Adsorptive removal of eight heavy metals from aqueous solution by unmodified and modified agricultural waste: tangerine peel. Int J Environ Sci Technol 15:2511–2518. https://doi.org/10.1007/s13762-018-1645-7

Abdou KA, Mohammed AN, El Rahman Moselhy WA, Farghali AA (2017) Assessment of modified rice husk and sawdust as bio-adsorbent for heavy metals removal using nanoparticles in fish farm. Asian J Anim Vet Adv 13:180–188. https://doi.org/10.3923/ajava.2018.180.188

Acharya J, Kumar U, Rafi PM (2018) Removal of heavy metal ions from wastewater by chemically modified agricultural waste material as potential adsorbent-a review. Int J Curr Eng Technol 8(3):526–530. https://doi.org/10.1016/j.biortech.2007.06.011

Achmad RT, Budiawan, Auerkari E (2017) Effects of chromium on human body. Annu Res Rev 17(2):1–8. https://doi.org/10.9734/ARRB/2017/33462

Ahmad T, Danish M (2018) Prospects of banana waste utilization in wastewater treatment. J Environ Manag 206:330–348. https://doi.org/10.1016/j.jenvman.2017.10.061

Ahmed MAA, Misganaw W (2019) Chemically modified banana peel as a potential bioadsorbent to remove chromium (VI) from effluent. Int J Res Eng Appl Manag 5(6). https://doi.org/10.35291/2454-9150.2019.0414

Alhumaimess MS, Alsohaimi IH, Alqadami AA, Kamel MM, Naushad M, Ahamad T, Alshammari H (2018) Synthesis of phosphorylated raw sawdust for the removal of toxic metal ions from aqueous medium: adsorption mechanism for clean approach. J Sol-Gel Sci Technol 80:602–615. https://doi.org/10.1007/s10971-018-4870-0

Amin MT, Alazba AA, Shafiq M (2017) Removal of copper and Lead using Banana biochar in batch adsorption systems: isotherms and kinctic studies. Arab J Sci Eng 43:5711–5722. https://doi.org/10.1007/s13369-017-2934-z

Babel S, Kurniawan TA (2004) Cr (VI) removal from synthetic wastewater using coconut shell charcoal and commercial activated carbon modified with oxidizing agents and/or chitosan. Chemosphere 54(7):951–967. https://doi.org/10.1016/j.chemosphere.2003.10.001

Baby Shaikh R, Saifullah B, Rehman FU (2018) Greener method for the removal of toxic metal ions from the wastewater by application of agricultural waste as an adsorbent. Water 10(10):1316. https://doi.org/10.3390/w10101316

Bayuo J, Pelig-Ba KB, Abukari MA (2018) Isotherm modeling of Lead (II) adsorption from aqueous solution using groundnut Shell as a low-cost adsorbent. IOSR J App Chem 11(11):18–23. https://doi.org/10.9790/5736-1111011823

Bayuo J, Abukari MA, Pelig-Ba KB (2019) Equilibrium isotherm studies for the sorption of hexavalent chromium (VI) onto groundnut shell. IOSR J App Chem 11(12):40–46. https://doi.org/10.9790/5736-1112014046

Beidokhti MZ, Naeeni STO, AbdiGhahroudi MS (2019) Biosorption of nickel (II) from aqueous solutions onto pistachio hull waste as a low-cost biosorbent. Civil Eng 5(2). https://doi.org/10.28991/cej-2019-03091259

Beni AA, Esmaeil A (2020) Biosorption, an efficient method for removing heavy metals from industrial effluents: a review. Environ Technol Innov 17. https://doi.org/10.1016/j.eti.2019.100503

Bibaj E, Lysigaki K, Nolan JW, Seyedsalehi M, Deliyanni EA, Mitropoulos AC, Kyzas GZ (2018) Activated carbons from banana peels for the removal of nickel ions. Int J Environ Sci Technol 16:667–680. https://doi.org/10.1007/s13762-018-1676-0

Bulgariu L, Bulgariu D (2018) Functionalized soy waste biomass - a novel environmental-friendly biosorbent for the removal of heavy metals from aqueous solution. J Clean Prod 197(1):875–885. https://doi.org/10.1016/j.jclepro.2018.06.261

Chen Q, Yao Y, Li X, Lu J, Zhou J, Huang Z (2018) Comparison of heavy metal removals from aqueous solutions by chemical precipitation and characteristics of precipitates. J Water Process Eng 26:289–300. https://doi.org/10.1016/j.jwpe.2018.11.003

Christopher D, Kaur S, Singh R (2012) Water quality status of river Yamuna in Delhi with reference to presence of heavy metals: a review. Int J Pharm Med Bio Sci 1(2):266–271

Cobbina SJ, Duwiejuah AB, Quainoo AK (2018) Single and simultaneous adsorption of heavy metals onto groundnut shell biochar produced under fast and slow pyrolysis. Int J Environ Sci Technol 16(7):3081–3090. https://doi.org/10.1007/s13762-018-1910-9

Correiaa IKDS, Santosa PF, Santanaa CS, Nerisa JB, Luzardoa FHM, Velascoa FG (2018) Application of coconut shell, banana peel, spent coffee grounds, eucalyptus bark, piassava (*Attalea funifera*) and water hyacinth (*Eichornia crassipes*) in the adsorption of Pb2+ and Ni2+ ions in water. J Environ Chem Eng 6(2):2319–2334. https://doi.org/10.1016/j.jece.2018.03.033

De Gisi S, Lofrano G, Grassi M, Notarnicola M (2016) Characteristics and adsorption capacities of low-cost sorbents for wastewater treatment: a review. Sustain Mater Technol 9:10–40. https://doi.org/10.1016/j.susmat.2016.06.002

Ehishan NS, Sapawe N (2018) Performance studies removal of chromium (Cr^{6+}) and lead (Pb^{2+}) by oil palm frond (OPF) adsorbent in aqueous solution. Mater Today 5(10):21897–21904. https://doi.org/10.1016/j.matpr.2018.07.048

El-Azazy M, El-Shafie AS, Issa AA, Al-Sulaiti M, Al-Yafie J, Shomar B, Al-Saad K (2019) Potato peels as an adsorbent for heavy metals from aqueous solutions: eco-structuring of a green adsorbent operating Plackett–Burman design. J Chem. https://doi.org/10.1155/2019/4926240

El-Maghraby, Lamiaa MM (2019) Removal of copper (II) and ferric (III) ions from aqueous solutions by adsorption using potato peel. Assiut J Agric Sci 50(3):98–111. https://doi.org/10.21608/ajas.2019.52737

Enniya I, Rghioui L, Jourani A (2018) Adsorption of hexavalent chromium in aqueous solution on activated carbon prepared from apple peels. Sustain Chem Pharm 7:9–16. https://doi.org/10.1016/j.scp.2017.11.003

Fergusson JE (1990) The heavy elements: chemistry, environmental impact and health effects. Pergamon Press, Oxford

Flores GS, Reyna SA, Elvir-Padilla LG, Mendoza-Castillo DI, Reynel-Avila HE, Bonilla-Petriciolet A (2018) Kinetics, thermodynamics, and competitive adsorption of heavy metals from water using Orange biomass. Water Environ Res 90(12):2114–2125. https://doi.org/10.2175/106143017X15131012188321

Gao X, Wu L, Xu Q, Tian W, Li Z, Kobayashi N (2018) Adsorption kinetics and mechanisms of copper ions on activated carbons derived from pinewood sawdust by fast H3PO4 activation. Environ Sci Pollut Res 25:7907–7915. https://doi.org/10.1007/s11356-017-1079-7

Gaur N, Kukreja A, Yadav M, Tiwari A (2018) Adsorptive removal of lead and arsenic from aqueous solution using soya bean as a novel biosorbent: equilibrium isotherm and thermal stability studies. Appl Water Sci 8(98). https://doi.org/10.1007/s13201-018-0743-5

Gough I (2017) Heat, greed and human need: climate change, capitalism and sustainable wellbeing. Edward Elgar Publishing

Gunatilake SK (2015) Methods of removing heavy metals from industrial wastewater. Methods 1(1):14

Guo H, Bi C, Zeng C, Ma W, Yan L, Li K, Wei K (2018) *Camellia oleifera* seed shell carbon as an efficient renewable bio-adsorbent for the adsorption removal of hexavalent chromium and methylene blue from aqueous solution. J Mol Liq 249:629–636. https://doi.org/10.1016/j.molliq.2017.11.096

He S, Li Y, Weng L, Wang J, He J, Liu Y, Zhang K, Wu Q, Zhang Y, Zhang Z (2018) Competitive adsorption of Cd2+, Pb2+ and Ni2+ onto Fe3+–modified argillaceous limestone: influence of pH, ionic strength and natural organic matters. Sci Total Environ 637–638:69–78. https://doi.org/10.1016/j.scitotenv.2018.04.300

Ibrahem S, Hassan M, Ibraheem Q, Arif K (2020) Genotoxic effect of Lead and cadmium on Workers at Wastewater Plant in Iraq. Int J Environ Res Public Health. https://doi.org/10.1155/2020/9171027

Iftekhar S, Ramasamy DL, Srivastava V, Asif MB, Sillanpää M (2018) Understanding the factors affecting the adsorption of Lanthanum using different adsorbents: a critical review. Chemosphere 204:413–430. https://doi.org/10.1016/j.chemosphere.2018.04.053

Ittrat P, Chacho T, Pholprayoon J, Suttiwarayanon N, Charoenpanich J (2014) Application of agriculture waste as a support for lipase immobilization. Biocatalysis 3(3):77–82. https://doi.org/10.1016/j.bcab.2014.02.002

Joseph L, Jun BM, Flora JRV, Park CM, Yoon Y (2019) Removal of heavy metals from water sources in the developing world using low-cost materials: a review. Chemosphere 229:142–159. https://doi.org/10.1016/j.chemosphere.2019.04.198

Kataria N, Garg VK (2018) Green synthesis of Fe3O4 nanoparticles loaded sawdust carbon for cadmium (II) removal from water: regeneration and mechanism. Chemosphere 208:818–828. https://doi.org/10.1016/j.chemosphere.2018.06.022

Khan F, Momtaz S, Abdollahi M (2019) The relationship between mercury exposure and epigenetic alterations regarding human health, risk assessment and diagnostic strategies. J Trace Elem Med Biol 52:37–47. https://doi.org/10.1016/j.jtemb.2018.11.006

Khatoon H, Rai JPN (2016) Agricultural waste materials as biosorbents for the removal of heavy metals and synthetic dyes-a review. Environ Res 4(3):208–229

Kołodyńska D, Bąk J (2018) Use of three types of magnetic biochar in the removal of copper (II) ions from wastewaters. Sep Sci Technol 53(7):1045–1057. https://doi.org/10.1080/01496395.2017.1345944

Kumar D, Pandey LK, Gaur JP (2016) Metal sorption by algal biomass: from batch to continuous system. Algal Res 18:95–109. https://doi.org/10.1016/j.algal.2016.05.026

Kyzas GZ, Bomis G, Kosheleva RI, Efthimiadou EK, Favvas EP, Kostoglou M, Mitropoulos AC (2019) Nanobubbles effect on heavy metal ions adsorption by activated carbon. Chem Eng J 356:91–97. https://doi.org/10.1016/j.cej.2018.09.019

Lee SY, Cho HJ (2018) Persimmon leaf bio-waste for adsorptive removal of heavy metals from aqueous solution. J Environ Manag 209:382–392. https://doi.org/10.1016/j.jenvman.2017.12.080

Lin S, Huang W, Yang H, Sun S, Yu J (2020) Recycling application of waste long-root Eichhornia crassipes in the heavy metal removal using oxidized biochar derived as adsorbents. Bioresour Technol 314:123749. https://doi.org/10.1016/j.biortech.2020.123749

Liu Y, Gao Q, Pu S, Wang H, Xia K, Han B, Zhou C (2019) Carboxyl-functionalized lotus seedpod: a highly efficient and reusable agricultural waste-based adsorbent for removal of toxic Pb2+ ions from aqueous solution. Colloid Surf A 568:391–401. https://doi.org/10.1016/j.colsurfa.2019.02.017

Mallampati R, Valiyaveettil S (2013) Apple peels-a versatile biomass for water purification? ACS Appl Mater Interfaces 5:4443–4449. https://doi.org/10.1021/am400901e

Malook K, Ul-Haquea I (2019) Investigation of aqueous Cr(VI) adsorption characteristics of Orange peels powder. Prot Met Phys Chem Surf 55(1):34–40. https://doi.org/10.1134/S2070205119010155

Mehrotra T, Shukla A, Singh R (2019) In vitro toxicological evaluation of domestic effluent treated by formulated synthetic autochthonous bacterial consortium. World J Microbiol Biotechnol 35(12):184. https://doi.org/10.1007/s11274-019-2756-0

Melia PM, Busquets R, Ray S, Cundy AB (2018) Agricultural wastes from wheat, barley, flax and grape for the efficient removal of Cd from contaminated water. RSC Adv 8(70):40378–40386. https://doi.org/10.1039/C8RA07877G

Mohubedu RP, Diagboya PN, Abasi CY, Dikio ED, Mtunzi F (2019) Magnetic valorization of biomass and biochar of a typical plant nuisance for toxic metals contaminated water treatment. J Clean Prod 209:1016–1024. https://doi.org/10.1016/j.jclepro.2018.10.215

Mullick A, Moulik S, Bhattacharjee S (2017) Removal of hexavalent chromium from aqueous solutions by low-cost Rice husk-based activated carbon: kinetic and thermodynamic studies. Indian Chem Eng 60:58–71. https://doi.org/10.1080/00194506.2017.1288173

Mutiara T, Rofiki I, Al Ghifari MAD (2018) Bio adsorbent from modified jackfruit wood sawdust for removal of Lead ions. Mater Sci Forum 934:159–164. https://doi.org/10.4028/www.scientific.net/MSF.934.159

Naoufal B, Szabolcs K, Zoltán P, Cecilia H (2018) Adsorption of nutrients using low-cost adsorbents from agricultural waste and by-products–review. Prog Agric Eng Sci 14(1):1–30. https://doi.org/10.1556/446.14.2018.1.1

Ngo HH, Guo W, Hang Nguyen TA, Surampalli RY, Zhang TC (2016) Agricultural by-products for phosphorous removal and recovery from water and wastewater: a green technology. Green Technol Sustain Water Manag. https://doi.org/10.1061/9780784414422.ch14

Nguyen HN, Pignatello JJ (2013) Laboratory tests of biochars as absorbents for use in recovery or containment of marine crude oil spills. Environ Eng Sci 30(7):374–380

Ojuederie OB, Babalola OO (2017) Microbial and plant-assisted bioremediation of heavy metal polluted environments: a review. Int J Environ Res Public Health 14(12):1504. https://doi.org/10.3390/ijerph14121504

Özsin G, Kılıç M, Varol EA, Pütün AE (2019) Chemically activated carbon production from agricultural waste of chickpea and its application for heavy metal adsorption: equilibrium, kinetic, and thermodynamic studies. Appl Water Sci 9(56). https://doi.org/10.1007/s13201-019-0942-8

Pandey R, Ansari NG, Prasad RL, Murthy RC (2014) Removal of cd(II) ions from simulated wastewater by HCl modified Cucumis sativus Peel: equilibrium and kinetic study. Air soil Water Res 7:93–101. https://doi.org/10.4137/ASWR.S16488

Prabha D (2018) Biosorption: a solution for removal of toxic metals. Int J Appl Eng Res 13(22):15895–15898

Prasad BB, Singh R, Lakshmi D (2006) Biosorption separation of 'Labile' fractions of copper and lead from river sediments and evaluation by differential pulse anodic stripping voltammetry. Int J Environ Pollut 27:179–190

Pyrzynska K (2019) Removal of cadmium from wastewaters with low-cost adsorbents. J Environ Chem Eng 7(1):102795. https://doi.org/10.1016/j.jece.2018.11.040

Qadeer HA, Mahomoodally MF, Nadeem F, Khanam A (2018) Wastewater treatment and dyes removal using electrocoagulation aided by natural biosorbents–a review. Int J Chem Biochem Sci 14:77–87

Rengganis AP, Arlin Wuriyudani H, Yuniarti N, Masturi N, Fianti N (2018) Adsorbent metal Chromium (Cr) in the liquid waste of batik made from Chitosan and Banana Peels. Adv Recycling Waste Manag 3(151):2

Rohmah DN, Saputro S, Masykuri M, Mahardiani L (2018) Combination of rice husk and coconut shell activated adsorbent to adsorb Pb (II) ionic metal and it's analysis using solid-phase spectrophotometry (sps). IOP Conf Ser Mater Sci Eng 333(1):012056

RoyChowdhury A, Datta R, Sarkar D (2018) Heavy metal pollution and remediation. Green Chem:359–373. https://doi.org/10.1016/B978-0-12-809270-5.00015-7

Shakoor MB, Niazi NK, Bibi I, Murtaza G, Kunhikrishnan A, Seshadri B, Shahid M, Ali S, Bolan NS, Ok YS, Abid M, Ali F (2016) Remediation of arsenic-contaminated water using agricultural wastes as biosorbents. Crit Rev Environ Sci Technol 46(5):467–499. https://doi.org/10.1080/10643389.2015.1109910

Sharma R, Sapna, Dhillon A, Kumar D (2018) Biosorbents from agricultural by-products: updates after 2000s. In: Bio-and nanosorbents from natural resources. Springer, Cham, pp 1–20. https://doi.org/10.1007/978-3-319-68708-7_1

Shrestha SL (2018) Study of the adsorption kinetics of Iron ion from wastewater using Banana Peel. Int J Adv Res Chem Sci 5(3):1–8

Shrivastava A, Ghosh D, Dash A, Bose S (2015) Arsenic contamination in soil and sediment in India: sources, effects, and remediation. Current Pollution Reports 1:35–46. https://doi.org/10.1007/s40726-015-0004-2

Shruthi KM, Pavithra MP (2018) A study on utilization of groundnut shell as biosorbent for heavy metals removal. Int J Eng Technol 4(3):411–415

Sidhu GPS, Singh HP, Batish DR, Kohli RK (2016) Effect of lead on oxidative stress, antioxidative response and metal accumulation in *Coronopus didymus*. Plant Physiol Biochem 105:290–296. https://doi.org/10.1016/j.plaphy.2016.05.019

Singh R (2012) Selective separation of Lead ions from waste water by immobilized Chlorella based biosorbents. Int J Chem Environ Eng 3(2):108–114

Singh R, Prasad BB (2000) Trace metal analysis: selective sample (copper II) enrichment on an AlgaSORB column. Process Biochem 35(9):897–905. https://doi.org/10.1016/S0032-9592(99)00150-8

Singh R, Sinha S (2013) Bioremediation of heavy metals by algae: a review on evaluation of low cost and high performance biosorbents. Indian J Agric Biochem 26(1):1–9

Singh NB, Nagpal G, Agrawal S, Rachna (2018) Water purification by using adsorbents: a review. Environ Technol Innov 11:187–240. https://doi.org/10.1016/j.eti.2018.05.006

Singh RJ, Martin CE, Barr D, Rosengren RJ (2019) Cucumber peel bead biosorbent for multi-ion decontamination of drinking water collected from a mine region in New Zealand. Environ Technol. https://doi.org/10.1080/09593330.2019.1703824

Sinha S, Nigam S, Singh R (2019a) Biosorption capacity of Cr (VI) on live and dead S. rubescens: kinetic, equilibrium and phytotoxicity studies. Indian J Agric Biochem 31(2):137–144. https://doi.org/10.5958/0974-4479.2018.00023.0

Sinha S, Srivastava A, Mehrotra T, Singh R (2019b) A review on the dairy industry waste water characteristics, its impact on environment and treatment possibilities. In: Emerging issues in ecology and environmental science. SpringerBriefs in Environmental Science. Springer, Cham. https://doi.org/10.1007/978-3-319-99398-0_6

Sinha S, Mehrotra T, Srivastava A, Srivastava A, Singh R (2020) Nanobioremediation technologies for potential application in environmental cleanup. In: Environmental biotechnology, vol 2. Springer, Cham, pp 53–73

Sirilert M, Maikrang K (2018) Adsorption isotherm of some heavy metals in water on unripe and ripe peel of banana. Sci Technol 26(1):128–141

Tan L, Wang X, Tan X, Mei H, Chen C, Hayat T, Alsaedi A, Wen T, Lu S, Wang X (2017) Bonding properties of humic acid with attapulgite and its influence on U (VI) sorption. Chem Geol 464:91–100. https://doi.org/10.1016/j.chemgeo.2017.01.024

Taralgatti PB (2019) Explore the potential of food waste in removal of heavy metal (Lead) from industrial wastewater. Techno-Soc:879–885. https://doi.org/10.1007/978-3-030-16848-3_79

Tatah VS, Ibrahim KLC, Ezeonu CS et al (2017) Biosorption kinetics of heavy metals from fertilizer industrial waste water using groundnut husk powder as an adsorbent. J Appl Biotechnol Bioeng 2(6):221–228. https://doi.org/10.15406/jabb.2017.02.00049

Tejada-Tovar C, González-Delgado Á, Villabona-Ortíz A (2018) Adsorption kinetics of orange peel biosorbents for Cr (VI) uptake from water. Contemporary Eng Sci 11(24):1185–1193

Vardhan KH, Kumar PS, Panda RC (2019) A review on heavy metal pollution, toxicity and remedial measures: current trends and future perspectives. J Mol Liq 209:111197. https://doi.org/10.1016/j.molliq.2019.111197

Villar da Gama BA, do Nascimento GE, Silva Sales DC, Rodríguez-Díaz JM, Bezerra de Menezes Barbosa CM, Menezes Bezerra Duarte MM (2018) Mono and binary component adsorption of phenol and cadmium using adsorbent derived from peanut shells. J Clean Prod 201:219–228. https://doi.org/10.1016/j.jclepro.2018.07.291

WHO (1996). https://www.who.int/water_sanitation_health/water-quality/guidelines/chemicals/chromium.pdf?ua=1. Accessed 3 July 2020

Xiao F, Cheng J, Cao W, Yang C, Chen J, Luo Z (2019) Removal of heavy metals from aqueous solution using chitosan-combined magnetic biochars. J Colloid Interface Sci 540:579–584. https://doi.org/10.1016/j.jcis.2019.01.068

Zhao L, Zheng W, Mašek O, Chen X, Gu B, Sharma BK, Cao X (2017) Roles of phosphoric acid in biochar formation: synchronously improving carbon retention and sorption capacity. J Environ Qual 46(2):393–401. https://doi.org/10.2134/jeq2016.09.0344

Zhu H, Tan X, Tan L, Chen C, Alharbi NS, Hayat T, Fang M, Wang X (2018) Biochar derived from sawdust embedded with molybdenum disulfide for highly selective removal of Pb2+. ACS Appl Nano Mater 1(6):2689–2698. https://doi.org/10.1021/acsanm.8b00388

Chapter 5
Current Trends and Emerging Technologies for Pest Control Management of Rice *(Oryza sativa)* Plants

Manjula Ramadass and Padma Thiagarajan

Abstract Rice (*Oryza sativa*), with a global production of around 770 million tons, is a widely consumed staple cereal in Asian countries. The Food and Agricultural Organization has declared it as a strategic crop for global food security. During their life cycle and development, these plants go through the vegetative, reproductive and ripening phases. Each of these phases is susceptible to attack by several kinds of bacterial, fungal, viral and insect pests, which ultimately lead to premature crop death and subsequent yield losses. Control of these pests is most effectively and quickly achieved by treating plants with synthetic chemicals and their agroformulations. But biological/herbal constituents and their combinations can also control specific types of pests over a longer period of time. Moreover, currently, smart techniques, technologies, and materials for pest control have overcome several constraints associated with conventional systems. Nanotechnology has especially proven to be a boon, and so have emulsion formulations that facilitate controlled and sustained release of loaded actives to achieve a better efficacy with minimal environmental impact. Genetically resistant plant varieties, which are naturally resistant to specific pests, and do not need additional treatments to prevent pathogenic attacks, are also being developed. All these agents have specific implications that include environmental policies, potency issues, necessitated time of action, repeated usage requirements, long-term efficacy, and pest resistance development. This review describes rice plants' developmental stages, pests that attack during these stages, and importantly also focuses on traditional and emerging pest control management strategies for achieving improved yields and economic sustainability.

Keywords *Oryza sativa* · Rice · *Xanthomonas oryzae* · Chemical pesticides · Agroformulations

---

M. Ramadass · P. Thiagarajan (✉)
School of Biosciences and Technology, Vellore Institute of Technology, Vellore, Tamil Nadu, India

K. M. Gothandam et al. (eds.), *Environmental Biotechnology Volume 4*, Environmental Chemistry for a Sustainable World 68,
https://doi.org/10.1007/978-3-030-77795-1_5

Abbreviations

AI	Active ingredient
EC	Emulsifiable concentration
EP	Emulsifiable powder
EW	Emulsion, oil in water
FS	Flowable concentration for seed treatment
GR	Granule
Ha	Hectare
PPM	Parts per million
RTD	Rice tungro disease
SC	Suspension concentration
SL	Soluble concentration
SP	Water soluble powder
WG	Water dispersible granules
WP	Wettable powder
Xoo	*Xanthomonas oryzae pv. oryzae*

5.1 Introduction

Rice plants serve as a global and versatile food crop. Rice is a seed of this monocot semi-aquatic annual grass plant. The global rice production is about 769.9 million tons and includes 510.6 million tons of milled rice (Food and Agricultural Organisation 2018). Its production and consumption are higher in Asian countries, wherein it is consumed by around 50% of the population. This plant belongs to the genus of *Oryza* and the family of *Poaceae* (Seck et al. 2012). It includes 22 species of this genus, out of which 20 are wild types. *Oryza sativa* and *Oryza glaberrima* are two important species that are fit for human consumption. *Oryza sativa* L. is its most widely grown variety and forms the staple food for over half the world's population. These plants are cultivated in more than a hundred countries in Asia, the European Union, the Middle East, Africa, North and South America, whereas *O. glaberrima* is confined to Africa, and is fast being replaced by *O. sativa* (Linares 2002). Rice varieties are classified as the long-grained, short/medium, and medium-grained varieties (Lu 1999). Cultivation of the long-grained variety is undertaken in tropical and subtropical Asia, whereas in northern China and Japan, the short or medium grain rice varieties are cultivated. Medium grained rice is grown in the Philippines and the mountain regions of Indonesia and Madagascar.

Rice is an essential dietary constituent for 3.5 billion people in the world (Consultative Group on International Agricultural Research 2016). Fifteen countries account for more than 90% of global rice production. These include Asian countries like Pakistan, China, India, Japan, Philippines, Sri Lanka, Thailand, Indonesia, South Korea, Vietnam, Bangladesh, Myanmar, Cambodia, and Nepal (Muthayya et al. 2012), and non-Asian countries like Brazil, Egypt, United States,

Nigeria, Madagascar, and Africa. Eight countries in Africa, nine in South and North America, and seventeen in the Pacific and Asian continents consume rice as a staple food. Based on the commercial production zones, *Oryza sativa* is divided into three significant subspecies, *viz., indica, javanica,* and *japonica*. In tropical and subtropical regions, *indica* is cultivated in South and Southeast Asia and Southern China, whereas *javanica* is cultivated in Indonesia. The *japonica* variety is grown in temperate countries like Japan, Korea, Nepal, and China (Gadal et al. 2019). Facts regarding global rice production, milling, supply chain, cost, consumption, market imports and exports have been reviewed (Gadal et al. 2019; Muthayya et al. 2014).

The total area for food grain production is 51.6%, out of which rice accounts for 42.5%, and it dominates the crop sectors of several countries (Ministry of Agricultural Development 2017). A diverse ecosystem supports rice cultivation, and they are classified mainly based on water and land systems such as rain-fed and irrigated water systems, and low and upland systems, respectively. The type of water used for rice production is classified as rainwater (green water), irrigated water (blue water), and polluted water. It also relates to the use of nitrogenous fertilizers in the rice field (greywater). Compared to the US and Pakistan, rice consumption's water footprint creates low stress on India's water resources. Estimation of evapo-transpiration and percolation flow in rice fields have been reported (Chapagain and Hoekstra 2011). Wetland and upland systems are two major systems used for rice production. About 85% of rice harvest is from wetland systems, out of which 75% is from irrigated wetlands (Bouman et al. 2007a).

Over the next few decades, Rice will feed billions of people worldwide, especially in the Asian and African countries (Seck et al. 2012). The production of red rice improves food security due to a genetic similarity with its commercial counterpart (Durand-Morat et al. 2018). The demand for rice is mainly from African countries. This is because, except for Egypt, 40% of rice consumed here is imported (Seck et al. 2010). This versatile crop can be grown in wetlands and under dry conditions at high and low altitudes. Its demand is due to population growth, rising incomes, and a shift in consumer preferences, mainly in the urban areas (Seck et al. 2012). Its export influences economic development also (Hegde and Hegde 2013). The nutrient composition of rice is based on environmental conditions and soil characteristics, along with types and levels of fertilizers used for its cultivation (Chandler 1979). Commonly, its grain consists of starch (80%), water (12%), protein (7.5%) and ash (0.5%). The levels of nitrogen fertilizers and climatic conditions affect the starch content of its varieties. In fields, it is found to contain 72–75% of starch, 1.5–2.6% of glucose, 1.6–2.1% of dextrin and 0.3–0.5% of sucrose. It is a good source of carbohydrates, protein, iron, amino acids, vitamin E, niacin, thiamine, and riboflavin (Grist 1986; Juliano 1993).

The green revolution (1964–1983) led to an increase in the production of rice. A structural shift that concerning technological changes, cost production, and marketing management occurred in the world rice market from 1950 to 2000 (Dawe 2002). Global food security was profoundly impacted by rice production from Asian countries, particularly from India and China, and hence policies to increase its export have recently been recommended (Bandumula 2018). Exports will serve to enhance income for farmers and increase employment in developing countries. Yield losses in rice occur due to microbial and insect attacks. The major biotic factors that affect these crops are bacterial leaf blight, fungal blast, and stem borer infections. Abiotic stresses like drought, salinity, and low temperature also affect yields. Environmental conditions like global warming, production of methane gas, and toxicity of sprayed pesticides may restrict its production in several cultivating areas (Van-Nguyen and Ferrero 2006). The application of phosphorus to crops generally increases grain yields and enhances its flow into the food chain (Ma et al. 2011). Rice functional genomics research aims to improve quality, increase resistance and nutrient contents, and yield from them. Its perspective view has been reported (Zhang et al. 2008a). This chapter describes the growth stages of rice plants starting from the seedling to harvest stages, pests that affect these plants, and diseases caused by them, along with treatment strategies for the control of such pests. The latter includes chemical and herbal formulations and smart methodologies and materials that focus on innovative technologies.

5.2 Life Cycle of Rice Plants and Their Developmental Phases

The life span of a rice plant (Fig. 5.1) from the germination stage to complete maturity depends on the variety and environment in which it is cultivated. Spans can be short (90–120 days), medium (120–140 days) or long (140–180 days). Their agronomic growth cycle is divided into the vegetative phase, which includes

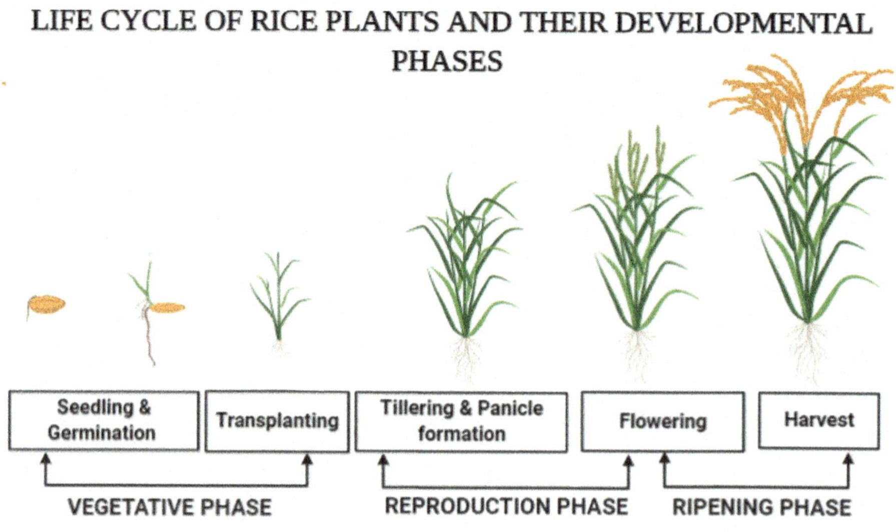

Fig. 5.1 Life cycle of rice plants and their developmental stages

germination to panicle initiation, the reproductive phase that consists of panicle initiation followed by flowering, and finally the ripening period, which involves later stages of flowering to maturity (Maclean et al. 2002).

5.2.1 Vegetative Phase

The rice grain or seed is a condensed form of the plant from which a vegetative stage begins. It mainly consists of a caryopsis and hull with several layers of differentiated thin tissue that encloses a potential embryo and endosperm. Under completely dehydrated conditions, a single grain weighs around 10–45 mg, which represents 20% of hull weight. The length, thickness, and width of seeds vary within their varieties. This phase's nascent seedling stage is initiated when seed dormancy is broken by a copious water supply at 25–30 °C. Initially, under anaerobic conditions, the coleoptile (protective sheath) first appears and can elongate under low anoxic stress. After that, the emergence of a radicle (embryonic shoot) from the seed necessitates aerobic conditions (Bouman et al. 2007b). Germinal and lateral roots then develop, followed by their growth and multiplication (Maclean et al. 2002). Nutrients are absorbed from the endosperm during this process. The seedling stage that commences during germination terminates after the development of the fifth leaf.

The transplanting stage, which is avoided indirectly seeded plants, involves uprooting seedlings after germination to full crop recovery. The transplanted plants undergo two or more growth processes, like growth in nurseries and transplantation in paddy fields. Seedling growth ceases after transplantation due to an environmental shock (Salam et al. 2001). The tillering stage starts with the development of the first tiller from the auxiliary bud. The rice stem consists of nodes and internodes. Upper nodes develop a leaf and a bud, which can grow into tillers. The number of nodes varies from 13 to 16 (Maclean et al. 2002). Primary tillers originate from lower nodes and give rise to secondary tillers followed by tertiary tillers, after which this stage terminates at around 30–40 days. Each tiller refers to an independent plant.

Plants start to grow in height at this stage, and tillers increase rapidly at optimum temperatures of 25–31 °C. Higher temperatures affect tiller numbers and negatively influence panicle and pollen development, bringing down crop yields (Yoshida 1981). This stage is also affected by environmental conditions like spacing, lighting, nutritional status, and cultural practices. The number of days required for the vegetative phase depends on crop varieties and may hover between 55 and 85 days.

5.2.2 Reproductive Phase

Here elongation of culms (jointed stem of rice, consisting of nodes and internodes) increases plant heights. It starts with bulging of leaf stems leading to panicle initiation (optimum temperature 15–30 °C). Base, primary and secondary branches, axis, pedicel, rudimentary glumes, and spikelets are structural features of panicles. Panicle tips develop in the booting stage and continue to grow into the heading stage for 10–14 days, during which 50% of panicles are extended. The flowering phase begins and continues until spikelets in panicles bloom in about 5–7 days at 35 °C (Matsui et al. 2000; Prasad et al. 2006). Pollination and fertilization then follow. Crops are more affected by heat stress at this stage rather than at the vegetative stage (Baker et al. 1992; Hall 1992). High nocturnal temperatures alter pollen germination and spikelet fertility, thus reducing yields (Mohammed and Tarpley 2009; Peng et al. 2004). The duration of time required for panicle initiation, booting, heading, and flowering stages depend on crop varieties.

5.2.3 Ripening Phase

After pollination and fertilization of rice grains, the ripening phase commences. It passes through milky, dough, yellow-ripe and maturity stages, based on colour and texture of growing grains, and typically takes 15–40 days at optimum temperatures of 20–25 °C. The ripening phase is affected by temperature. It extends to about 30 days in the tropics and 65 days in cooler regions, after which seeds are harvested (Yoshida 1981).

5.3 Diseases Caused in Rice Plants Due to Pest Infestations

Rice plants are infected by several pests (Table 5.1) at different stages of their life cycle, resulting in reduced crop yields and low-quality grains. Microbes, which include bacteria, fungi, and viruses, severely affect paddy cultivation. The major devastating diseases of rice include two bacterial infections, *viz.,* bacterial leaf blight and bacterial leaf streak, four fungal diseases, *viz.*, rice blast, sheath blight, sheath rot and brown spot, and a viral Rice Tungro Disease (Basso et al. 2011; Maclean et al. 2002). Early vegetative pests, root feeder, stem borer, leaf and planthoppers, defoliators, and grain sucking insects fall under the category of insect pests and these attack plants at all stages of their growth.

Table 5.1 Diseases caused in rice plants due to pest infestations

Pest category	Specific causative organism	Disease caused	References
Bacteria	*Xanthomonas oryzae pv. oryzae*	Bacterial leaf blight	Zhang and Wang (2013)
	Xanthomonas oryzae pv. oryzicola	Bacterial leaf streak	Tran et al. (2008)
Fungi	*Magnaporthe oryzae*	Rice blast	Sesma and Osbourn (2004)
	Rhizoctonia solani	Sheath blight	Groth (2008)
	Sclerotium oryzae	Sheath rot	Sakthivel (2001)
	Helminthosporium oryzae	Brown spot	Vidhyasekaran et al. (1990)
Virus	Rice tungro spherical virus RNA & Rice tungro bacilliform virus -RNA/ DNA	Rice Tungro disease	Jones et al. (1991)
Insects	*Lissorhoptrus oryzophilus*, *Leucopholis irrorata* *Gryllotalpa orientalis*	Root feeder	Pathak and Dhalival (1981), Matteson (2000)
	Chilo polychrysus *Chilo auricilius* *Chilo suppressalis* *Scirpophaga innotata* *Scirpophaga incertulas* *Sesamia inferens*	Stem borer	Zibaee et al. (2008), Cheng et al. (2010)
	Nephotettix malayanus *Nephotettix virescens*	Leaf hoppers	Cheng and Pathak (1972)
	Nilaparvata lugens *Sogatella furcifera*	Plant hoppers	Sujithra and Chander (2013)
	Cutworm (*Spodoptera litura* (Fabricius) Ear-cutting (*Mythimna separata* (Walker) & swarming caterpillars (*Spodoptera mauritia acronyctoides* (Guenee) *Dicladispa armigera* (Olivier) *Cnaphalocrocis medinalis*	Defoliators	Tanwar et al. (2010) Pathak and Dhalival (1981) Chakraborty and Deb (2012)
	Leptocorisa oratorius *Leptocorisa acuta*	Grain sucking insects	Torres et al. (2010)

5.3.1 Diseases Caused by Bacterial Infestations

Bacterial leaf blight is a primary common and most destructive disease caused by *Xanthomonas oryzae pv. oryzae* (*Xoo*) (Zhang and Wang 2013). It has an incidence of 70–80% and can lead to significant crop damage (Sere et al. 2005; Basso et al. 2011). *Xoo* may induce wilting of seedlings known as kresek that may be noticed 1–3 weeks after the transplantation process. The wilted greyish-green or yellow leaves roll up, and whole plants entirely wilt up and die. Infection is systemic and results in the death of whole seedlings (Nino-liu et al. 2006). *Xoo* also causes leaf blight, wherein yellowish stripes become visible on leaf blades that undergo marginal necrosis. Symptoms develop from leaf tips as wavy edges and spread to entire leaves resulting in their acute dropping, untimely plant death, and ultimate loss of crop yields (Mehar et al. 2009). The bacterial ooze resembles a milky or opaque dewdrop and is visible early in the morning on young lesions. It later dries up and forms small yellowish beads underneath leaves. In older plants, yellow-orange stripe lesions on leaf blades or leaf tips are visible, and lesions then progress to leaf bases. In susceptible rice varieties, lesions may develop in leaf sheaths. This infection, which occurs during booting stages, results in low and poor quality grains (Jonit et al. 2016; Zhang and Wang 2013).

A temperature of 25–34 °C and 70% of relative humidity favours infectious bacteria's growth in tropical and temperate regions. Infection is intense during continuous rainfall and windy weather as such conditions support the spread of bacteria and favours the development of lesions in plants. Infection can occur at different stages of plant growth and spread through water pores or wounds (Jonit et al. 2016). It is severe in rice fields that grow susceptible varieties wherein large amounts of nitrogenous fertilizers are sprayed.

Bacterial leaf streak, which is caused by *Xanthomonas oryzae pv. oryzicola,* occurs in plants cultivated in tropical and subtropical regions of Asia, Africa, Australia, and South America, wherein high humidity and temperatures prevail. Transmission of infection occurs through seeds, infected plant parts, and due to the presence of bacteria in fields after harvest. Leaf streaks develop from tillering to the panicle initiation stage. Symptoms initially are noted as small dark green, water-soaked streaks on inter-veins that later become translucent. Streaks spread longitudinally in veins and turn into orange or yellow-brown. The infection then develops as a large patch and covers the entire leaf surface (Tran et al. 2008). Tiny yellow or amber colour bacterial droplets then become visible. Infection in florets and seeds results in black or brown discoloration and deaths of the ovary and endosperm. Under severe conditions, plants show browning and drying of leaves with reduced grain weight. Yield loss is more during rains but depends on other climatic conditions and varieties of rice (International Rice Research Institute- Rice Knowledge Bank 2020a).

5.3.2 Diseases Caused by Fungal Infestations

Magnaporthe oryzae (formerly *Magnaporthe grisea*), causes the rice blast disease (Sesma and Osbourn 2004). It is an ascomycete fungus that produces sexual spores called asci (ascospores). Its asexual spore form is described as *Pyricularia oryzae.* It affects plant growth at all stages but is immensely damaging at the seedling stage (Park et al. 2009). It attacks all parts above the ground (leaf, collar, nodes, neck, panicle parts, pedicels, seeds, and sometimes leaf sheaths). Infected leaves project white or grey (initial), elliptical, or spindle or diamond (older lesions) shaped spots with brown margins. Lesions are usually elongated and pointed at ends. Spots enlarge and lead to drying. They can kill entire leaves and may extend into the sheath and produce spores on lesions. The nodal infection leads to a neck blast or rotten neck, which causes the failure of seeds to fill, or panicle blast, and plants die before grain filling. Branches may break at lesions. All plant parts die due to infected node rots (Wilson and Talbot 2009). Plants are killed during their seedling stages, and severe infection leads to yield losses (Skamnioti and Gurr 2009). Frequent rainfall and low temperatures favour this disease. The intensity of leaf blast depends on plant age, cultivator variety, and environmental conditions.

Rhizoctonia solani causes sheath blight. Initial symptoms are seen in leaf sheaths with oval or elliptical or irregular greenish-grey spots. As spots enlarge, centres become greyish white with irregular blackish brown or purple-brown border. The old sheaths (5–6 weeks) are more prone to develop infections. Lesions extend rapidly and cover entire tillers (water line to flag leaf), and the infection spreads to inner sheaths resulting in death of whole plants. Infection is higher in early heading and grain filling stages that lead to reduced grain filling. Rice sheath blight is noticed in rice cultivation regions, especially in intensive rice growing conditions with high nitrogen fertilizers. In highly humid subtropical areas, rice blast and sheath blight limit yields (International Rice Research Institute- Rice Knowledge Bank 2020b; Maclean et al. 2002). Symptoms occur throughout the tropical, subtropical, and temperate countries and start from the tillering to milk stage. According to an Internal Rice Research Institute study, a 6% yield loss due to this infection is seen in tropical Asia's lowland rice fields (Fahad et al. 2019). Under favourable conditions, it causes a 50% loss of yield (Bernardes-de-Assis et al. 2009; Groth 2008).

Sheath rot is caused by *Sclerotium oryzae*, which is found worldwide, notably in rain-fed areas. During heading to maturation stages, infections are high in nitrogen fertilizers, high humidity, and temperature conditions of 20–28 °C. Small black lesions start from the outer to inner sheaths and lead to tissue rotting. Culm collapse with a profusion of mycelial growth and an abundance of *Sclerotium oryzae* is seen during this infection in rotting tissues. It may also occur after harvest (Sakthivel 2001). Typical lesions start at the uppermost leaf sheath with irregular spots (reddish and brown margin). Whitish fungal mycelial growth is present on affected leaf sheaths. Emerged panicles and florets turn red to dark brown during infection. The disease arrests a panicle's emergence that leads to sterile panicles, unfilled seeds, and grain discoloration. It reduces grain yields and quality. Its incidence of occurrence is higher than sheath blight. It is a seed-borne disease that affects plants during growth stages (Bigirimana et al. 2015). Due to this disease, the annual loss in Japan is 16,000–35,000 tons, 80% in India and the Philippines, and 85% in Taiwan (International Rice Research Institute- Rice Knowledge Bank 2020c).

Helminthosporium oryzae causes brown spot disease. Typical symptoms are spots on the coleoptile, leaf blade, leaf sheath, spikelets, and glumes predominant under conditions of high humidity (86–100%), in nutrient-deficient soil, and unflooded regions with low temperature. This fungus is spread by air and present in seeds for more than 4 years. Seeds are the primary sources of infection. This disease starts from the seedling stage, and 10–58% mortality of seedlings results in low yield, low-quality grains, and reduced kernel weight (Pannu et al. 2006).

Circular, small brown or yellow spots are seen in infected seedlings that affect the coleoptile and collapse development of primary and secondary leaves. Spots become visible from the tillering stage. Initially, they are circular, small, and purple or brown. They later develop into oval or circular spots with reddish-brown margins due to the presence of fungal toxins (Vidhyasekaran et al. 1990). Around 14 mm long lesions occur in susceptible varieties resulting in leaf wilt. Brown pinhead size lesions are developed in resistant varieties of rice. Black to dark brown spots or discoloration are seen in infected panicle branches and glumes. Infection in floret and spikelets leads to discoloration or spots in grains, referred to as pecky rice the unfilled grains and grain damage results in reduced grain quality and yield (Chakrabarti 2001). It can affect plants at all stages of growth. However, maximum damage occurs from the tillering to ripening stages. Around 45% of crop damage is observed in severe infection (International Rice Research Institute- Rice Knowledge Bank 2020d), due to which huge yield losses of 90% have been reported (Padmanabhan 1973).

5.3.3 Diseases Caused by Viral Infestations

An association of two viruses causes rice tungro disease: rice tungro spherical virus, which is an RNA virus, and rice tungro bacilliform virus that is a RNA / DNA virus (Jones et al. 1991). Tungro virus is transmitted by six leafhoppers, among which five belong to the genus *Nephotettix*. The primary vector is *Nephotettix virescens*, which is monophagous and restricted to *Oryzae sativa* and some closely related wild rice species. This disease varies according to the type of virus, age of plants, rice varieties, and environmental conditions. It occurs at all stages of plant growth, especially in the vegetative phase, and mostly in the tillering stage. Yellow to orange-yellow discoloration starts from leaf tips and extends to the lower portion of leaves, whereas in the resistant varieties, mild or no discolorations are seen. Mottle or striped appearance with rust spots and necrosis of inter-veins are seen during infection. Plants show stunting symptoms, poor root development, low number of tillers, incomplete exertion of panicles, sterile panicles, partial grain filling, delays in the flowering stage that inhibits crop maturity, and yield losses (Muralidharan et al. 2003).

Disease severity depends on the virual source and vector. The virus transmits this disease to plants within 5–7 days. It also reacquires the infection when it feeds on an infected plant. The young infected leaves of tungro are efficient in transmitting this disease rapidly by vectors. Rice tungro spherical virus is restricted to phloem tissue (Tyagi et al. 2008), whereas Rice tungro bacilliform virus is localized in vascular bundles. It is a destructive disease seen in South and Southeast Asia (Dai and Beachy 2009). Severe infection during early growth stages results in 100% yield losses (International Rice Research Institute- Rice Knowledge Bank 2020e; Azzam and Chancellor 2002).

5.3.4 *Diseases Caused by Insect Infestations*

5.3.4.1 Early Vegetative Insect Pests

Pests like rice caseworm, gall midge, whorl maggot, and green semi looper feed on plants during their vegetative phase of growth. Caseworms (*Nymphula depunctalis*, Pyralidae, Lepidoptera), cut leaf tips to make leaf cases. They cut at right angles like a pair of scissors. Adults are nocturnal and attracted to light traps. Transplantation of young seedlings facilitates such pests' development (Matteson 2000; Pathak and Dhalival 1981; Vromant et al. 1998). Rice gall midge (*Orseolia oryzae*, Cecidomyiidae, Diptera) infection leads to the formation of a hollow cavity or tubular gall at the base of infected tillers. Infected tillers inhibit the growth of leaves, and deformation, wilting, rolling, and plant stunting are other symptoms. It causes elongation of leaf sheath that is referred to as silver shoot or onion leaf (Jagadeesha et al. 2009; Prasad 2011). Whorl maggots (*Hydrellia philippina*, Ferino, Ephyridae, Diptera), feed on rice and cause yellow, white, or transparent patches and pinholes. Winds easily break infected leaves that result in plant stunting. They are active during days, and symptoms disappear at the maximum tillering stage (Sain 2000). Green semiloopers (*Naranga aenescens*, Noctuidae, Lepidoptera) are minor pests, but plants can recover from the damage, and hence economic losses are rare. Mature larvae feed on leaf edges, and young larvae scrape tissues from leaf blades.

5.3.4.2 Root Feeders

These termites occur in patches, mostly during low rainfall and kill plants. In the irrigated field, the primary pests that affect plants are the water weevil (*Lissorhoptrus oryzophilus,* Curculionidae, Coleoptera). Larva feeds on roots and causes severe damage to the root system leading to poor or low yields (Chen et al. 2005). Root grubs (*Leucopholis irrorata*, Scarabaeidae, Coleoptera) feed on roots and lead to root loss. They can also cause wilting, plant discoloration and abnormal plant heights. Both larvae and adults feed on roots and leaves. Mole crickets (*Gryllotalpa orientalis*, Gryllotalpidae, Orthoptera) feed on seeds, tillers, and roots and occur in all environments. The nymphs feed on roots and damage crops. Symptoms like damaged roots, missing plants, loss of plant stand and death, or weak seedlings are caused by mole crickets (Matteson 2000; Pathak and Dhalival 1981).

5.3.4.3 Stem Borers

Stem borers (*Scirpophaga incertulas*) are the primary pests of rice plants and occur throughout India's rice-growing areas. This infection is higher in Southern states. They affect crops from tillering to mature stages, with temperatures lower than 28 °C and frequent rainfall leading to an outbreak. Adult moths lay eggs, and larvae bore stems leading to 'dead heart' (dead of central shoot system). Infections during reproductive stages result in whitish and unfilled grains in panicles. It is referred to as 'whiteheads' (Huang et al. 2008; Satpathi et al. 2012). Dark-headed, golden-fringed, striped, white, yellow, and pink are different types of stem borer diseases caused by *Chilo polychrysus, Chilo auricilius, Chilo suppressalis, Scirpophaga innotata, Scirpophaga incertulas,* Pyralidae, Lepidoptera, and *Sesamia inferens*, Noctuidae, and Lepidoptera, respectively. Pink stem borers are less important and prefer sugarcane to rice. Yellow stem borers are found in aquatic environments with continuous flooding. The second instar larvae enclose themselves in bodies and wrap leaves to make tubes detached from leaves to fall onto the water surface. They then attach themselves to tillers and bore stems. 20% loss in early planted crops and 80% damage in late-planted crops are observed. Striped stem borers are seen in temperate countries, and their final instar remains dormant in temperate areas. Their wide distribution and infection cause severe to 100% damage (Cheng et al. 2010; Zibaee et al. 2008). White stem borers cause an outbreak in wetland rice plants (Gomez and Bernardo 1974; International Rice Research Institute- Rice Knowledge Bank 2020f; Muralidharan and Pasalu 2006; Rahman et al. 2004).

5.3.4.4 Leaf and Planthoppers

Leafhoppers attack all aerial parts of plants, whereas plant hoppers attack basal portions. *Nephotettix malayanus* and *Nephotettix virescens* species can also spread tungro virus. Both nymphs and adults feed and extract plant saps. High concentrations of nitrogen encourage pest attacks. They feed on lateral leaves and the dorsal surface of leaf blades. They cause drying symptoms and lead to stunted plant growth with reduced tillers (Cheng and Pathak 1972). Brown (*Nilaparvata lugens*) and white-backed (*Sogatella furcifera*) Delphacidae, and Hemiptera plant-hoppers are two species that infect rice plants (Sujithra and Chander 2013). Feeding in large numbers results in complete crop dryness and is referred to as 'hopper-burn'. Cresent shaped white eggs, inserted into midribs or leaf sheaths, cause damage to tiller bases. Hoppers are common in irrigated and rain-fed areas (Dupo and Barrion 2009).

5.3.4.5 Defoliators and Other Pests

Several groups of insects feed on leaves and belong to different orders. They reduce yields by decreasing the photosynthetic capacity of plants. Plants at their active tillering stages can overcome this damage by compensating new tillers and preventing yield losses. Armyworms, rice hispa, and leaf folders are general defoliators that affect rice plants. Common rice black bugs, Malayan black bugs, and Japanese rice black bugs are three black bugs species (*Scotinophara coarctata*) that affect these plants. Bugs remove saps of plants, which result in browning of leaves, dead hearts, whiteheads, reduced tillers, plant stunting, and bug-burn. In severe cases, bugs prevent plants from producing seeds. Bugs are abundantly seen in densely planted fields and during dry seasons. Mealybug (*Brevennia rehi*, Pseudococcidae, Hemiptera) causes curling and wilting of plants. Both nymphs and adults feed and remove plant saps (Pathak and Dhalival 1981).

Armyworms are caterpillars that feed on rice crops. The most common species in Asia are common cutworm (*Spodoptera litura* (Fabricius), Noctuidae, Lepidoptera), ear-cutting (*Mythimna separata* (Walker), Noctuidae, Lepidoptera) and swarming caterpillars (*Spodoptera mauritia acronyctoides* (Guenee), Noctuidae, Lepidoptera). They affect all stages of growth in rice plants. The female worm lays 800–1000 eggs during its lifetime of about 1 week. It feeds on plants and cuts leaves, panicles, and young seedlings. The adult worm feeds, mates, and migrates during nights and rests in plants' bases during days while larvae feed on plants during nights or cloudy days. They mainly feed on rice canopy (upper portion of rice). Low temperature facilitates the survival of worms. In wetlands, they pupate, whereas, in dry-lands, they are found in plant bases and soil. A high population of worms leads to yield loss, and the spread of worms from one field to another leads to an outbreak (Tanwar et al. 2010).

Rice hispa, caused by *Dicladispa armigera* (Olivier), Chrysomelidae, Coleopteran, is abundant during the rainy season (Chakraborty and Deb 2012). The organism tunnels through leaf tissues and causes severe damage. The presence of grassy weeds in rice fields and heavy fertilizers encourages pest growth. It causes scraping of leaf blades, whitish, and membranous leaves and wilting of damaged leaves (Dutta and Nath 2003). Cnaphalocrocis medinalis, Pyralidae, Lepidoptera cause the rice leaf folder disease. It folds leaves around itself and feeds on leaves to cause a longitudinal white transparent streak on blades. Grassy weeds, high humidity, and high fertilizers favour its growth and development. The heavy infection results in many leaf foldings, and high feeding causes damage and yield loss (Shanmugam et al. 2006).

Other prevalent species of rice bugs include *Leptocorisa oratorius* and *Leptocorisa acuta*. They suck out grains' contents resulting in unfilled or empty grains (Torres et al. 2010). Both immature and adult bugs feed on grains leading to small or deformed spotted grains and empty grains with erect panicles. In severe cases, 30% of yield loss is seen. These bugs occur in all rice cultivating environments, and they prefer to infect during flowering or milky stages (Jahn et al. 2004; Hosamani et al. 2009).

5.4 Pest Control Management

Since a variety of microbial pests attack rice plants at various stages of growth, many synthetic chemicals, green constituents and formulations have been developed for pest control. The types of pests that infest seedlings and plants depend on host species' susceptibility, geographical locations of growth, soil conditions, fertilizers used therein, and the general environmental conditions. This has warranted the need for a wide range of pest control measures that focus not only on chemical and natural constituents but also on the novel and innovative materials, techniques and technologies (Table 5.2).

Table 5.2 Pest control strategies for microbial and insect infestations in *Oryza sativa*

Disease	Representative pest control strategies	References
Bacterial leaf blight	Plant based emulsion formulation with *Piper Sarmentosum*	Syed-Ab-Rahman et al. (2020)
	Bio-synthesied silver nanoparticles against *Xoo*	Ahmed et al. (2020), Ibrahim et al. (2019), Abdallah et al. (2019a), Ogunyemi et al. (2019)
	Green synthesis of magnesium oxide (MgO) nano-flowers and zinc oxide nanoparticles against *Xoo*	
	Seed treatment with *Kalanchoe pinnata*	Khoa et al. (2017)
	Biological control with *Pseudomonas aeruginosa, Paenibacillus polymyxa, Bacillus spp.,*	Yasmin et al. (2017), Abdallah et al. (2019b), Elshakh et al. (2016)
	Aspergillus against *Xoo*	Jiang et al. (2019)
Bacterial leaf streak	Biological control with *Bacillus amyloliquefaciens*	Zhang et al. (2012)
	Sulfone derivatives	Li et al. (2014)
	Treatment with chitosan	Li et al. (2013a)
Rice blast	*Cladosporium cladosporioides* as plant spray for seed treatment	Chaibub et al. (2020)
	Trichoderma harzianum	Chou et al. (2020)
	Burkholderia pyrrocinia & *Pseudomonas fluorescens*	De Sousa Oliveira (2020)
	Bacillus safensis, Bacillus tequilensis, Pseudomonas saponiphila, Pseudomonas koreensis & Stenotrophomonas rhizophila	Wei et al. (2020)
Sheath blight	Myco-silver nano particle from *Solanum nigrum*	Akther and Hemalatha (2019)
	Torulaspora indica & *Wickerhamomyces anomalus*	Into et al. (2020)
	Plant extracts (lemon grass, clove), bio fungicides (*Bacillus subtilis*) & new generation fungicides (Trifloxystrobin+Tebuconazole, Propineb, Cyclops (clove oil + cinnamon oil)	Persaud et al. (2019)
	Nitrogen fixing cyanobacteria *Nostoc piscinale* & *Anabaena variabilis*	Zhou et al. (2020)
	Fungicide molecule Difenoconazole 11.4% + Azoxystrobin 18.2% SC (1 ml/L)	Kumar (2020)
Sheath rot	Potassium fertilizer application	Zhang et al. (2020a)
	Trichoderma harzianum &Trichoderma viride	Selvaraj and Annamalai (2015)
	Bacillus subtilis, Pseudomonas fluorescens	Subramaniam et al. (2013)
Brown spot	Berberine (5 mg/ml)	Kokkrua et al. (2020)
	Trifloxystrobin 50% WP + Tebuconazole 25%, Hexaconazole 5% EC, Hexaconazole 25% WG + Zineb 68%	Monisha et al. (2019)
	Azadirachta indica (10%)	Kumar and Simon (2016)
	Trichoderma harzianum & *Trichoderma atroviride*	Khalili et al. (2012)
Rice tungro disease	Neem cake & 5% spray of neem seed kernel extract	Rajappan et al. (2000)

(continued)

Table 5.2 (continued)

Disease	Representative pest control strategies	References
Root feeder	Seed treatment with Chlorantraniliprole & Thiamethoxam	Kelly et al. (2020)
	Extracts of *Commelina communis L, Heteranthera limosa, Alternanthera philxeroides*	Sholl et al. (2020)
	Chlorantraniliprole seed treatment	Hummel et al. (2014)
Stem borer	Lepidopteran pests with *Trichogramma* spp.	Babendreier et al. (2020)
	Chlorantraniliprole	Meng et al. (2020)
	Carbon quantum dot, lipofectamine2000, chitosan	Wang et al. (2019a)
Leaf hopper	Malathion, alpha-cypermethrin, phosphamidon, clothianidin, thiamethoxam	Kumar et al. (2010a), Misra (2009), Sahithi and Misra (2006)
	Neem oil application	Kumar et al. (2010a)
	Essential oils	Longkumer and Misra (2020)
Plant hopper	Natural enemy spider	He et al. (2020)
	Sulfoxaflor-loaded natural polysaccharide microsphere	Yang et al. (2020)
	Insecticide imidacloprid and biological agent *Pseudomonas fluorescens*	Sangamithra et al. (2017)
	Triflumezopyrim	Zhang et al. (2020b)
Defoliators	Essential oils	Longkumer and Misra (2020)
	Insecticide dinotefuran and bio pesticide *Beauveria* spores	Sharma and Srivastava (2019)
	Trichogramma chilonis	Bharti et al. (2018)
	Fepronil & flubendamide	Soomro et al. (2020)
	Beauveria bassiana	Sharma et al. (2017)
	Cypermethrin & Fenvalarate	Bhattacharjee and Ray (2012)
Grain sucking insects	Neem oil	Prakash and Kunal (2020)
	Leaf extract of *Piper spp*	Nugroho et al. (2020)
Rice bug	Monocrotophos, profenophos & fenvalerate	Prakash and Kunal (2020)

5.4.1 *Synthetic Chemicals and Their Formulations*

Chemicals and their agro-formulations are best suited for pest control and disease management. However, effective chemical formulations for controlling rice bacterial blight are yet to be developed as microbial pathogens show extreme variability in their sensitivity towards antibiotics and chemicals. Disease management largely depends on factors like chemical availability, modes of action and access opportunity, pathogen susceptibility, economic value of crops, and market value of chemicals or bio-control agents. Strategies that deal with proper use of fertilizers, their controlled spraying, silica application, flooding of paddy fields, and utilization of improved chemicals like copper fungicide, organo-phosphorus, and organo-mercuric compounds, antibiotics like kasugamycin, blasticidin, and plant activators, need to be employed. Specific chemical insecticides are effective against pests at different geographical locations. Still, their indiscriminate usage would reduce natural enemies and cause environmental pollution and bioaccumulate their residues. Chemical methods of bacterial disease management have been reviewed (Naqvi 2019).

The development of strategies for managing bacterial plant diseases has traditionally remained a challenge for plant pathologists. Achieving effective disease control has been difficult as consumer preferences always have focused on disease-prone cultivars, and environmental factors have predominantly favoured pathogenic infestation. These challenges, along with achieved innovations and prospects of bacterial disease management, have been reviewed (Sundin et al. 2016).

Reasonable disease control has been achieved with foliar sprays of streptomycin sulfate. Compared to kasugamycin, its combinations with copper oxychloride, a fungicide-like mixture of tebuconazole and trifloxystrobin, and azoxystrobin and difenconazole, have produced better yields of rice at 92%, 86% and 87%. Such combinations have fared better than other fungicides that are used for pest management (Nasir et al. 2019). Broad-spectrum antibiotics like streptomycin, kanamycin, ampicillin, benzylpenicillin, chloramphenicol and sinobionic at different concentrations, are active against several virulent *Xoo* strains isolates from different district of Punjab, Eastern Pakistan, which have been deposited at Nuclear Institute for Agricultural and Biology, Faisalabad, Pakistan (Khan et al. 2012). Common plant pathogens like *Erwinia amylovora, Xanthomonas spp*., and *Pseudomonas spp* have been controlled by bactericide spray treatment (McManus et al. 2002).

The addition of a mixture of antibiotics has been used as an alternative procedure in some patho-systems because of antibiotic-resistance (Shtienberg et al. 2001). Traditionally, bacterial diseases have been controlled by spraying antibiotics in paddy fields (Singh et al. 1980; Khan et al. 2012). Combination sprays of streptomycin sulphate and tetracycline, along with copper oxychloride, serve best to control infection (Patel et al. 2009). If the infection persists, spraying may be repeated after 15 days. During the Kresek stage, the application of bleaching powder in irrigated water is recommended (Chand et al. 1979). Two sprays of copper hydroxide 77WP, 30DAP, and 45DAP are used for bacterial control. 100 ppm of agrimycin, along with 90% streptomycin sulphate and 10% of tetracycline hydrochloride, is currently used by farmers to control infections. Infection is reduced by treating seeds with zinc sulfate, ceresan, streptocyclin, agrimycin, and bleaching powder. Synthetic organic chemicals such as phenazine, phenazine N-oxide, nickel dimethyl dithiocarbamate, techlofthalam, and dithianone can also control bacteria in paddy (Gnanamanickam et al. 1999; Biswas et al. 2009).

Agrimycin, streptomycin, and tetramycin, control infection moderately. 20 ppm of chlorine minimizes field disease, whereas 30% reduces the frequency of infections in paddy fields. Better control has been achieved with 2-amino, 3, 4-thiadiazole. Subsequent application of Bordeaux mixture controls initial development of infection. Out of 11 chemicals like streptomycin, blasticidin, tricyclazole, sumithione, phytomycin, kasumin, celdion, mipcine, Saturn, hinosan and stem F-34, the first six inhibit bacterial growth *in vitro*, to produce good inhibition zones. But only blasticidin, streptomycin, and kasumin control this disease under *in vivo* conditions and that also, to a small extent (Ashrafuzzaman 1987). Agricterramycin 17, agrimycin 100, agrimycin 500, A.S. 50, vitavax 75, brestanol 45, fytolan, and streptocycline are used to control *Xoo*. Spraying combinations of bactericide and fungicide, such as agrimycin 100 and fytolan (Copper oxychloride) in ratios of 50:500 ppm effectively controls disease and prevents economic losses in susceptible varieties. After 10–12 days, repeated spraying becomes necessary (Singh et al. 1980).

Phenazine-1-carboxylic acid, an antibiotic that can be produced from *Pseudomonas spp*., is very effective against *Xoo*. It increases the accumulation of reactive oxygen species and, in during this process, decreases activities of antioxidant enzymes like superoxide dismutase and catalase. Phenazine-1-carboxylic acid also inhibits carbohydrate metabolism and reduces uptake of nutrients by this pathogen, thus successfully agitating its redox balance (Xu et al. 2015a). Oxytetracycline, along with copper oxychloride at different concentrations, controls bacterial pests in paddy. This chemical combination increases yields, panicle weights and lengths by 16.6%, 2.8 g, 27.6 cm, respectively, and decreases disease incidence by 51% (Singh et al. 2015a). This is followed by bactrinashak, bacterimycin and bronip (Patel et al. 2009). Ampicillin trihydrate, benzylpenicillin, kanamycin sulphate, and chloramphenicol are broad-spectrum antibiotics that effectively control blight disease (Khan et al. 2012). 200 ppm of streptomycin sulphate and 0.25% of copper oxychloride also reasonably achieve the same (Kumar et al. 2009). *In vitro* evaluation of 1000 ppm streptocyclin shows its effectiveness against bacterial rice blight. This is followed by 1000 ppm of streptocycline with copper oxychloride. *In vivo* conditions reveal the use of streptocycline, along with copper oxychloride, as the best measure to control infection by reducing disease indices and increasing grain yields (Thimmegowda et al. 2012).

Phenylacetic acid has been used to synthesize a series of 2, 5-substituted-1, 3, 4-oxadiazole/thiadiazoles. Compared to thiadiazole copper and bismerthiazol, a chemical derivative series, *viz.*, 2, 5-substituted-1, 3, 4-oxadiazole/thiadiazole, exhibits better activity against bacterial leaf blight under *in vivo* and *in vitro* conditions. A synthesized chemical compound 2-(methylsulfonyl)-5-(4-fluorobenzyl)-1, 3, 4-oxadiazole best inhibits bacterial blight and streaks (Li et al. 2014). 1, 3, 4-Oxadiazole/thiadiazole also shows antibacterial (Li et al. 2013b) and antifungal activities (Zhang et al. 2013). Sulfonate derivatives are reported to have antibacterial, antifungal (Xu et al. 2011) and herbicidal (Li et al. 2005) activities. A series of 2-(thioether/sulfone)-5-pyrazolyl-1, 3, 4-oxadiazole derivatives containing 1, 3, 4-oxadiazole, and pyrazole moiety show better control of bacterial and fungal diseases in paddy plants (Zheng et al. 2017). 2, 4-dichlorobenzoic acid is used as a starting material for the synthesis of pyridinium-tailored 2, 5-substituted-1, 3, 4-oxadiazole thioether/sulfoxide/ sulfonate derivatives. 1, 3, 4-oxadiazole derivatives with cinnamic acid moiety can be used for *Xoo* control. It serves as an alternative for the effective management of bacterial blight (Wang et al. 2016,2019b).

2-(methylsulfonyl)-5-(((4-fluorophenyl) sulfonyl) methyl)-1, 3, 4-oxadiazole, shows excellent activity against bacterial blight. These compounds largely inhibit bacterial growth and development from serving as promising agrochemicals for infection control (Li et al. 2018).

A series of 1, 3, 4-oxadiazole thioether derivatives containing a 6-fluoroquinazolinyl piperidinyl moiety is active against *Xoo*. Structure-activity relationships play a significant role in antibacterial activities against this pathogen. A strong electron-withdrawing group on the benzyl or benzene ring, cyano group at 2- or 3- position or mono-halogens like -Cl and -F on the phenyl ring, is found to be more beneficial for bacterial control in paddy. *In vitro* bioassay of these compounds shows better results than commercial bismerthiazol (Shi et al. 2016). Bismerthiazol (N, N-methylene-bis (2-amino-5-mercapto-1, 3, 4-thiadiazole) is a major bactericide used for blight control. Bismerthiazol resistant strains of *Xoo* is controlled by zinc thiazole (bis (2-amino-5-mercapto-1, 3, 4-thiadiazole). It is safe for non-target organisms. When assessed by growth inhibition studies, the baseline sensitivity of 109 strains of Xoo shows EC_{50} values between 0.53 and 9.62μg/ml, whereas complete inhibition of growth occurs between 5 and 40μg/ml. At 300μg/ml (1 and 2 days before or after inoculation) it exhibits greater than 88% control efficacy than bismerthiazol (Chen et al. 2015).

In the case of genetically diverse organisms isolated from several northern and southern districts of India, ciprofloxacin and niclosamide, from 300 to 1000 ppm are found to be more active for pathogenic control as compared to bionol, bactrinashak, chloramphenicol, streptomycin sulphate, and copper oxychloride (Praveen et al. 2019). 2.5% of copper hydroxide (kocide) 2000 DF and validamycin compound (Sheathmar 3 L) exhibit 6.4% and 55% greater control of bacterial blight as compared to control plants (Parthasarathy et al. 2014). Kasumin, kasuran, vitigran blue, copper oxychloride, chloramphenicol, streptomycin, and oxytetracycline have been evaluated against *Xoo*. The highest yields and best control have been observed with copper oxychloride followed by vitigran blue, kasumin, streptomycin, oxytetracycline, chloramphenicol, and streptomycin with copper oxychloride. Among these, copper oxychloride and vitigran blue show the best activities against *Xoo* with 43 and 48% disease incidence and highest yields of 3.63 and 3.58 t/ha, respectively, as compared to controls. Spraying of copper fungicide alternately with streptomycin 250 ppm may also be effective (Yasin et al. 2007).

A mixture of 2.5 kg copper sulphate and quick lime in 300 l/ha water (Bordeaux) is economical and also effective for disease control. Two or three sprays of this mixture control infection and increase grain yields of paddy. Spraying during the booting stages prevents infection. Bordeaux mixture, in combination with antibiotics like streptomycin and oxytetracycline, suppresses the development and spread of bacterial blight for about 10–15 days with disease incidence percentages of 41, 37 and 39 in research farms and 44, 40 and 43 in farmers' fields, respectively (Chaudhary et al. 2012).

Some drugs that are used for the treatment of human infections have a positive effect on blocking plant pathogens (McManus et al. 2002). Niclosamide and auranofin are two such drugs. The influence of niclosamide in inhibiting leaf blight caused by *Xoo* has been reported in three of its strains. After spraying, it travels as far as the distal tissues and increases salicylate levels. It has no deteriorative effects on the life cycle of plants and increases their yields. It prevents plants from wilting and lesion development. At a low concentration of 8μg/ml, it persists in plants for more than 4 days, thus blocking infections for longer periods of time (Kim et al. 2016a, b; Sahu et al. 2018). Auranofin is also reported to block bacterial growth in rice, and it works similarly to niclosamide. Lesion development is inhibited by the mediation of salicylate and jasmonate-dependent defense signalling pathways (Kim et al. 2018).

Systemic interference of fungicides with the electron transport chain reduces or inhibits the synthesis of new cellular materials required for biosynthesis processes, with cell structure disruption and increased permeability of cell membranes being the modes of action. The triazole groups of fungicides are demethylation inhibitors. Strobilurin is a quinone inhibitor. Carboxamide groups of fungicides are succinate-dehydrogenase inhibitors (Finch et al. 2014). Non-systemic fungicides like sulphur based ones disrupt normal cellular hydrogenation – dehydrogenation reactions and precipitate or inactivate proteins (enzymes of sulphydryl group), thus killing fungal spores. Mercury based fungicides destroy sulfhydryl group enzymes whereas quinone derivatives disturb the electron transport system. Tricyclazole (5-methyl-1, 2, 4-triazolo (3, 4-b) benzothiazole) effectively controls rice blast fungi by inhibiting melanization within appressoriums, thus reducing rigidity in appressorial walls (Peterson 1990).

As compared to *in-vitro* studies, *in-vivo* ones show excellent control of blast disease. Different modes of applications like foliar spraying, seed coating, soil drenching, and transplant bare-root soaking effectively inhibit fungal pathogens' growth with long-term control (Froyd et al. 1976). Ferric chloride, di-potassium hydrogen phosphate, and salicylic acid are reported to control rice blast. They significantly reduce disease severity when applied through foliar sprays and or through the soil. From seedling to the heading growth stage, foliar sprays need to be used twice for disease control. Ferric chloride alone increases grain yields significantly (Manandhar et al. 1998).

Phosphor thioate fungicides, edifenphos and iprobenfos, are mainly applied to the soil and have a systemic mode of action against blast fungi. They block the biosynthesis of choline, an essential constituent of fungal membranes. In India, copper fungicide is noted to be effective in controlling fungal infections. Dithiocarbamate, benomyl, carbendazim, tricyclazole, strobilurins, pyroquilon, bavistin, hinosan, carpropamid, fenoxalin, and tiadinil are systemic fungicides, melanin biosynthesis inhibitors, and plant activators used for controlling blast diseases in rice. The past, present and future research with respect to blast disease management have been reviewed (Pooja and Katoch 2014). The MTU7029 rice variety has been used for evaluation of different fungicides, *viz.*, 75% tricyclozole (Gain 75WP), 50% tebuconazole+25% trifloxystrobin (Native 75WG), 5% hexaconazole (5% Hexacon super SC), 25% difenoconazole (Score 250EC) and propiconazole (Tilt 25EC), at different concentrations, against rice blast disease. All of them are reported to control this disease. 75% Tricyclozole, 50% tebuconazole +25% trifloxystrobin and difenoconazole are effective in disease control with reduced disease severity of 12.9, 10.2 and 11.5%. Neck blast is effectively controlled by 50% tebuconazole +25% trifloxystrobin, propiconazole, and 25% difenoconazole with reduced disease percentages of 19.0, 14.39 and 11.6. Leaf blast is least effectively controlled by propiconazole (Dutta et al. 2012).

The effect of five chemical fungicides, *viz*., 0.025% of propiconazole and bitertanol, 0.045% tricyclazole, 0.05% carbendazim, and 0.1% cyproconazole has been evaluated for 3 years in native rice cultivar pankhali-203, which is highly susceptible to blast disease. All the fungicides effectively control leaf and neck blasts of paddy. They are sprayed twice during booting and flowering stages for effective disease control and increase grain yields. In comparison to a recommended fungicide Carbendazim 50 WP, tricyclazole exhibits superior control of leaves (62.9%) and neck blast (64.1%) disease and increases in grain yields (72.3) over control plants have been noted (Prajapati et al. 2004). Different fungicides like propiconazole, iprobenfos, tricyclazole, edifenphos, carbendazim, and mancozeb are effective against blast at concentrations of 500, 1000, and 1500 ppm. Mancozeb performs best at 1000, 2000, and 3000 ppm (Gohel et al. 2008). Tricyclazole is proven to be effective against leaf and neck blasts and increases grain weights, grain and straw yields (Gohel et al. 2009). Efficacy of isoprothiolane (Fuji-one 40E), at three different doses, has been tested against rice blast. Even low doses of 1.5 ml/l effectively control leaf (45.3%) and neck blast (16%) and increase yields (26.9%) (Raji and Louis 2007).

Kasugamycin, blasticidin, isoprothiolane, edifenphos, iprobenfos, ferimzone, and metominostrobin excellently inhibit blast growth causing fungi in paddy. Overuse of organophosphorus thiolate and kasugamycin leads to blast-resistance in fields. Due to fungicide resistance, two groups of non-fungicidal rice blast chemicals, *viz.,* melanin biosynthesis inhibitors (tricyclazole, carpropamid, phthalide, pyroquilon, fenoxanil, diclocymet) and plant defence activators (acibenzolar-s-methyl, probenazole, and tiadinil) are used for pest control. Chemicals that effectively control blast diseases have been reviewed (Yamaguchi 2004; Hirooka and Ishii 2013). Tebuconazole (25% WG) is effective in inhibiting rice blast and sheath blight. At 0.2% it significantly reduces disease severity by 17.2 and 10.2%, respectively and increases yields by 41.40 q/h (Hegde 2015). Jinggangmycin, a glucosaminidase glycoside antibiotic is the most widely used fungicide for the control of sheath blight in China. Its activity is similar to validamycin. It leads to 35 and 20% reductions in lesion lengths at 50 and 100 mg/l, respectively (Peng et al. 2014).

A sheath blight susceptible rice variety MTU7029 has been used to check the efficacy of fungicides. Pencycuron (Monceren 250SC) was found to be effective against sheath blight, and it has been evaluated for a period of 2 years under field conditions (Chowdhury and Sarkar 2006). Different fungicides like tebuconazole (Folicur 250EW), validamycin (Rhizocin 3 L), O,O-dimethyl-O(2,6-dichloro-4-methyl-phenyl)-phosphorothioate (Rhizolex 50WP), kresoxim-methyl (RIL 010/F1 25 SC, RIL 010/F1 50 SC), tebuconazole (Folicur 250EW), pencycuron (Monceren 250 SC), hexaconzole (Contaf 5EC) and propiconazole (Tilt 25EC) are known to control sheath blight fungi in rice. Fungicides like pencycuron (Monceren 250SC) exhibit superior disease control. Least control is reported with 2.62% copper sulphate (shield 2.62SC) under field conditions (Hunjan et al. 2011).

Upon comparison of six fungicides, *viz.*, captan 50 WP, mancozeb 75 WP, copper oxychloride 50WP, dodine 65WP, zineb 80 EP and propineb 70 WP, at 50–1000μg/ml of active ingredients, 95.3% of mean mycelial growth inhibition of sheath blight fungi has been observed in mancozeb, followed by 93.8 and 93.7% in captan and dodine respectively. Copper oxychloride was the least effective with 44.1%. Complete inhibition of sheath blight is seen with 500μg/ml of the active ingredient of mancozeb 75 WP, captan 50 WP, dodine 65 WP and propineb 70 WP. 100% of mycelial growth inhibition is observed with 1000μg/ml of active ingredient, and the least inhibitions of 95% and 84.4% are seen with zineb and copper oxychloride, respectively. All six fungicides reduce disease incidence as well as intensity and increase grain yields. Mancozeb shows superior control, followed by captan, dodine, propineb, zineb, and copper oxychloride (Mughal et al. 2017).

The efficacy of eight fungicides, *viz.*, rhizocin 3 L (Validamycin), tilt 25EC (Propiconazole), kitazin 48EC (Iprobenphos), baycor 25WP (Bitertanol), bavistin 50WP (Carbendazim), contaf 5EC (hexaconazole), saaf 75WP (Carbendazim-mancozeb) and dithane Z-78 (Zineb) has been compared against three rice fungal pathogens, *viz.*, sheath blight, sheath rot, and brown spot, under both *in-vitro* and *in vivo* conditions. 0.1% of tilt 25EC excellently controls all three diseases, whereas reduced severity of infection is seen with regard to sheath blight (15.7%), brown spot (9%), and sheath rot (6.7%) as compared to controls. 0.1% of bavistin 50WP is effective against sheath blight (18.1%) and sheath rot (10.5%), followed by contaf 5EC. Baycor 25WP, saaf 75WP, kitazin 48EC, and dithane Z-78 are least effective against all three fungal diseases. Bavistin 50WP, contaf 5EC, and tilt 25EC are highly effective against grain discoloration (Hunjan et al. 2011). In comparison to captan 50%WP and hexaconazole 5%SC, TAQAT 75%WP, which consists of captan 70% along with hexaconazole 5%, effectively controls sheath blight and sheath rot fungus in paddy. The lowest disease index of 14.1 and 14.2% is reported in sheath blight and sheath rot, respectively. Lowest disease index and increased grain yield is observed with TAQAT 75% WP (Pramesh et al. 2017).

The effectiveness of several other fungicides, *viz.*, trifloxystrobin along with tebuconazole (Nativo 75WG), thifluzamide (Spencer 24SC), propiconazole (Tilt 25EC), tebuconazole (Folicur 25EC) and pencycuron (Monceren 250SC) have been evaluated against rice sheath blight and brown spot fungus under both *in vitro* and *in vivo* conditions. 0.04% trifloxystrobin along with tebuconazole and 0.1% of tebuconazole is effective against both fungi followed by 0.1% propiconazole. They reduce severity of sheath blight infection and brown spots in paddy. Pencycuron is least effective against brown spot disease but effective against sheath blight. Higher levels of disease control in sheath blight and brown spots are seen with trifloxystrobin, tebuconazole, tebuconazole, and propiconazole. Thifluzamide is the least effective under both *in vivo* and *in vitro* conditions for brown spot and sheath blight (Hunjan et al. 2011). The combination of zineb and hexaconazole has been evaluated for control of brown spot and sheath blight in paddy. The foliar spray of zineb along with hexaconazole (2 g/l) is useful here. This spray is found to be more effective 30 and 45 days after transplantation. Reduced disease severity of 36.5% is seen in brown spot condition and 21.1% in sheath blight. It also increases grain yields of paddy with a maximum yield of 4.49 t/ha (Dinakaran et al. 2012).

Brown spot disease of paddy has been reviewed (Sunder et al. 2014). Fungicides like mancozeb birational molecule, with 2.62% copper sulphate, tridemorph, triadimenol, ridomil, bitoxazol, edifenphos, and oxalyl chloride are highly effective in controlling mycelial growth of fungi that cause brown spot disease. Hexaconazole, propiconazole, iprobenphos, iprodione, strobilurins, azoxystrobin, kresoxim-methyl, and trifloxystrobin are strong inhibitors of brown spot disease. Application of 20% azoxystrobin along with 12.5% difenoconazole from leaf booting stage to milky grain stage shows that this process, when undertaken during different growth stages of the rice cultivar IRGA 424, reduces disease severity and the incidence of brown spot along with increasing grain yields (Barua et al. 2019). Benzoic and salicylic acids also completely inhibit fungal growth at 9 Mm (*in-vitro*). *In vivo* experiments reveal that 20 mM benzoic acid effectively controls disease severity and increases grain yields. A significant increase in photosynthetic pigments, total carbohydrates and proteins in grains has also been recorded (Shabana et al. 2008).

Chemical structures of insecticides generally define their target sites and modes of action. Most insecticides are neurotoxins except for a few commonly used ones like insect growth regulators and some with different active ingredients like energy inhibitors, borates, and dehydrating dust. Neurological target sites of neurotoxins include modulation and blockage of sodium channels, inhibition of acetylcholinesterase enzyme, acetylcholine receptor stimulation, Gamma-aminobutyric acid receptor blockage and glutamate receptor stimulation. Pyrethrins and pyrethroids (cyfluthrin, beta-cyfluthrin, cypermethrin, pyrethrins, permethrin, deltamethrin, and lambda-cyhalothrin), oxadiazines (Indoxacarb), and semicarbazone (Metaflumizone) block sodium channels.

Chemical class of organophosphates (Malathion, profenophos, acephate, monocrotophos, diazinon, chlorpyrifos (Dursban), dichlorvos, and propetamphos) and carbamate (Carbofuran, cartap, carbaryl, oxamyl, methomyl, fenobucarb, and methiocarb) insecticides are acetyl-cholinesterase inhibitors. Neonicotinoids (Imidacloprid, thiamethoxam, dinotefuran, clothianidin, and acetamiprid) and spinosyns (Chemicals produced by soil bacteria) bind to acetylcholine receptors on post synapse nerve cells. Phenylpyrazole class of insecticides, like fipronil, bind and block Gamma-aminobutyric acid receptors (blocking chloride channel). Avermectins (Benzoate, ivermectin) bind to glutamate receptors.

A diamide insect growth regulator, amidinohydrazone, pyrrole, fumigants, borates, and dehydrating dust are insecticides that do not target insect nervous systems. Diamides (chlorantraniliprole) technically are neurotoxins, but they act on muscular calcium channels. Juvenile hormone analogs (mimic juvenile hormones) and chitin synthesis inhibitors (blocks chitin formation) are insect growth regulators. Amidinohydrazones (hydramethylnon), pyrroles (chlorfenapyr), and fumigants (sulfuryl fluoride) inhibit energy production. Non-specific metabolic disruption has been reported in borates (Borax, boric acid, and borates). Dehydrating dust, *viz.*, silica gel, act *via* desiccation (Adsorption of cuticular wax layer).

A synthetic insecticide cartap hydrochloride 4G and imidacloprid 17.8 SL have been evaluated against yellow stem borers and leaf folder infection in paddy. Both insecticides reduce disease incidence in paddy (Singh et al. 2008). The use of insecticide fipronil 5SC at a concentration of 50 g active ingredient (a.i)/ha, lowers disease incidence of stem borer and increases yields. This is followed by 300 g a.i/ha of cartap hydrochloride 50 SP and 750 g a.i/ha of cartap hydrochloride 4G. Cartap hydrochloride 50SP is reported to be more economical than others with the highest cost: benefit ratio, followed by fipronil 5SC and cartap hydrochloride 4G (Mishra et al. 2012).

The effectiveness of insecticides, *viz.*, profenophos 50EC (0.08%), cartap hydrochloride 4G (0.20%), fipronil 0.3G (0.01%), spinosad 45SC (0.01%) has been evaluated against yellow stem borerS and yield performance in paddy. Minimum damage has been recorded in fipronil 0.3G and spinosad 45SC, and they were on par with each other. Profenophos 50EC and cartap hydrochloride 4G show significantly lower damage over controls and are on par with each other. All insecticides increase grain yields over controls. The highest yields of 59% are found in fipronil, followed by spinosad (57.2), profenophos (51.71), and cartap (50.3) (Kakde and Patel 2019).

Application of various insecticides, *viz.*, flubendiamide 48% SC at 50 ml/ha, monocrotophos 36SL at 850 ml/ha, and dinotefuran 20SG at 200 g/ha has been evaluated against white stem borers. All of these significantly control disease and increase yields. The minimum dead heart and white ears are seen with flubendiamide and rynaxypyr. For effective management of stem borers, a very low dose of flubendiamide is effective and economical (Tandon and Srivastava 2018). Observation of a rice ecosystem from 2013 to 15, shows that several insect pests are associated with rice crops. The efficacy of different insecticides has been tested against yellow, pink, dark-headed, white, and striped stem borers. Minimum and maximum disease incidence due to all of these stem borers, occurs during grain hardening and tillering stages, respectively. The percentage of disease incidence is higher with yellow stem borer (24.7) followed by pink (8.1), white (4.9%), darkhead (3.3) and stripped stem borers (1.8%). Disease incidence is highest due to yellow stem borer (2.66 individuals/sweep) followed by dark-headed (1.54 individuals/sweep), pink (1.26 individuals/sweep), white (0.68 individual/sweep), and stripped stem borers (0.47 individuals/sweep). Infection incidence differs considerably during different growth stages of plants. But throughout their growth, the incidence due to the white stemborer is low.

Insecticide fipronil 0.3G exhibits reasonable control against stem borer disease. It shows the reduction of whiteheads (65.3%) and dead heart (56%). Dursban 20EC results in a 40% reduction in whiteheads and a 30.0% reduction in dead heart. Both insecticides control stem borers and increase grain yields. Yields of 3.18 and 3.04 t/ha have been reported in fipronil and Dursban, respectively (Mondal and Chakraborty 2016). A survey of disease incidence, the distribution pattern of yellow, gold-fringed, and pink stem borers and the effects of insecticides on the management of stemborer on three rice cultivars (ASD 16, TPS 3 and TPS5) has been reported. Five classes of pesticides, *viz.*, synthetic pyrethroid (λ-cyhalothrin), organophosphates (Profenofos, chlorpyriphos), anthranilic diamide (chlorantraniliprole), phenyl pyrazole (fipronil) and oxadiazine (indoxacarb) have been evaluated against stem borer infection. Among them, chlorantraniliprole 0.4GR was superior for controlling stem borers. It reduces disease incidence and increases the yield of paddy. The maximum grain yield was reported in rice treated with chlorantraniliprole 0.4GR, chlorantraniliprole 20SC, and profenofos 50 EC (Justin and Preetha 2014).

The bioefficacy of imidacloprid 600FS for seed treatment against brown planthoppers has been studied. Different doses of imidacloprid (1.5–2.5 ml/g kg −1 seed) have been tested over a standard control thiamethoxam 300 FS (3 ml/g kg −1 seed). Imidacloprid at 2.5 ml/g kg^{-1} seed exhibits superior control, with a 66% disease reduction as compared to control plants. At 2 ml/g and 1.5 ml/g kg, Imidacloprid exhibits 62.6 and 51.8% reduction over controls (Sangamithra et al. 2017). A new formulation fipronil 0.8G has been evaluated against brown planthopper. Its efficacy is observed from January to April (first season) and October to January (second season). The mean population reduction of planthoppers was noted at concentrations of 40, 50, and 75 g a.i/ha and a reduction was found to be 77.4, 82.7 and 86.3% in the first season, and 56.3, 79.8 and 78.0% in the second season. A granular application of insecticide was given at 21 and 45 days, after transplantation by broadcasting method (Guruprasath and Ayyasamy 2019).

Efficacy of insecticides like Dursban (chlorpyrifos 20% EC), fame 480 SC (flubendiamide 39.35%), mifpro-G (fipronil 0.3% G), mahveer GR (fipronil 0.3% G), regent (fipronil 0.3% G), miftap (cartap hydrochloride 4% G), nidan (cartap hydrochloride 4% G), marktriazo (triazophos 40% EC) and sutathion (triazophos 40% EC) has been evaluated against leaf folders and stem borers on basmati rice. All of them effectively controlled stem borer infestation and leaf folder incidence. They reduced dead hearts and white ears, and increased grain yields of basmati rice variety. Fame 480 SC at 50 ml/ha was superior for disease control. The equally effective Dursban (2500 ml/ha), sutathion (875 ml/ha) and marktriazo also reduced dead heart and leaf folder incidence in rice plants. Three brands of fipronil (15 kg/ha) and two of cartap (25 kg/ha) were found to be equally effective in controlling both diseases. Compared to other insecticides, a lower dose is reported for fame 480 SC, and it works excellently for controlling thIS disease in paddy fields. Application of fame hence results in low pesticide usage (Sandhu and Dhaliwal 2016).

New granular insecticides like chlorpyrifos, cartap hydrochloride, phorate, fipronil 50 DAT, quinalphos, ethoprophos, and isazofos, carbosulfan, and carbofuran have been evaluated against significant pests of rice like stem borer, leaf folder, planthopper, and gall midge. Reduced dead heart infestation is seen in rice treated with chlorpyrifos and isazofos, and found to be efficacious during nursery applications. Application of fipronil 50 DAT, ethoprophos, isazofos, chlorpyrifos, and carbofuran are effective against stem borer whiteheads. Fipronil, chlorpyrifos, and quinalphos effectively control leaf folder incidence. The population of plant-hoppers can be reduced only by carbofuran and isazofos insecticides. Carbofuran followed by isazofos, increase grain yields. The silver shoots caused by gall midge are effectively controlled by carbosulfan, carbofuran, and isazofos granules. Cartap hydrochloride and phorate are the commonly recommended insecticide granules, but they are moderately ineffective in reducing insect pests and increasing grain yields (Sontakke and Dash 2000).

The efficacy of newer insecticides has been evaluated against major pest stem borer. Insecticides like acephate, flubendiamide, cartap hydrochloride, monocrotophos, chlorpyrifos, imidacloprid, and fibronil, when sprayed twice on plants, reduce disease infestation and increase yields. Among tested insecticides, disease infestation is maximally reduced by monocrotophos, followed by chlorpyrifos, flubendiamide, acephate, cartap hydrochloride, imidacloprid, and fibronil. Monocrotophos is effective and shows an increase in yield and economy, followed by chlorpyrifos, flubendiamide, cartap hydrochloride, fibronil, imidacloprid and acephate, as compared to controls (Sharanappa et al. 2017).

A new granular insecticide thiocyclam hydrogen oxalate 4G, has been evaluated against stem borers and leaf folders at different doses of 300–500 g a.i/ha and compared with fipronil (45 g a.i/ha) and cartap hydrochloride (1000 g a.i/ha) in basmati rice variety. At 500 g a.i/ha, it was effective in controlling disease infestation (Dhawan et al. 2010). Effect of eight insecticides like cartap hydrochloride 4G, monocrotophos 36 SL, chlorpyrifos 20 EC, imidacloprid 200 SL, endosulfan 35 EC, triazophos 40 EC, quinalphos 25 EC and methyl parathion 50 EC was checked against white-backed planthopper, stemborer and leaf folder disease of paddy. Plant-hopper population was high during the tillering and booting stages. During the tillering phase, this population is effectively controlled by chlorpyrifos, endosulfan, imidacloprid and cartap. At the booting stage, imidacloprid, endosulfan, cartap, and quinalphos reduce white-backed plant-hopper population.

Chloropyriphos, cartap, and triazophos effectively control white ear infestation. Cartap, monocrotophos, and chlorpyrifos exhibit excellent activity against leaf folders. Cartap hydrochloride treated rice results in higher grain yield, followed by monocrotophos and chlorpyrifos. Among others, methyl parathion acts as the least effective insecticide (Sarao and Mahal 2008). Sucking pests of rice, *viz.*, plant and leafhoppers are controlled by 40% ethiprole along with 40% imidacloprid −80 WG. Different dose concentrations have been evaluated against this pest. Ethiprole 40% + imidacloprid 40% – 80 WG, at a concentration of 125 g/ha exhibit 35% reduction in disease (Kumar et al. 2010b).

5.4.2 *Biological Agents, Green Constituents and Their Formulations*

The use of biological agents for pest control is gaining importance as this presents a safe and eco-friendly approach for the same. Such agents also reduce environmental toxicity and mostly not hazardous to non-target organisms. Beneficial bacterial isolates associated with plants also play a broader role in disease control and support plant growth. *Bacillus* and *Pseudomonas spp.* are commonly used for bacterial leaf blight control. Rhizospheric *Pseudomonas* has been documented as a beneficial bio-control agent against bacterial leaf blight caused by *Xoo* (Yasmin et al. 2016). Formulation of botanicals, along with biological agents, minimizes environmental toxicity. Plant extracts are used to control infections in rice. Biological control by phenolic compounds, like those present in natural agents, has been reviewed (Naqvi 2019).

Management and efficacy of seeds treated with plant growth-promoting bacteria *Bacillus spp*, for control of seed-borne bacterial blight in rice, through induced systemic response, has been reported. Several *Bacillus* strains, *viz., Bacillus pumilus* INR7, T4 and SE34, *Bacillus subtilis* GBO3 and IN937B, *Bacillus amyloliquefaciens* IN937 and *Brevibacillus brevis* IPC11, are used for seed-borne blight control. Fresh culture suspension, sodium alginate, and talc formulations under both *in vivo* and *in-vitro* conditions have been evaluated. Fresh suspensions of *Bacillus subtilis* GBO3 and *Bacillus pumilus* SE34 exhibit 86 and 85% of germination and high vigour indices of 1374 and 1323 respectively, over controls. Increased accumulation of peroxidase, phenyl oxidase, and phenylalanine ammonia-lyase are reported in treated seeds. Seeds treated with fresh suspensions show 71% protection with *Bacillus pumilus* SE34 followed by 58 and 52% of protection with *Bacillus subtilis* GBO3 and *Bacillus pumilus* T4, respectively (Chithrashree et al. 2011).

Out of 512 bacterial isolates obtained from different rice plant sites, 79 show antagonistic activity against different *Xoo* strains. *Pseudomonas aeruginosa* BRP3 produces indole acetic acid (IAA) and siderophores, and causes phosphorous solubilization. This isolate effectively controls disease and increases straw and grain yields by 55 and 51%, respectively. By producing siderophores (1-hydroxyphenazine, pyochelin, and pyocyanin), rhamnolipids, and due to presence of 1, 2, 3, 4-tetrahydroxy-2-alkylquinolines, 2, 3, 4-trihydroxy-2-alkylquinolines and 4-hydroxy-2-alkylquinolines, it contributes to effective disease control. Secondary metabolites produced by rhizospheric bacteria *Pseudomonas aeruginosa* BRP3, suppresses bacterial blight effectively in rice. It also supports host plants' growth by regulation of biotic and abiotic stresses (Singh and Jha 2016; Yasmin et al. 2017).

From 2005 to 10, a collection of 2690 actinomycete strains have been preserved at Vietnam type culture collection (Vietnam National University Hanoi). Ten dominant *Xoo* strains occurring in Japan and Northern Vietnam have been employed for disease control studies using actinomycetes. Upon screening for identification of actinomycetes strains, which can inhibit all ten dominant *Xoo* strains without causing any adverse effect on beneficial microbes, seventeen of them were able to do so. *Streptomyces toxytricini* VN08-A-12 strain, isolated from soil and leaf-litter samples, effectively reduced lesion length by 38.3%. Moreover, it was able to reduce yield losses to the extent of 43.2% in infected rice cultivars. In fact, it facilitated increased yields in healthy cultivars of rice too (Van Hop et al. 2014). Plant-associated bacteria, *Pseudomonas fluorescence,* is well known to produce an antibiotic 2,4-diacetylphloroglucinol with antimicrobial, antifungal and other beneficial biological properties (Notz et al. 2001). In a study, 27 strains from this subpopulation were screened for presence of 2, 4-diacetylphloroglucinol. Out of these, seven strains, *viz.,* PTB9 (64.5%), IMV14 (56.7%), MDR7 (54.4%), KAD7 (53.9%), PDY7 (51.9), VGP13 (51.2%) and VEL 17 (50.8%), controlled 50% of disease infection. Both *in vitro* and *in vivo* studies revealed 2, 4-diacetylphloroglucinol, which suppresses bacterial growth by 59–64% and controls lesion length (Velusamy et al. 2006).

Pseudomonas fluorescence PDY7 has been evaluated for bacterial blight control in paddy. 2, 4-diacetylphloroglucinol and indole acetic acid were responsible for controlling infection and promoting the growth of plants. It has been reported that PDY7 could be an excellent agent for controlling *Xoo* infection (Velusamy et al. 2013). *Pseudomonas fluorescens Pf1* effectively controls *Xoo* in seeds and plants by induction of systemic resistance. Seed treatment and foliar sprays of a powder formulation of *Pseudomonas fluorescence* also effectively control bacterial blight and increases yields (Vidhyasekaran et al. 2001). Seven isolates of fluorescent *Pseudomonads* (FLP2, FLP3, FLP28, FLP84, FLP85, FLP88, and FLP90) and *Pseudomonas fluorescent* 83 (*Pf*83) have been used against bacterial blight. Fluorescent *Pseudomonas* significantly reduces disease severity, with a maximum reduction of 62% being seen with *Pseudomonas fluorescens* 83 and FLP85, followed by 60.8% with FLP 90 and 58.7% with FLP88. Lowest disease control of49.5% is observed with FLP 20. All isolates of fluorescent *Pseudomonas* effectively increase grain yields. FLP 88 is found to be the best with 60.5%, followed by FLP84 (52.3%) and *Pf* 83 (50.7%). However, a maximum increase of 25–26% is reported for FLP88, *Pf83*, and FLP 84. Rice leaf isolate *Pseudomonas fluorescens* 83 (*Pf83*), and fluorescent *Pseudomonas* (FLP 84, 85, 88, and FLP90) have been reported to be useful for disease control (Gangwar and Sinha 2012).

It has been observed that several bacteria and fungi control bacterial leaf blight in paddy. *Aspergillus niger, Aspergillus flavus, Penicillium oxalicum, Curvularia clavata, Chetomimum spp, Trichoderma viride, Trichoderma hamatum, Trichoderma harzanium, Trichoderma pseudokunigii, Glomerella cingulate, Pseudomonas fluorescence* and *Bacillus subtillis* are representative some such organisms that are effective. *Pseudomonas fluorescence* effectively controls bacterial growth and 67.2% of inhibition over controls has been observed against *Xoo*. In contrast, 45.5, 51.8, 54.2, 56.1 and 59.6% inhibition is shown by *Trichoderma pseudokunigii, Trichoderma hamatum, Bacillus subtillis, Trichoderma harzanium,* and *Trichoderma viride*, respectively. 1% *Pseudomonas fluorescence* and 300 ppm of nimbicidin (*Azadirachtin*) at 7.5 ml/l is least effective against blight bacteria (Parthasarathy et al. 2014).

Rhizospheric bacteria from rice plants effectively control bacterial blight infection in paddy. *Lysobacter antibioticus* strain 13-1 has been used as a biocontrol agent against 38 isolates of *Xoo* strains wherein 32 isolates were found to be effectively controlled by 13-1 strain. Disease severity and its suppression depend on *Xoo* isolates and cultivars tested (Ji et al. 2008). 1000μl/ml of salicylic acid (natural signal) is used to trigger resistance in rice plants by induction of systemic phenolics, and the effect persists in susceptible cultivars for at least 3 days before *Xoo* infection (Babu et al. 2003). The formation of biofilms in paddy fields helps bacteria acquire environmental stress-resistance and supports the spread of infection.

Anthranilamide, a secondary metabolite produced from *Streptomycetes spp.*, controls this biofilm formation by *Xoo,* thus suppressing blight infection and increasing yields (Ham and Kim 2018). Bacterial antagonism can be achieved by direct suppression of plant pathogens through the production of antibiotics, enzymes like proteases, chitinases, glucanases, and siderophores or through indirect mechanisms in which antagonistic bacteria compete with pathogens for nutrient sites (Bardin et al. 2015). Depending on bacterial strains and rice varieties, *Bacillus and Pseudomonas spp.*, effectively control rice plants' disease up 90% (Montano et al. 2013).

Antibacterial activity of 33 aqueous plant extracts from different plant parts and plant families were studied against *Xanthomonas oryzae*. Six extracts, *viz., Allium sativum, Curcuma domestica, Datura metel, Eucalyptus globulus, Ocimum sanctum,* and *Syzygium aromaticum* effectively control infection. Among them, five plant extracts, *viz., Allium sativum, Curcuma domestica, Datura metal, Ocimum sanctum,* and *Syzygium aromaticum* were evaluated under field conditions. They showed increases in yields (513.3, 406.7, 436.3, 392, 457 gm/m^2 respectively) when compared to controls (315.3 gm/m^2). Extracts of *Datura metel* (Leaf), *Ocimum sanctum* (leaf), and *Curcuma domestica* (stem) show excellent activity against bacterial blight by reducing lesion lengths and by increasing grain yields (Yugander et al. 2015). Ametalin B in *Datura metel* is responsible for antibacterial activity against these pathogens (Meena et al. 2013). Sulphur and allicin in *Allium sativum* show maximum inhibition against *Xoo*, and curcumin in *Curcuma domestica* is responsible for its antibacterial activity (Sunder et al. 2005; Gurjar et al. 2012).

Trichoderma polysporum, Trichoderma harzianum, Trichoderma pseudokoningii, Paecilomyces variotii, Paecilomyces lilacinus, and *Glicladium virens* are the six bio-control agents used to control rice blast. Maximum growth inhibition has been reported in *Paecilomyces lilacinus*, followed by *Trichoderma pseudokoningii, Trichoderma polysporum, Trichoderma harzianum, Glicladium virens,* and *Paecilomyces variotti* (Khanzada and Shah 2012). Bacteria isolated from the rhizosphere and endorhizosphere have been evaluated for controlling rice blast disease. Out of 35 isolates, *Bacillus pumilus* and *Pseudomonas pseudoalcaligenes* were selected due to their growth-promoting ability and production of secondary metabolites. Both isolates solubilized phosphorous, produced siderophores, gibberellic acid, and indole acetic acid. Higher levels of production were seen in *Pseudomonas pseudoalcaligenes*. This bacterium supports plants in overcoming the deleterious effects of fungal pathogens. It suppresses fungal growth by producing β1–3, glucanase and chitinase, and it enhances plant heights, root lengths and dry weights (Jha and Subramanian 2014).

The efficacy of seed treatment of different *Trichoderma spp.* has been evaluated against rice blast disease. Species of *Trichoderma spp.* were isolated from upland rice systems. Isolate Th-3 treated seeds of samba mahsuri showed a maximum plant height of 57%, followed by 44% in plants treated with Tv-12 isolate as compared to controls. *Trichoderma* treated seeds exhibited low disease intensity, an increase in the number of leaves, tillers and panicles. It reduced disease severity by 10–25%. *Trichoderma spp* has emerged as a potential biological control agent against many plant pathogenic fungi, and it also supports plant growth (Aravindhan et al. 2012). Five different bio-agents, *viz., Bacillus subtilis, Pseudomonas fluorescens,*

Trichoderma viride, *Trichoderma virens,* and *Trichoderma harzianum* were evaluated against rice fungal diseases such as blast and sheath blight. The highest inhibition of mycelial growth with a mean of 68.5% was observed in *Trichoderma virens*, followed by *Trichoderma viride* (62%), (Kulmitra et al. 2017). Three bio-control agents, *viz., Pseudomonas fluorescens, Trichoderma viride*, and *Trichoderma harzianum* were evaluated against blast of rice. Minimum disease severity of 22% was seen in *Pseudomonas fluorescens,* followed by *Trichoderma viride* (24.5%) and *Trichoderma harzianum* (23.1%), over controls. In comparison with *Trichoderma spp., Pseudomonas fluorescens* exhibits good control against rice blast disease and supports plant growth. It increases the number of tillers, root weight (fresh and dry), and shoots lengths, followed by *Trichoderma harzianum* and *Trichoderma viride* (Kumar et al. 2017).

Antagonism of *Trichoderma harzianum*/ GPO80, *Chaetomium globosum*/N76-1, *Trichothecium roseum* / T372, *Micromonospora spp* / PS 6-2, *Gliocladium roseum, Chaetomium cochlioides,* and *Chaetomium cuniculorum* have been evaluated against rice blast. The highest percentage of mycelial growth inhibition was reported with *Trichoderma harzianum*/GPO80 and *Chaetomium globosum*/N76–1. Inhibition of mycelial growth (71%) and conidial germination (88%) was seen with *Trichoderma harzianum*/GPO80. 68% of mycelial growth inhibition and 76% inhibition in conidial germination were observed with *Chaetomium globosum*/N76-1. Least inhibition was noted with *Trichothecium roseum*/T372 and *Micromonospora spp*/PS 6-2 (Gouramanis 1995). The efficacy of antimicrobial metabolites produced from *Pseudomonas fluorescens* has been evaluated against rice fungal pathogens, *viz.,* blast, sheath blight, sheath rot, and brown spot disease. Totally, one hundred and twenty soil samples were used for isolation of *Pseudomonas fluorescens,* collected from rhizosphere region of rice. Among these, ten isolates were tested against fungal pathogens. One isolate, *Pseudomonas fluorescens* 8 (*Pf* 8), effectively inhibited mycelial growth (50–85%) of all rice fungi and produced higher amounts of siderophores, salicylic acid, and hydrogen cyanide. Isolates *Pf* 3, *Pf* 7, and *Pf* 8 exhibited good control of blast and sheath blight disease and isolates *Pf* 3, *Pf* 6, and *Pf* 8 controlled sheath rot fungi. *Pf* 1, *Pf* 3 *Pf* 5, *Pf* 6, and *Pf* 8 isolates were effective against paddy brown spot disease (Reddy et al. 2008).

An antifungal metabolite produced from *Streptomycetes sp*., PM5, was evaluated against blast and sheath blight pathogens. SPM5C-1 and SPM5C-2 with a lactone and ketone carbonyl unit are the two antifungal aliphatic compounds. Excellent inhibition of mycelial growth was reported with SPM5C-1. Complete inhibition of blast and sheath blight fungus was observed at 25–100μg/ml. Spraying of 500μg/ml on rice plants significantly reduced infection of sheath blight (82.3%) and blast (76.1%), and increased grain yields when compared to controls (Prabavathy et al. 2006). Twenty-five strains of *Pseudomonas fluorescens*, five strains of *Trichoderma viride*, and five strains of *Trichoderma harzianum* were evaluated against sheath blight disease. All strains were isolated from rhizosphere soil of plants like rice, wheat, tomato, tea, sugarcane, and chickpea. Among them, six strains, *viz., Pseudomonas fluorescens* PF-08, *Pseudomonas fluorescens* PF-10, *Trichoderma harzianum* UBSTH-501, *Trichoderma viride* UBSTV-05, *Trichoderma viride*

UBSTV-10, and *Trichoderma harzianum* UBSTH-101 were found to be effective. Co-inoculation of *Pseudomonas fluorescens* PF-08 and *Trichoderma harzianum* UBSTH-501 showed good control of sheath blight disease. Increases in defense-related enzymes and root and shoot biomass were seen along with stimulation of the production of indole acetic acid, which results in enhanced root exudation (Singh et al. 2016). A positive effect of *Pseudomonas fluorescens* and *Trichoderma harzianum* in the rhizosphere of many crop ecosystems has been reported (Harman et al. 2004).

Bacteria associated with up and low land rhizospheric soils are effective against sheath blight disease in rice. Fluorescens and non-fluorescens bacteria detected from sheath blight infected plants control this disease. Totally, 50 isolates of sheath blight associated bacteria have been isolated for antagonism tests and screened against sheath blight fungi. Eleven isolates exhibited inhibition activity, and out of these, three isolates are effective in controlling this disease under both *in vivo* and *in vitro* conditions. Both mycelial growth and sclerotial germination are inhibited. When seeds are soaked with antagonistic bacteria, 100% control is observed. Biological methods of control minimize disease incidence and serve as alternatives to chemical control methods (Bashar et al. 2010). Antagonistic activity of rice associated bacteria has been evaluated against rice sheath blight. During a phylogenetic analysis study, five different strains of *Bacillus amyloliquefaciens* (RAB6, RAB9, RAB16, RAB17S, and RAB18) were evaluated against sheath blight. All bacteria had the ability to control this disease. Complete inhibition of sclerotia germination and restricted development of sheath blight lesions were reported. Among these isolates, RAB9 was excellent for disease control (Shrestha et al. 2016).

Plant growth-promoting bacteria were isolated from different agroecosystems of Tamil Nadu, India, and they were tested against rice sheath rot pathogen. Strains Pf1, TDK1, and PY15 were effective against this pathogen. Further, a combination of these three strains was more effective in controlling sheath rot under greenhouse and field conditions. When compared to individual strain treatment, a strain combination increases the accumulation of defense-related enzymes in plants (Saravanakumar et al. 2009). *Trichoderma harzianum and Trichoderma viride* were effective against sheath rot fungi, wherein they are able to achieve a disease inhibition of 96% and 86%, respectively (Selvaraj and Annamalai 2015).

A bio-fungicide from *Trichoderma harzianum* has been evaluated against sheath blight, sheath rot, and brown spot fungus. An increase in yield of 22.2% was seen with 3% bio-fungicide. It effectively controls disease severity and also protects seeds from infection. Excellent control has been reported for brown spot disease (Mahmud et al. 2016). *In vivo* and *in vitro* control of brown spot disease of rice has been studied using *Trichoderma spp*. Forty-five isolates of *Trichoderma spp*., have been isolated from phyllosphere and soil samples in rice fields. Out of this, 31 isolates belonged to *Trichoderma harzianum*, whereas eleven and three isolates belonged to *Trichoderma virens* and *Trichoderma atroviride* respectively. These isolates significantly reduce mycelial growth of fungi by producing volatile and non-volatile metabolites. Two isolates of *Trichoderma harzianum* control brown spot disease significantly. *Trichoderma atroviride* significantly increases seedling growth of paddy (Khalili et al. 2012).

Extracts of neem (*Azadirachta indica L*) and garlic cloves (*Allium sativum L*) have been evaluated against sheath blight, sheath rot, and brown spot fungus. Aqueous extracts were prepared at 1% and 2% concentrations. Garlic extract (2%) exhibited reasonable control against the disease and also increased yields and number of panicles (Mahmud et al. 2016). Berberine effectively controls rice blast and brown spot disease. At 5 mg/ml, it suppresses brown spot disease. 49.8% reduction in disease severity is seen in case of rice blast fungi at 10 mg/ml concentration (Kokkrua et al. 2020). A foliar spray of 10% *Azadirachta indica* is effective against brown spot fungi. It supports plant growth at all growth phases, increases plant height and also the number of leaves. This neem formulation shows superior activity over controls and increases in the number of buds, flowers, and shoot height are seen. It also increases yields, the number of pods and seed weights (Kumar and Simon 2016).

Crude extracts of botanicals, *viz.*, ginger, garlic, onion, chili, neem, aak, turmeric, datura, and eucalyptus were evaluated to control rice blast fungi. In the case of leaf blast disease, ginger and garlic extracts were superior, followed by onion, neem, aak, and chili. Aak and chili were most effective against neck blast followed by garlic, onion, turmeric and ginger (Shafaullah 2016). Ethanol and acetone solvent extracts of maruthani (*Lawsonia inermis*) and kodukkapuli (*Pithecolobium dulce*) were effective against brown spot disease. Both extracts exhibited 76% disease reduction over controls. This was followed by 48% jamoon (*Syzygium cumini*), 46.5% vasambu (*Acorus calamus*), 41.6% eucalyptus (*Eucalyptus globulus*) and 34.8% turmeric (*Curcuma longa*) (Akila and Mini 2020).

Biopesticides serve as essential ingredients for crop protection. Microbes have been used to control various plant pathogens and pestiferous insects. The most commonly used biopesticide bacterium is *Bacillus thuringiensis*, an insect pathogenic bacterium. It produces δ-endotoxin, which is host specific and causes death of host within 48 h. It lyses the gut cell wall when consumed by susceptible insects, is safe for the environment and for other beneficial microbes of the system. Biopesticides used for biological pest management, and their commercialization with future prospects, have been reviewed (Chandler et al. 2011). Biopesticides have been used to control leaf folder and stem borer infections in paddy. Usage of *Bacillus thuringiensis* at 1.5 kg/ha, *Beauveria bassiana* at 2.5 kg/ha, and eggs of *Trichogramma japonicum* at 1,00,000 eggs/ha has been effective against leaf folder and stem borer infection in paddy. Neem gold at 5 ml/l and 3 ml/l, along with 75,000 eggs/ha of *Trichogramma japonicum* is effective against yellow stem borers (Singh et al. 2008).

In comparison to bio-control agents, *viz.*, *Bacillus thuringiensis*, *Trichogramma japonicum*, multi-neem extract, and synthetic insecticides are active against yellow stemborers of paddy. An egg parasite, *Trichogramma japonicum*, exhibits 11.75 and 11.50% of white ear head and dead heart, respectively, with an increased grain yield of 31 q/ha. This is on par with synthetic chemical insecticides (Bhushan et al. 2012). Neem oil and an egg parasite *Trichogramma japonicum* showed a reduction of 20 and 13.3% in the white ear and 44.1 and 27.6% in dead heart, respectively, in rice cultivar ASD16 (Justin and Preetha 2014). Foliar sprays of *Pseudomonas fluorescens* and insecticide imidacloprid are also effective in controlling brown planthopper populations (Sangamithra et al. 2017). Botanicals like neemarin and biopesticide *Bacillus thuringiensis* reduce leaf folders in paddy (Dhaka et al. 2011).

The efficacy of egg parasite *Trichogramma japonicum* and *Trichogramma chilonis* has been evaluated against yellow stem borer of rice. Recent findings indicate that the use of egg parasite *Trichogramma spp.,* effectively controls striped stem borers and leaf folders. Application of 200,000 eggs of *Trichogramma japonicum* increases yields by 12% in the paddy field over controls (Tang et al. 2017). Use of *Trichogramma chilonis* at 5 card/ha has been evaluated along with synthetic pesticides against leaf folder and stem borer diseases in paddy. It effectively reduces disease incidence and increases grain yields. The effectiveness of synthetic pesticides and bio-pesticides under farmer's field conditions has been evaluated, and yields are comparable with synthetic pesticides (Bharti et al. 2018). As a biological control agent in agro-ecosystems, the effectiveness of parasitoid wasps has been reviewed (Wang et al. 2019c). Different doses of egg parasitoids, *viz., Trichogramma japonicum*, and *T. chilonis* have been evaluated against rice leaf folders and yellow stem borers. The release of 50,000/ha of egg parasitoids in fields reduces tiller damage caused by leaf folders and stem borers. The reduction range is from 64 to 75.5% in leaf folders, followed by 50–61% in stem borers. A reduction of 73–82 and 78–82% has been reported in leaf folders and stem borers, respectively, at one lakh/ha of egg parasitoids. All doses of parasitoids were effective in disease management, wherein superior control has been reported with 1, 00,000 /ha of eggs over lower doses and controls (Kumar and Khan 2005). *Bacillus thuringiensis* suppress the feeding and larval growth of rice pests. It reduces damages caused by leaf folders (Nathan et al. 2005).

High activity against insect pests of rice has been reported with *Empedobacter brevis* and *Beauveria bassiana. Beauveria bassiana* reduces the density of green leafhopper by 77–81% for the overwintering generation and 73–83% for the second generation. Application of *Paecilomyces carnrus* leads to 66–91% death of planthoppers in fields from 3 to 7 days. Leaf folders can be controlled by *Cnaphalocrocis medinalis granulovirus*. It persists for more than 30 days and increases their mortality by around 20%. Biological control procedures for rice insect pests have been reviewed (Liu et al. 2013; Lou et al. 2014).

Four species of *Trichogramma, viz., T. chilonis, T. japonicum, T. dendrolimi*, and *T. ostriniae* have been used to control stripped stemborers in paddy. They have been evaluated at four different humidity conditions (90, 70, 50, and 30%) and five different temperatures (34, 30, 26, 22, 18 °C). Humidity and temperature were found to affect the abilities of all species of *Trichogramma. T. japonicum, T. dendrolimi*, and *T. ostriniae* performed best at a wide range of temperatures. At 26 °C, *T. chilonis* parasitized more eggs, and *T. ostriniae* parasitized fewer host eggs at 30–70% of relative humidity. At 70% of relative humidity, more number of hosts was parasitized. At 30–34 °C, the egg parasite *T. japonicum* effectively reduces the population of small brown hoppers, and at 18–26 °C, *T.dendrolimi* works best against hoppers. *T. japonicum* and *T.dendrolimi* performed better than other species. Significant differences in all *Trichogramma* species were reported at tested temperature and humidity conditions at 30 °C and 50%, respectively (Yuan et al. 2012). *T. japonicum, T.dendrolimi*, and *T. australicum* are the most commonly used species, and currently, *T. japonicum* is recommended as a superior *Trichogramma* species against lepidopteran rice pests (Chen et al. 2010).

The highest control of plant hopper population, *viz.*, brown and black-headed, and rice leaf folders have been achieved by a biological insecticide 12α-hydroxy rotenone EW (Mo et al. 2014). 1.5 kg/ha *Bacillus thuringiensis* reduces dead hearts and white ears in yellow stem borers. After 21 days of second spraying, the disease control percentage was 6.9, and the cost: benefit ratio was 1:2.73. Grain yield percentage was 39.07% q/ha (Singh et al. 2015b). The bio-efficacy of two biopesticides, *viz.*, 5% eupatorium and 5% Melia at a concentration of 2.5 l/ha, has been evaluated against white stem borers. Eupatorium shows a minimum disease infestation of 12.78%, and it was on par with a synthetic chemical pesticide. A minimum of 9.6% and 10.9% were recorded for eupatorium and Melia respectively. Yields of 34 and 35% have been reported with 5% of Melia and eupatorium (Tandon and Srivastava 2018).

Bacillus thuringiensis and *Metarhizium anisopliae* have been evaluated against stem borer and leaf folder induced diseases of rice under both *in vitro* and *in vivo* conditions. Significant reductions in disease incidence were seen under both conditions. *Bacillus thuringiensis*, *Metarhizium anisopliae*, and their combination effectively control stem borer and leaf folder damages over controls. The reported bacteria and fungi cause an equal rate of mortality in the insect population. Superior control was reported with a combination of both bacteria and fungi, resulting in increased mortality rates. After 48 and 72 h, mortality rates were 65 and 75%, respectively, whereas 100% mortality was observed after 96 h (Shahid et al. 2003). Insecticides from botanical sources may serve as alternatives to synthetic chemical pesticides (Magierowicz et al. 2020). In greenhouse crops, biological sources play a central role in the production. Insect repellent activities of certain plant extracts and essential oils serve as potent controls against crops' insect pests. The toxicity of synthetic chemical pesticides can be overcome by using natural green-based botanical pesticides.

Efficacies of three plant extracts, *viz.*, neem (*Azadirachta indica*), akando (*Calotropis procera*), and tobacco (*Nicotiana tabacum*) have been evaluated against stem borers. Azaridactin, calotropin, and nicotine are active substances present in respective plant extracts. 15 ml/L sprays of all three plant extracts (dose applied per hectare) controlled stem borers. 38.3% of dead hearts and 48.1% of white hearts were reduced by spraying the neem extract. Leaf and stem extracts of akando showed suppression of 16.9% of white hearts and 15.3% of dead hearts followed by 24.3%white hearts and 22.2% dead hearts in leaf extract of tobacco. Yields of 3.07, 2.86, and 2.68 t/ha were reported in neem, tobacco, and akando extracts. Neem extracts controlled stem borers in a similar way as synthetic chemicals (Mondal and Chakraborty 2016).

Neemarin (1500 ppm) was effective against yellow stem borers. The percentage of dead hearts and white ears were 5.6 after second spraying that was carried out at

21 days. The cost: benefit ratio was 1:0.25 (Singh et al. 2015b). The use of such botanical pesticides in fields counteracts the toxicity of chemical pesticides, and they are found to be harmless. They reduce disease infestation and increase plant yields (Rahaman et al. 2014). Neem oil has been used to control stem borer disease in rice cultivar ASD16. This oil reduces disease incidence of dead hearts and white ears by 44.1 and 20%, respectively, over controls at a concentration of 3 ml/l. In the TPS 3 rice variety, the use of neem oil at 3 ml/l reduces incidence by 25.44% in dead heart and by 13.94% in white ear compared to controls.

In the TPS 5 rice variety, decreases in disease incidence were reported to be 43.2 and 36.7% in dead hearts and white ears, respectively (Justin and Preetha 2014). The application of a neem-based formulation, multi-neem (1500 ppm) at 3.75 l/ha, has been evaluated against rice stem borers. It reduces disease incidence by 21.80% over controls and increases mean yield by 27.8% (Bhushan et al. 2012).

5.4.3 Novel Techniques, Technologies and Materials for Pest Control

Microbes frequently acquire resistance against synthetic chemicals, which poses severe constraints in developing novel pesticide molecules. Also, environmental toxicity that arises due to bio-accumulation and bio-magnification of synthetic chemicals, repeatedly sprayed for pest control, is a matter of concern. Adopting smart techniques, technologies, and materials can control pests efficaciously and prove economical in the process.

5.4.3.1 Smart Farming Techniques

Adopting specific procedures for ploughing, mechanical harvesting, and irrigation can control the yellow, stripped, and pink stem borers in rice. Approximately 70% of overwintering stem borers may be killed by ploughing and irrigation (Guo et al. 2013). Delaying or postponing the process of sowing and transplantation has a significant effect on reducing the population of small brown plant-hoppers and preventing plants from acquiring viral infections (Zhu et al. 2011). Light traps in fields can attract adult pests, which can control rice stem borers and plant-hopper populations (Xu et al. 2015b). Using resistant rice varieties for cultivation decreases the insect pest population in them. Cultivation of genetically modified rice varieties and conventional breeding of insect-resistant types are also being successfully attempted (Chen et al. 2011).

Balanced application of nitrogen, potassium, phosphorous and other elements in rice fields protect plants from insect attacks. Smart use of fertilizers decreases pest infections. Through ecological engineering, the frequency of pest attacks and their population sizes can be controlled. Intercropping methods have also been beneficial

(Zhang et al. 2008b; Peng et al. 2002). The frequency of pest outbreaks can be effectively inhibited by developing such methods (Chen et al. 2016). Employing natural enemies and bio-trapping of pests are smart ways to control pests and pathogenic diseases. They also reduce the use of chemical pesticides. Sustainable management of insect pests by such non-chemical methods has been reviewed (Hong-Xing et al. 2017). Hence attacks by predominant insects, toxicity issues of chemical pesticides, and environmental pollution can be prevented or controlled by nifty management techniques.

5.4.3.2 Nano-Technological Solutions

Smart nano-based pesticide formulations are used to control insect pests of crops. Compared to conventional pesticides, they offer greater efficacy and controlled release. Nanocarriers like lipids, clays, polymers, nanomaterials, organic metal frameworks, and green formulations are used for developing smart pesticides. A new trend in the synthesis of smart formulations involves incorporating nanoparticles as active ingredients to target pathogens. Smart nano pesticides offer a wide range of benefits like enhanced permeability, controlled release kinetics, stability, solubility, use of low concentrations of active ingredients with extended durations of control, and prevention of degradation of active ingredients.

Emerging opportunities that concern nano-based smart pesticides have been reviewed (Kumar et al. 2019; Medina-Perez et al. 2019). Nanotechnology enhances agrochemicals' controlled release and ensures site-targeted delivery of various macromolecules to impart resistance to plants against diseases. The potency of nanomaterials in this process and nanomaterial delivery mechanisms to plants have been reviewed (Remya et al. 2010; Fraceto et al. 2016).

Current and future advances of nanotechnology in agricultural crop management focus on three significant concepts, viz., improvements in conventional pesticide formulations with this technology, development of pesticide delivery systems, and nanoparticles as active ingredients and carriers (Athanassiou et al. 2018). Use of ceramic devices, nano-based agrochemicals, filters, and laminating methods more efficiently alter conventional agro practices. Challenges and applications of nanotechnology in sustainable Indian agriculture, outlining present and future possibilities, have been reviewed along with their use in crop development and promotion of plant growth in general (Pandey 2018; Shang et al. 2019).

Silver nanoparticles, prepared by green technology, can control bacterial leaf blight caused by *Xanthomonas oryzae* (Mankad et al. 2018). Copper nanoparticles also control bacterial leaf blight in plants. CuNP-3 shows good antibacterial activity, and it produces reactive oxygen species (ROS) that cause membrane disintegration and DNA breakdown of pathogenic bacteria. Due to ROS production, there is an increased activity against pathogens as compared to commercial antibiotics (Majumdar et al. 2019). Halloysite nanotube (a clay mineral nanomaterial), surface modified with a surfactant, has been used to control phytopathogens. Surfactant modified SM-Halloysite nanotubes effectively controls *Xanthomonas oryzae* and

Ralstonia solanacearum. It also increases ROS production in phyto-pathogens (Abhinayaa et al. 2019).

Myco-nanoparticles have emerged as outstanding solutions for the control of various agricultural pests. They act as bio-fertilizers and pesticides. They attack targets effectively, are easy to maintain, environmental-friendly, and are cost-effective. The synthesis of metal nanoparticles mediated by fungi is non-toxic, clean, and eco-friendly (Gautam and Avasthi 2020). A nanocomposite of chitosan/ TiO_2, in a ratio of 1:5, exhibits potent antibacterial activity against *Xoo* under both light and dark conditions (Li et al. 2016). Bio-conjugated nano molecules are also used to enhance the antibacterial activities of pathogens. Conjugation is achieved by bio-molecular interaction of silver clusters with 3-dichloro-5, 6-dicyano-1, 4-benzoquinone. They effectively control a wide range of pathogenic microbes and different species of *Xanthomonas*. They serve as alternative tools to combat drug-resistant pathogenic strains (Baker et al. 2020).

5.4.3.3 Emulsion Based Formulations

The use of emulsions for delivering actives to target pests is a safe and eco-friendly approach for pest control. Emulsions reduce environmental toxicity and are efficient delivery systems. The controlled release of pesticide molecules with active spreading is feasible through emulsion formulations. Delivery of an essential gaseous plant growth regulator, 1-methyl cyclopropane, as a microcapsule-based oil dispersion, for controlled release has been reported with reasonable release rates and stability (Guo et al. 2019). Glycolipids can form a stable emulsion in oil and water systems and are useful in loading both hydrophobic and hydrophilic bioactive molecules. They also facilitate the controlled release of actives and prevent the degradation of active molecules. Bioactive molecules like Vitamin B2, C, E, and curcumin can be loaded by direct solid dispersion methods.

Alkyl polyglycoside (APG), polyoxyethylene 3-lauryl ether, and methyl laurate as oil phase have been used for the synthesis of biocompatible oil in water emulsion for pesticide delivery. β-cypermethrin, a water-insoluble pesticide incorporated into this emulsion, is stable and spreadable (Du et al. 2016). A hydrophobic pesticide lambda-cyhalothrin with aromatic hydrocarbon as an oil phase, along with a non-ionic surfactant, has been used to synthesize oil in water emulsion, a low energy emulsification method. The impact of surfactant concentration and its types on the physical stability of oil in water emulsion loaded with pesticides has been reported (Feng et al. 2018).

As a norcantharidin microemulsion, a new bio-pesticide formulation controls insect pests of crops effectively (Shao et al. 2018). Plant-based emulsion formulations have been developed to control bacterial leaf blight and sheath rot fungi in rice plants. A methanol crude extract of *Piper sarmentosum* emulsion has been synthesized using a non-ionic and ethoxylated surfactant. *In vitro* and *in vivo* studies reveal that this emulsion can effectively control bacterial leaf blight and sheath rot disease.

Disease reduction, to the extent of 81 and 55–61%, is noted in bacterial leaf blight and sheath rot, respectively (Syed-Ab-Rahman et al. 2020).

5.4.3.4 Carbohydrates Polymer-Based Materials

Polymers are a class of soft materials with excellent mechanical and chemical properties. Due to their versatile applications, they are used in agrochemicals to improve soil quality and control pesticides' release. They act as bio-sorbents and super-absorbents. They are environmentally friendly, economically sustainable materials. It is vital to use an agrochemical combined with a polymer to avoid toxicity and contamination.

Use of polymers in agriculture and their applications have been reviewed (Milani et al. 2017). Carbohydrate polymers have been used as controlled release matrices for pesticide delivery. Natural polymers, *viz.*, alginates, starch and its derivatives, chitosan, cyclodextrin, and carboxymethyl cellulose or ethyl cellulose, are polysaccharides used for this purpose. The advantages and disadvantages of polymers, efficacy, toxicity, and safety of formulations have been reviewed (Neri-badang and Chakraborty 2019).

5.4.3.5 Development of Pest-Resistant Varieties of Plants

A new approach based on DNA methylation can be adopted to improve the disease resistance of crops. Here, plants reprogram their transcriptome and manage their genome stability in a precise manner to increase their ability to cope with a dynamic environment. DNA methylation pattern changes during pathogenic attacks, improving resistance by regulating defence responses (Tirnaz and Batley 2019). Immunity and starvation present two new opportunities to increase disease resistance in crops. Blocking pathogenic access to nutrients and resisting mobilization of sugars to prevent microbial colonization are promising strategies for disease control. A different control strategy model has been proposed based on the pathogenic ability to deal with host immunity or starvation (Oliva and Quibod 2017).

Targeted genomic editing is a new way to develop or improve resistance to pathogenic diseases and also for controlling insect pests in crops. Currently, the significant potential of genomic editing through engineered nuclease has been established in plants. So far, inclusive of rice, it has been demonstrated for more than 50 plant species. *Cas9/CRISPR*- based two-component genomic editing system has been successful with broader applicability. Novel propositions of developing pathogen and insect resistance by genomic editing have been reviewed (Bisht et al. 2019).

Cas9/CRISPR in gene editing and crop improvement, with highlights of progression through *CRISPR* legacy, information on traditional delivery method of *Cas9*-gRNA complex in plant cells, and incorporation of advent *CRISPR* ribo-nucleoproteins are significant advances in the field of pest control research.

Such methods have come up with solutions for various limitations (Arora and Narula 2017). Many cloned genes are currently used in agriculture and are useful and beneficial in controlling plant diseases. They also provide increased yield, resistance to abiotic and biotic stresses, and produce good quality grains. Research on rice functional genomics, their progress, and implications on crop genetic improvements have been reviewed (Jiang et al. 2012).

Development and cultivation of resistant rice varieties can serve to overcome yield losses caused by bacterial leaf blight disease and reduce the use of synthetic chemicals or antibiotics. A broad-spectrum resistance gene *Xa21*, derived from *O. longistaminata*, is extensively used for developing bacterial leaf blight resistance in rice crops cultivated in Asian countries (Cao et al. 2005). Minghui 63 is highly susceptible to bacterial leaf blight and is most widely used in China's hybrid rice production. *Xa21* gene was introduced in a breeding program in this variety by molecular marker-assisted selection to enhance bacterial leaf blight resistance.

Plants were found to develop resistance against six pathogenic *Xoo* strains (Chen et al. 2000). In India, *Xa21* gene has been introduced in PR106 rice variety to gain resistance against this disease. In Punjab, *Xa21* has effectively acquired resistance against 17 pathogenic *Xoo* strains (Singh et al. 2001). *Xa21* gene is considered to be most effective in inducing resistance against bacterial leaf blight. In India, it is 88% resistant, followed by *Xa13* (84%), *Xa8* (64%), *Xa5* (30%), *Xa7* (17%) and *Xa4* (14%) (Mishra et al. 2013).

Genes *Xa4, Xa5, Xa13,* and *Xa21*, are used during plant breeding for developing bacterial leaf blight resistance. Lines carrying genes *Xa4* and *Xa13* are more resistant than those carrying either *Xa4* or *Xa13* gene. A higher level of resistance, seen in lines carrying more genes, is due to an additive effect, or gene interaction (Huang et al. 1997). A rice cultivar LT2 has been developed by a marker-assisted selection of *Xa21* gene to confer resistance against bacterial leaf blight. The marker-assisted selection that accompanies phenotypic selection could be a reliable strategy in backcross breeding (Nguyen et al. 2018). Different genes can be cloned in one rice variety to gain resistance against multiple diseases in a rice field. *Xa21, Bt* fusion gene and chitinase gene are resistance genes of *Xoo*, insect pest, and sheath blight, respectively. All three genes are combined in a single rice line, and they are found to be effective against respective diseases and help in increasing yields (Datta et al. 2002).

A novel *Bph32* gene, which encodes for an unknown SCR domain-containing protein, exhibits brown plant-hopper resistance in paddy. Transgenic introgression of this gene in a susceptible rice variety imparts plant resistance towards brown plant-hoppers. It is highly expressed in leaf sheath, where hoppers settle and feed. *Pph32* may thus inhibit feeding by brown plant-hopper (Ren et al. 2016). *Bph14, Bph17, Bph26,* and *Bph29* are genes that cause resistance towards brown plant-hoppers. Recent progress in genetic and molecular breeding of rice plants to acquire resistance against brown plant-hoppers, has been reviewed (Hu et al. 2016). Genotypes with R resistant leucine-rich repeat genes control specific races of parasites. Insufficient field resistance has been seen in temperature regions.

Pyramiding of R resistant genes, *viz., Pib, Pik, Piz, Piz-t,* and *Pita2* japonica material has been undertaken to produce two lines of *SJKT-2* and *SJKK*, which have

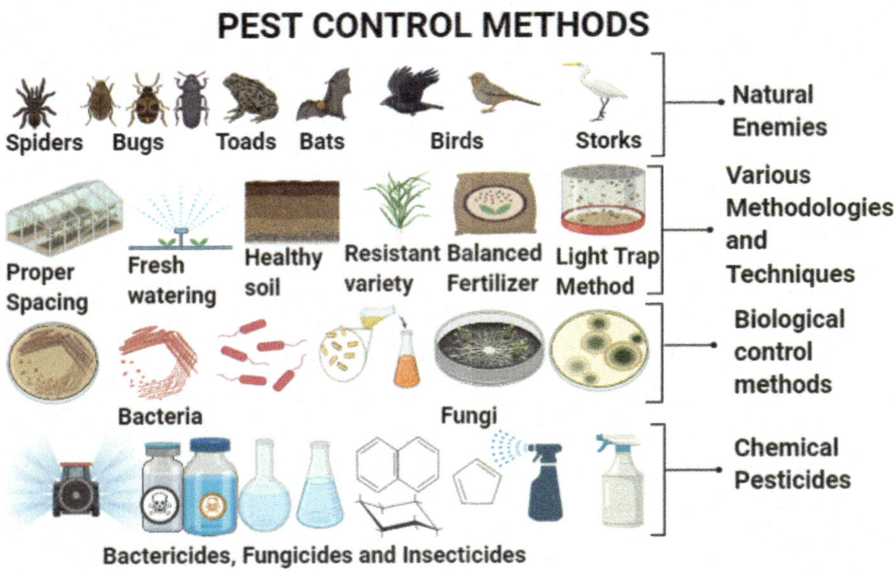

Fig. 5.2 Consolidated view of various pest control methods in *Oryza sativa*

four pyramiding genes. When tested in greenhouse and field conditions, it exhibits excellent resistance against blast (Orasen et al. 2020). *Pi54r,* cloned from wild species of rice, confers broad-spectrum resistance against blast. It exhibits a high degree of resistance to seven pathogenic strains in different geographical locations of India. It can be used in a breeding program to develop resistance (Das et al. 2014). Four simple sequence repeat (SSR) markers, *viz., RM413, RM1233, RM5961*, and *RM8225* are significantly associated with blast resistance. They account for about 20% of phenotypic variation (Ashkani et al. 2011). Figure 5.2 gives a consolidated view of the various methods of pest control.

5.5 Conclusion

Rice is a staple food for populations across the globe. Rice plants are susceptible to attack by several pathogenic microbes and insects that cause yield and economic losses. Several chemical and green constituents and their formulations have been developed for pest control based on their types, efficacy and environmental implications. However, resistance acquired by pests towards these constituents pose a major concern and propels a continuous search for new compounds that may be synthetic or natural. Adopting smart techniques and technologies and employing smart materials is currently serving to open up new and exciting frontiers for counteracting deleterious effects of these pathogens and rendering holistic protection to rice plants to obtain better quality high yields, along with economic benefits.

Acknowledgement The authors are grateful to Vellore Institute of Technology, Vellore, India, for supporting the preparation of this chapter.

Conflict of Interest The authors report no conflict of interest.

References

Abdallah Y, Ogunyemi SO, Abdelazez A, Zhang M, Hong X, Ibrahim E, Hossain A, Fouad H, Li B, Chen J (2019a) The green synthesis of MgO nano-flowers using *Rosmarinus officinalis L.(Rosemary)* and the antibacterial activities against *Xanthomonas oryzae pv. oryzae*. Biomed Res Int 2019:1–8. https://doi.org/10.1155/2019/5620989

Abdallah Y, Yang M, Zhang M, Masum MM, Ogunyemi SO, Hossain A, An Q, Yan C, Li B (2019b) Plant growth promotion and suppression of bacterial leaf blight in rice by *Paenibacillus polymyxa* Sx3. Lett Appl Microbiol 68(5):423–429. https://doi.org/10.1111/lam.13117

Abhinayaa R, Jeevitha G, Mangalaraj D, Ponpandian N, Meena P (2019) Toxic influence of pristine and surfactant modified halloysite nanotubes on phytopathogenic bacteria. Appl Clay Sci 174:57–68. https://doi.org/10.1016/j.clay.2019.03.022

Ahmed T, Shahid M, Noman M, Niazi MB, Mahmood F, Manzoor I, Zhang Y, Li B, Yang Y, Yan C, Chen J (2020) Silver nanoparticles synthesized by using *Bacillus cereus* SZT1 ameliorated the damage of bacterial leaf blight pathogen in rice. Pathogens 9(3):1–17. https://doi.org/10.3390/pathogens9030160

Akila R, Mini ML (2020) Solvent extraction and antifungal assay of *Lawsonia inermis Linn.* against the brown spot fungus *Bipolaris oryzae*. J Pharmacogn Phytochem 9(1):5–8

Akther T, Hemalatha S (2019) Mycosilver nanoparticles: synthesis, characterization and its efficacy against plant pathogenic fungi. BioNanoScience 9(2):296–301. https://doi.org/10.1007/s12668-019-0607-y

Aravindan S, Yadav MK, Sharma P (2012) Biological control of rice blast disease with *Trichoderma spp.* under upland rice system. ORYZA- Int J Rice 53(2):167–173

Arora L, Narula A (2017) Gene editing and crop improvement using CRISPR-Cas9 system. Front Plant Sci 8:1–21. https://doi.org/10.3389/fpls.2017.01932

Ashkani S, Rafii MY, Sariah M, Akmar AS, Rusli I, Rahim HA, Latif MA (2011) Analysis of simple sequence repeat markers linked with blast disease resistance genes in a segregating population of rice (*Oryza sativa*). Genet Mol Res 10(3):1345–1355. https://doi.org/10.4238/vol10-3gmr1331

Ashrafuzzaman H (1987) Chemical control of bacterial leaf blight of paddy caused by *Xanthomonas campestris pv. oryzae*. Curr Plant Sci Biotechnol Agric 4:955–958. https://doi.org/10.1007/978-94-009-3555-6_205

Athanassiou CG, Kavallieratos NG, Benelli G, Losic D, Rani PU, Desneux N (2018) Nanoparticles for pest control: current status and future perspectives. J Pest Sci 91(1):1–5. https://doi.org/10.1007/s10340-017-0898-0

Azzam O, Chancellor TC (2002) The biology, epidemiology, and management of rice tungro disease in Asia. Plant Dis 86(2):88–100. https://doi.org/10.1094/PDIS.2002.86.2.88

Babendreier D, Hou M, Tang R, Zhang F, Vongsabouth T, Win KK, Kang M, Peng H, Song K, Annamalai S, Horgan FG (2020) Biological control of lepidopteran pests in rice: a multi-nation case study from Asia. J Integr Pest Manag 11(1):1–11. https://doi.org/10.1093/jipm/pmaa002

Babu RM, Sajeena A, Samundeeswari AV, Sreedhar A, Vidhyasekaran P, Seetharaman K, Reddy MS (2003) Induction of systemic resistance to *Xanthomonas oryzae pv. oryzae* by salicylic acid in *Oryza sativa* (*L.*). J Plant Dis Prot 110(5):419–431. https://doi.org/10.1007/BF03356119

Baker JT, Allen JLH, Boote KJ (1992) Response of rice to carbon dioxide and temperature. Agric For Meteorol 60:153–166. https://doi.org/10.1016/0168-1923(92)90035-3

Baker S, Prasad MN, Chouhan RS, Kumar KM, Satish S (2020) Development of bio conjugated nano-molecules against targeted microbial pathogens for enhanced bactericidal activity. Mater Chem Phys 242:1–9. https://doi.org/10.1016/j.matchemphys.2019.122292

Bandumula N (2018) Rice production in Asia: key to global food security. Proc Natl Acad Sci, India Sect B: Biol Sci 88(4):1323–1328. https://doi.org/10.1007/s40011-017-0867-7

Bardin M, Ajouz S, Comby M, Lopez-Ferber M, Graillot B, Siegwart M, Philippe C (2015) Is the efficacy of biological control against plant diseases likely to be more durable than that of chemical pesticides? Front Plant Sci 6:1–15. https://doi.org/10.3389/fpls.2015.00566

Barua M, Quintana L, Ortiz A (2019) Chemical control of rice brown spot (*Bipolaris oryzae*) in Paraguay. J Trop Plant Res 6(1):148–151. https://doi.org/10.22271/tpr.2019.v6.i1.021

Bashar MA, Hossain MA, Rahman MM, Uddin MN, Begum MN (2010) Biological control of sheath blight disease of rice by using antagonistic bacteria. Bangladesh J Sci Ind Res 45(3):225–232. https://doi.org/10.3329/bjsir.v45i3.6529

Basso A, Onasanya A, Issaka S, Sido AY, Haougui A (2011) The brown rot of rice in Niger: pathological diversity of isolates collected from the irrigated areas. J Appl Biosci 38:2551–2563

Bernardes-de-Assis J, Storari M, Zala M, Wang W, Jiang D, Shidong L, Jin M, McDonald BA, Ceresini PC (2009) Genetic structure of populations of the rice-infecting pathogen *Rhizoctonia solani* AG-1 1A from China. Phytopathology 99:1090–1099. https://doi.org/10.1094/PHYTO-99-9-1090

Bharti N, Singh UK, Kumar V, Kumar A (2018) Evaluation of insecticides and biopesticides against rice stem borer *Scirpophaga incertulas* (Walker) and leaf folder (*Cnaphalocrosis medinalis*) in kharif rice under farmers' field condition. Int J Curr Microbiol App Sci 7:3436–3441

Bhattacharjee PP, Ray DC (2012) Bioefficacy of insecticides against rice hispa, *Dicladispa armigera* (Olivier) (Coleoptera: Chrysomelidae) on paddy. J Entomol Res 36(2):151–155

Bhushan S, Singh RP, Shankar R (2012) Biopesticidal management of yellow stem borer (*Scirpophaga incertulas* Walker) in rice. Bioscan 7(2):317–319

Bigirimana VD, Hua GK, Nyamangyoku OI, Hofte M (2015) Rice sheath rot: an emerging ubiquitous destructive disease complex. Front Plant Sci 6:1–16. https://doi.org/10.3389/fpls.2015.01066

Bisht DS, Bhatia V, Bhattacharya R (2019) Improving plant-resistance to insect-pests and pathogens: the new opportunities through targeted genome editing. Semin Cell Dev Biol Acad Press 96:65–76. https://doi.org/10.1016/j.semcdb.2019.04.008

Biswas SK, Rai M, Srivastava SS (2009) Evaluation of antibiotics and their suitable use against bacterial blight of paddy [*Xanthomonas oryzae pv. oryzae* (Ishiyama) dye]. Indian Phytopathol 62(1):126–128

Bouman BAM, Humphreys E, Tuong TP, Barker R (2007a) Rice and water. Adv Agron 92:187–237. https://doi.org/10.1016/S0065-2113(04)92004-4

Bouman B, Barker R, Humphreys E, Tuong TP, Atlin G, Bennett J, Dawe D, Dittert K, Dobermann A, Facon T, Fujimoto N, Gupta R, Haefele S, Hosen Y, Ismail A, Johnson D, Johnson S, Khan S, Shan L, Ilyas M, Matsuno Y, Pandey S, Peng S, Muthukumarisami T, Wassman R (2007b) Rice: feeding the billions. In Molden D (ed) Water for food, water for life: a comprehensive assessment of water management in agriculture. Earthscan/International Water Management Institute (IWMI), London/Colombo, pp 515–549

Cao LY, Zhan XD, Zhuang J, Cheng SH (2005) Breeding of Indica hybrid rice Guodao 1 with good quality, high yield and resistance to bacterial leaf blight by marker-assisted selection technique. Hybrid Rice 20(3):16–18. (in Chinese with English abstract)

Chaibub AA, de Sousa TP, De Oliveira MI, Arriel-Elias MT, De Araujo LG, De Filippi MC (2020) Efficacy of *Cladosporium cladosporioides* C24G as a multifunctional agent in upland rice in agroecological systems. Int J Plant Prod 14(10):463–474. https://doi.org/10.1007/s42106-020-00097-2

Chakrabarti NK (2001) Epidemiology and disease management of brown spot of rice in India. In: Sreenivasaprasad S, Johnson R (eds) Major fungal diseases of rice – recent advances. Kluwer Academic Publishers, Springer, Dordrecht, pp 293–306. https://doi.org/10.1007/978-94-017-2157-8_21

Chakraborty K, Deb DC (2012) Incidence of rice hispa, *Dicladispa armigera* (Coleoptera: *Chrysomelidae*) on kharif paddy in the agro climatic conditions of the northern parts of West Bengal, India. Glob J Sci Front Res Biol Sci 12:53–61

Chand T, Sing N, Sing H, Thind BS (1979) Field efficacy of stable bleaching powder to control bacterial blight of rice. Int Rice Res Newsl 4(4):12–13

Chandler RF (1979) Rice in the tropics: a guide to the development of national programs. CRC Press 2019, New York, pp 9–21

Chandler D, Bailey AS, Tatchell GM, Davidson G, Greaves J, Grant WP (2011) The development, regulation and use of biopesticides for integrated pest management. Philos Trans R Soc B: Biol Sci 366(1573):1987–1998. https://doi.org/10.1098/rstb.2010.0390

Chapagain AK, Hoekstra AY (2011) The blue, green and grey water footprint of rice from production and consumption perspectives. Ecol Econ 70(4):749–758. https://doi.org/10.1016/j.ecolecon.2010.11.012

Chaudhary SU, Iqbal J, Hussain M (2012) Effectiveness of different fungicides and antibiotics against bacterial leaf blight in rice. J Agric Res (Pak) 50(1):109–117

Chen S, Lin XH, Xu CG, Zhang Q (2000) Improvement of bacterial blight resistance of 'Minghui 63', an elite restorer line of hybrid rice, by molecular marker assisted selection. Crop Sci 40(1):239–244. https://doi.org/10.2135/cropsci2000.401239x

Chen H, Chen Z, Zhou Y (2005) Rice water weevil (Coleoptera: *Curculionidae*) in mainland China: invasion, spread and control. Crop Prot 24:695–702. https://doi.org/10.1016/j.cropro.2004.12.005

Chen HF, Huang SS, Zhang YZ, Zeng X, Huang ZH (2010) Control efficacy of *Trichogramma japonicum* against *Chilo suppressalis* and *Chilaraea auricilia*. J Appl Ecol 21(3):743–748

Chen M, Shelton A, Ye GY (2011) Insect-resistant genetically modified rice in China: from research to commercialization. Annu Rev Entomol 56:81–101. https://doi.org/10.1146/annurev-ento-120709-144810

Chen Y, Yang X, Gu CY, Zhang AF, Zhang Y, Wang WX, Gao TC, Yao J, Yuan SK (2015) Activity of a novel bactericide, zinc thiazole against *Xanthomonas oryzae pv. oryzae* in Anhui province of China. Ann Appl Biol 166(1):129–135. https://doi.org/10.1111/aab.12170

Chen GH, Zhu PY, Zheng XS, Yao XM, Zhang FC, Sheng XQ, Xu HX, Lu ZX (2016) The practice of ecological engineering on the control of rice insect pests in Jinhua China. Plant Prot 36(1):31–36

Cheng CH, Pathak MD (1972) Resistance to *Nephotettix virescens* in rice varieties. J Econ Entomol 65(4):1148–1153. https://doi.org/10.1093/jee/65.4.1148

Cheng X, Chang C, Dai SM (2010) Responses of striped stem borer, *Chilo suppressalis* (Lepidoptera: *Pyralidae*) from Taiwan to a range of insecticides. Pest Manag Sci 66(7):762–766. https://doi.org/10.1002/ps.1939

Chithrashree, Udayashankar AC, Chandra NS, Reddy MS, Srinivas C (2011) Plant growth-promoting rhizobacteria mediate induced systemic resistance in rice against bacterial leaf blight caused by *Xanthomonas oryzae pv. oryzae*. Biol Control 59:114–122. https://doi.org/10.1016/j.biocontrol.2011.06.010

Chou C, Castilla N, Hadi B, Tanaka T, Chiba S, Sato I (2020) Rice blast management in Cambodian rice fields using *Trichoderma harzianum* and a resistant variety. Crop Prot 135:1–7. https://doi.org/10.1016/j.cropro.2019.104864

Chowdhury AK, Sarkar SC (2006) Performance of different fungicides on the management of sheath blight of rice. Pestic Res J 18(2):157–158

Consultative Group on International Agricultural Research (CGIAR) (2016) The global staple. http://ricepedia.org/rice-asfood/the-global-stable-rice-consumers

Dai S, Beachy RN (2009) Genetic engineering of rice to resist rice tungro disease. In Vitro Cell Dev Biol Plant 45(5):517–524. https://doi.org/10.1007/s11627-009-9241-7

Das B, Sengupta S, Prasad M, Ghose TK (2014) Genetic diversity of the conserved motifs of six bacterial leaf blight resistance genes in a set of rice landraces. BMC Genet 15(1):1–15. https://doi.org/10.1186/1471-2156-15-82

Datta K, Baisakh N, Thet KM, Tu J, Datta S (2002) Pyramiding transgenes for multiple resistance in rice against bacterial blight, yellow stem borer and sheath blight. Theor Appl Genet 106(1):1–8. https://doi.org/10.1007/s00122-002-1014-1

Dawe D (2002) The changing structure of the world rice market, 1950–2000. Food Policy 27(4):355–370. https://doi.org/10.1016/S0306-9192(02)00038-6

De Sousa Oliveira MI, Chaibub AA, Sousa TP, Cortes MV, de Souza AC, da Conceicao EC, De Filippi MC (2020) Formulations of *Pseudomonas fluorescens* and *Burkholderia pyrrocinia* control rice blast of upland rice cultivated under no-tillage system. Biol Control 144:1–6. https://doi.org/10.1016/j.biocontrol.2019.104153

Dhaka SS, Prajapati CR, Singh DV, Singh R (2011) Field evaluation of insecticides and biopesticides against rice leaf folder, *Cnaphalocrocis Medinalis* (Guenee). Ann Plant Prot Sci 19(2):324–326

Dhawan AK, Mahal MS, Sarao PS, Virk JS, Suri KS, Paramjit S, Ravinder S (2010) Relative efficacy of new granular formulation against stem borers and leaf folder on basmati rice. J Insect Sci 23(4):414–419

Dinakaran D, Mathiyazhagan S, Thiruvudainambi S, Gajendran G, Kathiresan G (2012) Efficacy of Zineb+ Hexaconazole on the management of sheath blight, brown spot and grain discoloration in rice. ORYZA-Int J Rice 49(2):144–146

Du Z, Wang C, Tai X, Wang G, Liu X (2016) Optimization and characterization of biocompatible oil-in-water nanoemulsion for pesticide delivery. ACS Sustain Chem Eng 4(3):983–991. https://doi.org/10.1021/5b01058

Dupo ALB, Barrion AT (2009) Taxonomy and general biology of delphacid planthoppers in rice agroecosystems. In: Heong KL, Hardy B (eds) Planthoppers: new threats to the sustainability of intensive rice production systems in Asia. International Rice Research Institute, Los Banos, pp 3–155

Durand-Morat A, Nalley LL, Thoma G (2018) The implications of red rice on food security. Glob Food Sec 18:62–75. https://doi.org/10.1016/j.gfs.2018.08.004

Dutta BC, Nath RK (2003) Seasonality of rice hispa, *Dicladispa armigera* (*Olivier*) (Coleoptera: *Chrysomelidae*). Res Crop 4:258–262

Dutta D, Saha S, Ray DP, Bag MK (2012) Effect of different active fungicides molecules on the management of rice blast disease. Int J Agric Environ Biotechnol 5(3):247–251

Elshakh AS, Anjum SI, Qiu W, Almoneafy AA, Li W, Yang Z, Cui ZQ, Li B, Sun GC, Xie GL (2016) Controlling and defence-related mechanisms of *Bacillus* strains against bacterial leaf blight of rice. J Phytopathol 164(7–8):534–546. https://doi.org/10.1111/jph.12479

Fahad S, Adnan M, Noor M, Arif M, Alam M, Khan IA, Ullah H, Wahid F, Mian IA, Jamal Y, Basir A (2019) Major constraints for global rice production. In: Advances in rice research for abiotic stress tolerance. Woodhead Publishing, Duxford, pp 1–22. https://doi.org/10.1016/B978-0-12-814332-2.00001-0

Feng J, Chen Q, Wu X, Jafari SM, McClements DJ (2018) Formulation of oil-in-water emulsions for pesticide applications: impact of surfactant type and concentration on physical stability. Environ Sci Pollut Res 25(22):21742–21751. https://doi.org/10.1007/s11356-018-2183-z

Finch S, Samuel A, Lane GP (2014) Lockhart and Wiseman's crop husbandry including grassland, 9th edn. Woodhead Publishing, Cambridge. https://doi.org/10.1533/9781782423928.1.119

Food and Agricultural Organisation (FAO) (2018) Rice market monitor. http://www.fao.org/economic/est/publications/rice-publications/rice-market-monitorrmm/en/

Fraceto LF, Grillo R, de Medeiros GA, Scognamiglio V, Rea G, Bartolucci C (2016) Nanotechnology in agriculture: which innovation potential does it have? Front Environ Sci 4(20):1–5. https://doi.org/10.3389/fenvs.2016.00020

Froyd JD, Paget CJ, Guse LR, Dreikorn BA, Pafford JL (1976) Tricyclazole: a new systemic fungicide for control of *Pyricularia oryzae* on rice. Phytopathology 66(1):1135–1139

Gadal N, Shrestha J, Poudel MN, Pokharel B (2019) A review on production status and growing environments of rice in Nepal and in the world. Arch Agric Environ Sci 4(1):83–87. https://doi.org/10.26832/24566632.2019.0401013

Gangwar GP, Sinha AP (2012) Evaluation of *Trichoderma spp* and *fluorescent pseudomonads* for the management of bacterial leaf blight of rice. Indian Phytopathol 65(1):89–91

Gautam AK, Avasthi S (2020) Myconanoparticles as potential pest control agents. In: Thangadurai D, Sangeetha J, Prasad R (eds) Nanotechnology for food, agriculture, and environment, Nanotechnology life science. Springer, Cham, pp 189–226. https://doi.org/10.1007/978-3-030-31938-0_10

Gnanamanickam S, Priyadarisini V, Narayanan N, Vasudevan P, Kavitha S (1999) An overview of bacterial blight disease of rice and strategies for its management. Curr Sci 77(11):1435–1443. https://www.jstor.org/stable/24105230

Gohel NM, Chauhan HL, Mehta AN (2008) Bio-efficacy of fungicides against *Pyricularia oryzae* the incident of rice blast. J Plant Dis Sci 3(2):189–192

Gohel NM, Chauhan HL, Chaudhary PP, Mehta AN (2009) Field evaluation of fungicides, bioagent and botanicals against blast of rice. J Plant Dis Sci 4(2):164–166

Gomez KA, Bernardo RC (1974) Estimation of stem borer damage in rice fields. J Econ Entomol 67(4):509–513. https://doi.org/10.1093/jee/67.4.509

Gouramanis GD (1995) Biological and chemical control of rice blast disease (*Pyricularia oryzae*) in northern Greece. Cah Options Mediterr 15(3):61–68. http://om.ciheam.org/article.php?IDPDF=CI011017

Grist DH (1986) Characteristics of the plant rice, Tropical agriculture series, 6th edn. Longman Press, London and New York, pp 69–98

Groth DE (2008) Effects of cultivar resistance and single fungicide application on rice sheath blight, yield and quality. Crop Prot 27:1125–1130. https://doi.org/10.1016/j.cropro.2008.01.010

Guo R, Han M, Shu F (2013) Strategies and measures of green control of rice pests based on reduction of pesticides application in paddy field China. Plant Prot 33(10):38–41

Guo X, Zhao R, Zhang J, Du YJ, Yang LG, Chen L, Pang S, Xu Y, Zhang Z, Wu X (2019) A microcapsule oil dispersion for the controlled release of 1-methylcyclopropene in an open environment. RSC Adv 9(41):23465–23473. https://doi.org/10.1039/c9ra03762d

Gurjar MS, Ali S, Akhtar M, Singh KS (2012) Efficacy of plant extracts in plant disease management. Agric Sci 3(3):425–433. https://doi.org/10.4236/as.2012.33050

Guruprasath S, Ayyasamy R (2019) Bioefficacy of newer formulation of fipronil 0.8 G against brown plant-hopper of rice *Nilaparvata lugens* (*Stal.*). J Entomol Res 43(3):283–287. https://doi.org/10.5958/0974-4576.2019.00053.7

Hall AE (1992) Breeding for heat tolerance. Plant Breed Rev 10:129–168

Ham Y, Kim TJ (2018) Anthranilamide from *Streptomyces spp.* inhibited *Xanthomonas oryzae* biofilm formation without affecting cell growth. Appl Biol Chem 61(6):673–680. https://doi.org/10.1007/s13765-018-0405-1

Harman GE, Howell CR, Viterbo A, Chet I, Lorito M (2004) *Trichoderma* species opportunistic, avirulent plant symbionts. Nat Rev Microbiol 2(1):43–56. https://doi.org/10.1038/nrmicro797

He X, Qiao Y, Sigsgaard L, Wu X (2020) The spider diversity and plant hopper control potential in the long-term organic paddy fields in sub-tropical area, China. Agric Ecosyst Environ 295:1–8. https://doi.org/10.1016/j.agee.2020.106921

Hegde GM (2015) Efficacy of tebuconazole 25% WG against blast and sheath blight diseases of rice in central Western Ghats of Uttar Kannada district. ORYZA- Int J Rice 52(3):222–226

Hegde S, Hegde V (2013) Assessment of global rice production and export opportunity for economic development in Ethiopia. Int J Sci Res 2(6):2319–7064. https://doi.org/10.1.1.681.5988

Hirooka T, Ishii HJ (2013) Chemical control of plant diseases. J Gen Plant Pathol 79(6):390–401. https://doi.org/10.1007/s10327-013-0470-6

Hong-xing XU, Ya-jun Y, Yan-hui LU, Xu-song Z, Jun-ce T, Feng-xiang L, Qiang FU, Zhong-xian L (2017) Sustainable management of rice insect pests by non-chemical-insecticide technologies in China. Rice Sci 24(2):61–72. https://doi.org/10.1016/j.rsci.2017.01.001

Hosamani V, Pradeep S, Sridhara S, Kalleshwaraswamy CM (2009) Biological studies on paddy ear head bug *Leptocorisa oratorius Fabricius* (Hemiptera: *Alydidae*). Acad J Entomol 2(2):52–55

Hu J, Xiao C, He Y (2016) Recent progress on the genetics and molecular breeding of brown planthopper resistance in rice. Rice 9(1):1–12. https://doi.org/10.1186/s12284-016-0099-0

Huang N, Angeles ER, Domingo J, Magpantay G, Singh S, Zhang G, Kumaravadivel N, Bennett J, Khush GS (1997) Pyramiding of bacterial blight resistance genes in rice: marker-assisted selection using RFLP and PCR. Theor Appl Genet 95(3):313–320. https://doi.org/10.1007/s001220050565

Huang SH, Cheng CH, Yang HY (2008) Investigation on the occurrence of rice stem borers in Central Taiwan. Plant Prot Bull 50:37–46

Hummel NA, Meszaros A, Ring DR, Beuzelin JM, Stout MJ (2014) Evaluation of seed treatment insecticides for management of the rice water weevil, *Lissorhoptrus oryzophilus* Kuschel (Coleoptera: *Curculionidae*), in commercial rice fields in Louisiana. Crop Prot 65:37–42. https://doi.org/10.1016/j.cropro.2014.06.025

Hunjan MS, Lore JS, Pannu PPS, Thind TS (2011) Performance of some new fungicides against sheath blight and brown spot of rice. Plant Dis Res 26(1):61–67

Ibrahim E, Fouad H, Zhang M, Zhang Y, Qiu W, Yan C, Li B, Mo J, Chen J (2019) Biosynthesis of silver nanoparticles using endophytic bacteria and their role in inhibition of rice pathogenic bacteria and plant growth promotion. RSC Adv 9(50):29293–29299. https://doi.org/10.1039/C9RA04246F

International Rice Research Institute- Rice Knowledge Bank (2020a) Crop health, disease: bacterial leaf streak. http://www.knowledgebank.irri.org/training/fact-sheets/pest-management/diseases/item/bacterial-leaf-streak

International Rice Research Institute- Rice Knowledge Bank (2020b) Crop health, disease: sheath blight. http://www.knowledgebank.irri.org/training/fact-sheets/pest-management/diseases/item/sheath-blight

International Rice Research Institute- Rice Knowledge Bank (2020c) Crop health, disease: sheath rot. http://www.knowledgebank.irri.org/training/fact-sheets/pest-management/diseases/item/sheath-rot

International Rice Research Institute- Rice Knowledge Bank (2020d) Crop health, disease: brown spot. http://www.knowledgebank.irri.org/training/fact-sheets/pest-management/diseases/item/brown-spot

International Rice Research Institute- Rice Knowledge Bank (2020e) Crop health, disease: Tungro. http://www.knowledgebank.irri.org/training/fact-sheets/pest-management/diseases/item/tungro

International Rice Research Institute- Rice Knowledge Bank (2020f) Crop health, disease: stem borer. http://www.knowledgebank.irri.org/training/fact-sheets/pest-management/diseases/item/stem-borer

Into P, Khunnamwong P, Jindamoragot S, Am-in S, Intanoo W, Limtong S (2020) Yeast associated with rice phylloplane and their contribution to control of rice sheath blight disease. Microorganisms 8(3):1–21. https://doi.org/10.3390/microorganisms8030362

Jagadeesha KBD, Chakravarthy AK, Doddabasappa B, Basavaraju BS (2009) Biology of the rice gall midge, *Orseolia oryzae* (Wood-Mason) in southern Karnataka. Karnataka J Agric Sci 22(3):535–537

Jahn GC, Domingo I, Liberty M, Almazan P, Pacia J (2004) Effect of rice bug *Leptocorisa oratorius* (Hemiptera: *Alydidae*) on rice yield, grain quality, and seed viability. J Econ Entomol 97(6):1923–1927. https://doi.org/10.1603/0022-0493-97.6.1923

Jha Y, Subramanian RB (2014) Characterization of root-associated bacteria from paddy and its growth-promotion efficacy. 3 Biotech 4:325–330. https://doi.org/10.1007/s13205-013-0158-9

Ji GH, Wei LF, He YQ, Wu YP, Bai XH (2008) Biological control of rice bacterial blight by *Lysobacter antibioticus* strain 13-1. Biol Control 45(3):288–296. https://doi.org/10.1016/j.biocontrol.2008.01.004

Jiang Y, Cai Z, Xie W, Long T, Yu H, Zhang Q (2012) Rice functional genomics research: progress and implications for crop genetic improvement. Biotechnol Adv 30(5):1059–1070. https://doi.org/10.1016/j.biotechadv.2011.08.013

Jiang B, Wang Z, Xu C, Liu W, Jiang D (2019) Screening and identification of *Aspergillus* activity against *Xanthomonas oryzae pv. oryzae* and analysis of antimicrobial components. J Microbiol 57(7):597–605. https://doi.org/10.1007/s12275-019-8330-5

Jones MC, Gough IK, Dasgupta I, Rao BS, Cliffe J, Qu R, Shen P, Kaniewska M, Blakebrough M, Davies JW, Beachy RN (1991) Rice tungro disease is caused by an RNA and a DNA virus. J Gen Virol 72(4):757–761. https://doi.org/10.1099/0022-1317-72-4-757

Jonit NQ, Low YC, Tan GH (2016) *Xanthomonas oryzae pv. oryzae*, biochemical tests, rice (*Oryza sativa*), bacterial leaf blight (BLB) disease, Sekinchan. J Appl Environ Microbiol 4:63–69. https://doi.org/10.12691/jaem-4-3-3

Juliano BO (1993) Rice in human nutrition. FAO Food Nutr Ser 26:163

Justin CGL, Preetha G (2014) Survey on the occurrence, distribution pattern and management of stem borers on rice in Kanyakumari District, Tamil Nadu. J Entomol Zool Stud 2(67):86–90

Kakde AM, Patel KG (2019) Yield performance of different insecticides against rice yellow stem borer, Scirpophaga incertulas walker. J Rice Res 7(1):1–5. https://doi.org/10.4172/2375-4338.1000203

Kelly FR, Gore J, Cook DR, Catchot AL, Golden BR, Krutz LJ, Crow WD, Towles TB, Bond JA (2020) Evaluation of flood removal in combination with insecticide seed treatment for rice water weevil (Coleoptera: *Curculionidae*) larval management in rice. J Econ Entomol 113:1–6. https://doi.org/10.1093/jee/toaa158

Khalili E, Sadravi M, Naeimi S, Khosravi V (2012) Biological control of rice brown spot with native isolates of three *Trichoderma* species. Braz J Microbiol 43(1):297–305. https://doi.org/10.1590/S1517-83822012000100035

Khan JA, Siddiq R, Arshad HMI, Anwar HS, Saleem K, Jamil FF (2012) Chemical control of bacterial leaf blight of rice caused by *Xanthomonas oryzae pv. oryzae*. Pak J Phytopathol 24(2):97–100

Khanzada M, Shah GS (2012) In-vitro evaluation of fungicides, plant extracts and bio-control agents against rice blast pathogen *Magnaporthe oryzae* couch. Pak J Bot 44(5):1775–1778

Khoa ND, Xa TV, Hao LT (2017) Disease-reducing effects of aqueous leaf extract of *Kalanchoe pinnata* on rice bacterial leaf blight caused by *Xanthomonas oryzae pv. oryzae* involve induced resistance. Physiol Mol Plant Pathol 100:57–66. https://doi.org/10.1016/j.pmpp.2017.06.005

Kim SI, Kwak JS, Song JT, Seo HS (2016a) Long-term effect of niclosamide on inhibition of bacterial leaf blight in rice. J Plant Prot Res 56(4):323–327. https://doi.org/10.1515/jppr20160051

Kim SI, Song JT, Jeong JY, Seo HS (2016b) Niclosamide inhibits leaf blight caused by *Xanthomonas oryzae* in rice. Sci Rep 6:1–13. https://doi.org/10.1038/srep21209

Kim SI, Lee HY, Song JT, Seo HS (2018) The human drug auranofin inhibits the growth of *Xanthomonas oryzae pv. oryzae* and *Magnaporthe oryzae*, which cause rice leaf blight and blast. Plant Breed Biotechnol 6(2):119–124. https://doi.org/10.9787/PBB.2018.6.2.119

Kokkrua S, Ismail SI, Mazlan N, Dethoup T (2020) Efficacy of berberine in controlling foliar rice diseases. Eur J Plant Pathol 156(1):147–158. https://doi.org/10.1007/s10658-019-01871-3

Kulmitra AK, Sahu N, Kumar VB, Thejesha AG, Ghosh A, Gulnaz Y (2017) *In vitro* evaluation of bio-agents against *Pyricularia oryzae* (Cav.) causing rice blast disease. Agric Sci Digest- Res J 37(3):247–248. https://doi.org/10.18805/asd.v37i03.8989

Kumar B (2020) Efficacy of modern combination fungicide molecules against sheath blight of rice. Indian Phytopathol 73:1–5. https://doi.org/10.1007/s42360-020-00273-4

Kumar S, Khan MA (2005) Bio-efficacy of *Trichogramma spp.* against yellow stem borer and leaf folder in rice ecosystem. Ann Plant Prot Sci 13(1):97–99

Kumar M, Simon S (2016) Efficacy of certain botanical extracts in the management of brown leaf spot of rice cause by *Helminthosporium Oryzae*. Biosci Biotechnol Res Asia 13(4):2015–2018. https://doi.org/10.13005/bbra/2358

Kumar M, Parate RL, Ninawe BN (2009) Effect of botanicals, bioagents and some chemicals against *Xanthomonas oryzae pv. oryzae*. J Plant Dis Sci 4(1):60–63

Kumar V, Shanker K, Kumar A, Kumar M (2010a) Comparative effect of insecticides against green leaf hopper (*Nephotettix virescens* distant) on rice crop. Progress Agric 10(3):83–86

Kumar BV, Boomathi N, Kumaran N, Kuttalam S (2010b) Non target effect of ethiprole+ imidacloprid 80 WG on predators of rice plant-hoppers. Madras Agric J 97(4/6):153–156

Kumar S, Lal AA, Kumar N, Jaiswal S, Kumar H, Kumar A, Kumar M (2017) Effect of bio control agents and botanicals against blast of paddy caused by *Pyricularia oryzae*. Int J Chem Stud 5(1):314–318

Kumar S, Nehra M, Dilbaghi N, Marrazza G, Hassan AA, Kim KH (2019) Nano-based smart pesticide formulations: emerging opportunities for agriculture. J Control Release 294:131–153. https://doi.org/10.1016/j.jconrel.2018.12.012

Li YX, Wang YM, Liu B, Wang SH, Ming Li Z (2005) Synthesis and herbicidal activity of 2-alkyl (aryl)-3-methylsulfonyl (sulfinyl) pyrano-[4, 3-c] pyrazol-4 (2H)-ones. Heteroat Chem 16(4):255–258. https://doi.org/10.1002/hc.20067

Li B, Liu B, Shan C, Ibrahim M, Lou Y, Wang Y, Xie G, Li HY, Sun G (2013a) Antibacterial activity of two chitosan solutions and their effect on rice bacterial leaf blight and leaf streak. Pest Manag Sci 69(2):312–320. https://doi.org/10.1002/ps.3399

Li P, Yin J, Xu W, Wu J, He M, Hu D, Yang S, Song B (2013b) Synthesis, antibacterial activities, and 3 D-QSAR of sulfone derivatives containing 1, 3, 4-oxadiazole moiety. Chem Biol Drug Des 82(5):546–556. https://doi.org/10.1111/cbdd.12181

Li P, Shi L, Yang X, Yang L, Chen XW, Wu F, Shi QC, Xu WM, He M, Hu DY, Song BA (2014) Design, synthesis, and antibacterial activity against rice bacterial leaf blight and leaf streak of 2, 5-substituted-1, 3, 4-oxadiazole/thiadiazole sulfone derivative. Bioorg Med Chem Lett 24(7):1677–1680. https://doi.org/10.1016/j.bmcl.2014.02.060

Li B, Zhang Y, Yang Y, Qiu W, Wang X, Liu B, Wang Y, Sun G (2016) Synthesis, characterization, and antibacterial activity of chitosan/TiO2 nanocomposite against *Xanthomonas oryzae pv. oryzae*. Carbohydr Polym 152:825–831. https://doi.org/10.1016/j.carbpol.2016.07.070

Li P, Hu D, Xie D, Chen J, Jin L, Song B (2018) Design, synthesis, and evaluation of new sulfone derivatives containing a 1, 3, 4-oxadiazole moiety as active antibacterial agents. J Agric Food Chem 66(12):3093–3100. https://doi.org/10.1021/acs.jafc.7b06061

Linares OF (2002) African rice (*Oryza glaberrima*): history and future potential. Proc Natl Acad Sci 99(25):16360–16365. https://doi.org/10.1073/pnas.252604599

Liu Q, Xu J, Wang Y, Li C, Han G, Qi J, Sun J, Ma T (2013) Synergism of CmGV and *Bacillus thuringiensis* against larvae of *Cnaphalocrocis medinalis* Guenee. J Yangzhou Univ Agric Life Sci 34(4):89–93

Longkumer IY, Misra AK (2020) Efficacy of essential oils against rice insect pest in Pusa, Bihar. J Entomol Zool Stud 8(3):1980–1983

Lou YG, Zhang GR, Zhang WQ, Hu Y, Zhang J (2014) Reprint of: biological control of rice insect pests in China. Biol Control 68:103–116. https://doi.org/10.1016/j.biocontrol.2013.09.018

Lu BR (1999) Taxonomy of the genus *Oryza* (*poaceae*): historical perspective and current status. Int Rice Res Notes 24:4–8

Ma W, Ma L, Li J, Wang F, Sisak I, Zhang F (2011) Phosphorus flows and use efficiencies in production and consumption of wheat, rice, and maize in China. Chemosphere 84(6):814–821. https://doi.org/10.1016/j2011.04.055

Maclean JL, Dawe DC, Hettel GP, Hardy B (2002) Rice almanac: source book for the most important economic activity on earth. International Rice Research Institute, Los Baños

Magierowicz K, Gorska-Drabik E, Golan K (2020) Effects of plant extracts and essential oils on the behavior of *Acrobasis advenella* (Zinck.) caterpillars and females. J Plant Dis Prot 127(1):63–71. https://doi.org/10.1007/s41348-019-00275-z

Mahmud H, Hossain I, Dey P, Hasan AK, Afsar S (2016) Effects of BAU-biofungicide and proud on disease severity, seed quality, vigour index and health status of seed against brown spot disease of rice. Afr J Agric Res 11(50):5092–5099. https://doi.org/10.5897/AJAR2016.11134

Majumdar TD, Singh M, Thapa M, Dutta M, Mukherjee A, Ghosh CK (2019) Size-dependent antibacterial activity of copper nanoparticles against *Xanthomonas oryzae pv. oryzae* – a synthetic and mechanistic approach. Colloid Interf Sci Commun 32:1–10. https://doi.org/10.1016/j.colcom.2019.100190

Manandhar HK, Jorgensen HL, Mathur SB, Smedegaard-Petersen V (1998) Resistance to rice blast induced by ferric chloride, di-potassium hydrogen phosphate and salicylic acid. Crop Prot 17(4):323–329. https://doi.org/10.1016/S0261-2194(98)00020-9

Mankad M, Patil G, Patel D, Patel P, Patel A (2018) Comparative studies of sunlight mediated green synthesis of silver nanoparaticles from Azadirachta indica leaf extract and its antibacterial effect on Xanthomonas oryzae pv. oryzae. Arab J Chem 13(1):1–8. https://doi.org/10.1016/j.arabjc.2018.07.016

Matsui T, Omasa K, Horie T (2000) High temperature at flowering inhibits swelling of pollen grains, a driving force for thecae dehiscence in rice (*Oryza sativa L*). Plant Prod Sci 3(4):430–434. https://doi.org/10.1626/pps.3.430

Matteson PC (2000) Insect pest management in tropical Asian irrigated rice. Annu Rev Entomol 45(1):549–574. https://doi.org/10.1146/annurev.ento.45.1.549

McManus PS, Stockwell VO, Sundin GW, Jones AL (2002) Antibiotic use in plant agriculture. Annu Rev Phytopathol 40:443–465. https://doi.org/10.1146/annurev.phyto.40.120301.093927

Medina-Perez G, Fernandez-Luqueno F, Campos-Montiel RG, Sanchez-Lopez KB, Afanador-Barajas LN, Prince L (2019) Nanotechnology in crop protection: status and future trends. In: Nano-biopesticides today and future perspectives. Academic, Cambridge, pp 17–45. https://doi.org/10.1016/B978-0-12-815829-6.00002-4

Meena C, Gopalakrishnan J, Dureja P (2013) Antibacterial with ametelin B from *Datura metel* against *Xanthomonas oryzae pv. oryzae*. Biopestic Int 9(1):31–37

Mehar S, Shah A, Abassi FM, Akhtar MA, Rafi A (2009) Resistance characterization of wild relatives of rice in response to bacterial blight. Pak J Bot 41:917–925

Meng X, Zhang N, Yang X, Miao L, Jiang H, Ji C, Xu B, Qian K, Wang J (2020) Sublethal effects of chlorantraniliprole on molting hormone levels and mRNA expressions of three Halloween genes in the rice stem borer, *Chilo suppressalis*. Chemosphere 238:1–7. https://doi.org/10.1016/j.chemosphere.2019.124676

Milani P, França D, Balieiro AG, Faez R (2017) Polymers and its applications in agriculture. Polimeros 27(3):256–266. https://doi.org/10.1590/0104-1428.09316

Ministry of Agricultural Development (MOAD) (2017) Statistical information on Nepalese agriculture, 2016/17. Published by Ministry of Agricultural Development, Singh Durbar, Kathmandu

Mishra MK, Singh RB, Dwivedi JL, Ali S (2012) Efficacy of insecticides against *Scirpophaga incertulas* (Walker) on basmati rice. Ann Plant Prot Sci 20(2):310–313

Mishra D, Vishnupriya MR, Anil MG, Konda K, Raj Y, Sonti RV (2013) Pathotype and genetic diversity amongst Indian isolates of *Xanthomonas oryzae pv. oryzae*. PLoS One 8(11):1–11. https://doi.org/10.1371/journal.pone.0081996

Misra HP (2009) New insecticides for management of rice green leafhopper, *Nephotettix virescens* Dist. Ann Plant Prot Sci 17(1):5–8

Mo J, Ding WB, Li YZ, Li GH (2014) Control effect of 9% 12 α-hydroxy-rotenone EW against three pests of rice. Hunan Agric Sci Technol 10:40–42

Mohammed AR, Tarpley L (2009) High night time temperatures affect rice productivity through altered pollen germination and spikelet fertility. Agric For Meteorol 149(6–7):999–1008. https://doi.org/10.1016/j.agrformet.2008.12.003

Mondal IH, Chakraborty K (2016) Relative efficiency of some selected insecticide formulations on yellow stemborer, *Scirpophaga incertulas* (walk) in rice field at Murshidabad, West Bengal, India. J Entomol Zool Stud 4(6):471–477

Monisha S, Praveen NM, Ramanathan A (2019) Isolation, characterization and management of brown spot disease of rice. J Pharmacogn Phytochem 8(3):4539–4545

Montano FP, Alias-Villegas C, Bellogin RA, Cerro PD, Espuny MR, Guerrero IJ, Lopez-Baena FJ, Ollero FJ, Cubo T (2013) Plant growth promotion in cereal and leguminous agricultural important plants: from microorganism capacities to crop production. Microbiol Res 169(5–6):325–336. https://doi.org/10.1016/j.micres.2013.09.011

Mughal MN, Ahmed M, Bashir S, Anwar A (2017) Comparative efficacy of some non-systemic fungicides against rice sheath blight (*Thanatephorus cucumeris*). Int J Curr Microbiol App Sci 6(9):1824–1832. https://doi.org/10.20546/ijcmas.2017.609.225

Muralidharan K, Pasalu IC (2006) Assessments of crop losses in rice ecosystems due to stem borer damage (Lepidoptera: *Pyralidae*). Crop Prot 25(5):409–417. https://doi.org/10.1016/j.cropro.2005.06.007

Muralidharan K, Krishnaveni D, Rajarajeswari NV, Prasad AS (2003) Tungro epidemics and yield losses in paddy fields in India. Curr Sci 85(8):1143–1147. www.jstor.org/stable/24108612

Muthayya S, Hall J, Bagriansky J, Sugimoto J, Gundry D, Matthias D, Prigge S, Hindle P, Moench-Pfanner R, Maberly G (2012) Rice fortification: an emerging opportunity to contribute to the elimination of vitamin and mineral deficiency worldwide. Food Nutr Bull 33(4):296–307. https://doi.org/10.1177/156482651203300410

Muthayya S, Sugimoto JD, Montgomery S, Maberly GF (2014) An overview of global rice production, supply, trade, and consumption. Ann N Y Acad Sci 1324(1):7–14

Naqvi SA (2019) Bacterial leaf blight of rice: an overview of epidemiology and management with special reference to Indian sub-continent. Pak J Agric Res 32(2):359–380. https://doi.org/10.17582/journal.pjar/2019/32.2.359.380

Nasir M, Iqbal B, Hussain M, Mustafa A, Ayub M (2019) Chemical management of bacterial leaf blight disease in rice. J Agric Res 57(2):93–98

Nathan SS, Chung PG, Murugan K (2005) Effect of biopesticides applied separately or together on nutritional indices of the rice leaffolder *Cnaphalocrocis medinalis*. Phytoparasitica 33(2):187–195. https://doi.org/10.1007/BF03029978

Neri-Badang MC, Chakraborty S (2019) Carbohydrate polymers as controlled release devices for pesticides. J Carbohydr Chem 38(1):67–85. https://doi.org/10.1080/07328303.2019.1568449

Nguyen HT, Vu QH, Van Mai T, Nguyen TT, Vu LD, Nguyen TT, Nguyen LV, Vu HT, Nong HT, Dinh TN, Toshitsugu N (2018) Marker-assisted selection of *XA21* conferring resistance to bacterial leaf blight in *Indica* rice cultivar LT2. Rice Sci 25(1):52–56. https://doi.org/10.1016/j.rsci.2017.08.004

Nino-liu DO, Ronald PC, Bogdanove AJ (2006) *Xanthomonas oryzae* pathovars: model pathogens of a model crop. Mol Plant Pathol 5:303–324. https://doi.org/10.1111/J.1364-3703.2006.00344.X

Notz R, Maurhofer M, Schnider-Keel U, Duffy B, Haas D, Defago G (2001) Biotic factors affecting expression of the 2, 4-diacetylphloroglucinol biosynthesis gene *phlA* in *Pseudomonas fluorescens* biocontrol strain CHA0 in the rhizosphere. Phytopathology 91:873–881. https://doi.org/10.1094/PHYTO.2001.91.9.873

Nugroho LH, Pratiwi R, Soesilohadi RH, Subin ER, Wahyuni S, Hartini YS, Lailaty IQ (2020) Repellent activity of *Piper spp.* leaves extracts on rice ear bugs (*Leptocorisa oratorius* Fabricius) and the characters of its volatile compounds. Annu Res Rev Biol 35(9):34–45. https://doi.org/10.9734/ARRB/2020/v35i930000

Ogunyemi SO, Abdallah Y, Zhang M, Fouad H, Hong X, Ibrahim E, Masum MM, Hossain A, Mo J, Li B (2019) Green synthesis of zinc oxide nanoparticles using different plant extracts and their antibacterial activity against *Xanthomonas oryzae pv. oryzae*. Artif Cells Nanomed Biotechnol 47(1):341–352. https://doi.org/10.1080/21691401.2018.1557671

Oliva R, Quibod IL (2017) Immunity and starvation: new opportunities to elevate disease resistance in crops. Curr Opin Plant Biol 38:84–91. https://doi.org/10.1016/j.pbi.2017.04.020

Orasen G, Greco R, Puja E, Pozzi C, Stile MR (2020) Blast resistance R genes pyramiding in temperate japonica rice. Euphytica 216(3):1–10. https://doi.org/10.1007/s10681-020-2575-2

Padmanabhan SY (1973) The Bengal famine. Annu Rev Phytopathol 11:11–26. https://doi.org/10.1146/annurev.py.11.090173.000303

Pandey G (2018) Challenges and future prospects of agri-nanotechnology for sustainable agriculture in India. Environ Technol Innov 11:299–307. https://doi.org/10.1016/j.eti.2018.06.012

Pannu PPS, Chahal SS, Sharma VK, Mandeep K, Bagga PS (2006) Occurrence of brown leaf spot of rice in Punjab, its effect on grain yield and its control. Indian Phytopathol 59:190–193

Park JY, Jin J, Lee YW, Kang S, Lee YH (2009) Rice blast fungus (*Magnaporthe oryzae*) infects Arabidopsis via a mechanism distinct from that required for the infection of rice. Plant Physiol 149(1):474–486. www.plantphysiol.org/cgi/doi/10.1104/pp.108.129536

Parthasarathy R, Indira J, Manikandan A (2014) To study the antagonistic organisms and new chemicals against the bacterial blight pathogen (*Xanthomonas oryzae pv. oryzae*) of rice. Int J Res Biotechnol Biochem 4(1):6–11

Patel SJ, Chauhan BL, Mehta AN, Gohil NM (2009) Management of bacterial leaf blight of rice with chemicals, botanicals and antagonists. J Plant Dis Sci 4(2):208–211

Pathak MD, Dhalival GS (1981) Trends and strategies for rice insect problems in tropical Asia, Res Pap Ser 64. International Rice Research Institute, Los Banos, pp 1–15

Peng SB, Huang JL, Zhong XH, Yang JC, Wang GH, Zhou YB, Zhang FS, Zhu QS, Bures R, Witt R (2002) Research strategy in improving fertilizer nitrogen use efficiency of irrigated rice in China. Agric Sci China 35(9):1095–1103

Peng S, Huang J, Sheehy JE, Laza RC, Visperas RM, Zhong X, Centeno GS, Khush GS, Cassman KG (2004) Rice yields decline with higher night temperature from global warming. Proc Natl Acad Sci U S A 101(27):9971–9975. https://doi.org/10.1073/pnas.0403720101

Peng D, Li S, Wang J, Chen C, Zhou M (2014) Integrated biological and chemical control of rice sheath blight by *Bacillus subtilis* NJ-18 and jinggangmycin. Pest Manag Sci 70(2):258–263. https://doi.org/10.1002/ps.3551

Persaud R, Khan A, Isaac WA, Ganpat W, Saravanakumar D (2019) Plant extracts, bioagents and new generation fungicides in the control of rice sheath blight in Guyana. Crop Prot 119:30–37. https://doi.org/10.1016/j.cropro.2019.01.008

Peterson LG (1990) Tricyclazole for control of Pyriculariaoryzae on rice: the relationship of the mode of action and disease occurrence and development. In: Pest management in rice. Springer, Dordrecht, pp 122–130. https://doi.org/10.1007/978-94-009-0775-18

Pooja K, Katoch A (2014) Past, present and future of rice blast management. Plant Sci Today 1(3):165–173. https://doi.org/10.14719/pst.2014.1.3.24

Prabavathy VR, Mathivanan N, Murugesan K (2006) Control of blast and sheath blight diseases of rice using antifungal metabolites produced by *Streptomyces sp.* PM5. Biol Control 39(3):313–319. https://doi.org/10.1016/j.biocontrol.2006.07.011

Prajapati KS, Patel RC, Pathak AR (2004) Field evaluation of new fungicides against blast of rice. Pestic Res J 16(2):26–28

Prakash R, Kunal G (2020) Evaluation of efficacy of insecticides and bio pesticide against paddy Earhead bug, *Leptocorisa oratorius* F. J Pharmacogn Phytochem 9(2):2021–2023. https://doi.org/10.22271/phyto.2020.v9.i2ag.11150

Pramesh D, Alase S, Muniraju KM, Kumara MK (2017) A combination fungicide for the management of sheath blight, sheath rot and stem rot diseases of paddy. Int J Curr Microbiol App Sci 6:3500–3509. https://doi.org/10.20546/ijcmas.2017.609.430

Prasad R (2011) Status of the rice gall midge (*Orseolia oryzae* W.M) in the state of Jharkhand. J Rice Res 4:19–21

Prasad PVV, Boote KJ, Allen JLH, Sheehy JE, Thomas JMG (2006) Species, ecotype and cultivar differences in spikelet fertility and harvest index of rice in response to high temperature stress. Field Crop Res 95:398–411. https://doi.org/10.1016/j.fcr.2005.04.008

Praveen NM, Ramanathan A, Monisha S (2019) Isolation, characterization and chemical management of bacterial leaf blight pathogen of rice. J Pharmacogn Phytochem 8(3):3320–3323

Rahaman MM, Islam KS, Jahan M, Mamun MA (2014) Relative abundance of stem borer species and natural enemies in rice ecosystem at Madhupur, Tangail, Bangladesh. J Bangladesh Agric Univ 12:267–272. https://doi.org/10.22004/ag.econ.211239

Rahman MT, Khalequzzaman M, Khan MA (2004) Assessment of infestation and yield loss by stem borers on variety of rice. J Asia Pac Entomol 7(1):89–95. https://doi.org/10.1016/S1226-8615(08)60203-4

Rajappan K, Ushamalini C, Subramanian N, Narasimhan V, Kareem AA (2000) Effect of botanicals on the population dynamics of *Nephotettix virescens*, rice tungro disease incidence and yield of rice. Phytoparasitica 28(2):109–113. https://doi.org/10.1007/BF02981739

Raji P, Louis V (2007) Isoprothiolane-a new fungicide for the management of rice blast. ORYZA-Int J Rice 44(4):366–367

Reddy BP, Reddy KR, Rao MS, Rao KS (2008) Efficacy of antimicrobial metabolites of *Pseudomonas fluorescens* against rice fungal pathogens. Curr Trends Biotechnol Pharm 2(1):178–182

Remya N, Saino HV, Baiyu G, Maekawa T, Yashida Y, Sakthi KI (2010) Nano particular material delivery to plants. Plant Sci 523:341–351. https://doi.org/10.1016/j.plantsci.2010.04.012

Ren J, Gao F, Wu X, Lu X, Zeng L, Lv J, Su X, Luo H, Ren G (2016) *Bph32*, a novel gene encoding an unknown SCR domain-containing protein, confers resistance against the brown planthopper in rice. Sci Rep 6:1–14. https://doi.org/10.1038/srep37645

Sahithi S, Misra HP (2006) Control of rice green leaf hoppers, *Nephotettix virescens* (dist.) by the use of insecticides. Ann Plant Prot Sci 14(1):80–82

Sahu SK, Zheng P, Yao N (2018) Niclosamide blocks rice leaf blight by inhibiting biofilm formation of *Xanthomonas oryzae*. Front Plant Sci 9:1–16. https://doi.org/10.3389/fpls.2018.00408

Sain M (2000) Bionomics and management of rice whorl maggot *Hydrellia spp*. (Diptera: *Ephydridae*) – a review. Agric Rev 21:110–115

Sakthivel N (2001) Sheath rot disease of rice: current status and control strategies. In: Major fungal diseases of rice recent advances. Kluwer Academic Publishers, Boston, pp 271–283. https://doi.org/10.1007/978-94-017-2157-8_19

Salam MU, Jones JW, Kobayashi K (2001) Predicting nursery growth and transplanting shock in rice. Exp Agric 37(1):65–81. https://doi.org/10.1017/S0014479701001016

Sandhu GS, Dhaliwal NS (2016) Evaluation of different insecticides against major insect pests of rice in Punjab. Int J Plant Prot 9(1):187–192. https://doi.org/10.15740/HAS/IJPP/9.1

Sangamithra S, Samiayyan K, Chandrasekaran S (2017) Effect of Imidacloprid 600 FS as seed treatment against brown plant-hopper in rice. Trends Biosci 10(4):1225–1227

Sarao PS, Mahal MS (2008) Comparative efficacy of insecticides against major insect pests of rice in Punjab. Pestic Res J 20(1):52–58

Saravanakumar D, Lavanya N, Muthumeena K, Raguchander T, Samiyappan R (2009) *Fluorescent pseudomonad* mixtures mediate disease resistance in rice plants against sheath rot (*Sarocladium oryzae*) disease. BioControl 54(2):273–286. https://doi.org/10.1007/s10526-008-9166-9

Satpathi CR, Chakraborty K, Shikari D, Acharjee P (2012) Consequences of feeding by yellow stem borer (*Scirpophaga incertulas Walk*) on rice cultivar Swarna mashuri (MTU 7029). World Appl Sci J 17:532–539

Seck PA, Tollens E, Wopereis MC, Diagne A, Bamba I (2010) Rising trends and variability of rice prices: threats and opportunities for sub-Saharan Africa. Food Policy 35(5):403–411. https://doi.org/10.1016/j.foodpol.2010.05.003

Seck PA, Diagne A, Mohanty S, Wopereis MC (2012) Crops that feed the world 7: rice. Food Sec 4(1):7–24. https://doi.org/10.1007/s12571-012-0168-1

Selvaraj K, Annamalai P (2015) *In vitro* antagonism of indigenous trichoderma isolates against phytopathogen causing sheath rot disease in paddy. World J Pharm Res 4(4):429–437

Sere Y, Onasanya A, Verdier V, Akator K, Ouedraogo LS (2005) Rice bacterial leaf blight in West Africa: preliminary studies on disease in farmers' fields and screening released varieties for resistance to the bacteria. Asian J Plant Sci 4:577–579. https://doi.org/10.3923/ajps.2005.577.579

Sesma A, Osbourn AE (2004) The rice leaf blast pathogen undergoes developmental processes typical of root-infecting fungi. Nature 431(7008):582–586. https://doi.org/10.1038/nature02880

Shabana YM, Abdel-Fattah GM, Ismail AE, Rashad YM (2008) Control of brown spot pathogen of rice (*Bipolaris oryzae*) using some phenolic antioxidants. Braz J Microbiol 39(3):438–444. https://doi.org/10.1590/S1517-83822008000300006

Shafaullah KM (2016) Management of *Pyricularia grisea*, the rice blast pathogen through botanical pesticides. Int J Sci Res 5:973–976

Shahid AA, Nasir IA, Zafar AU, Sumrin A, Chaudhry B, Riazuddin S (2003) The use of CAMB biopesticides to control pests of rice (*Oryza sativa*). Asian J Plant Sci 2(15–16):1079–1082

Shang Y, Hasan M, Ahammed GJ, Li M, Yin H, Zhou J (2019) Applications of nanotechnology in plant growth and crop protection: a review. Molecules 24(14):1–23. https://doi.org/10.3390/molecules24142558

Shanmugam TR, Sendhil R, Thirumalvalavan V (2006) Quantification and prioritization of constraints causing yield loss in rice (*Oryza sativa*) in India. Agric Tropica Subtrop 39(3):194–204

Shao H, Xi N, Zhang Y (2018) Microemulsion formulation of a new biopesticide to control the diamondback moth (Lepidoptera: *Plutellidae*). Sci Rep 8(1):1–9. https://doi.org/10.1038/s41598-018-28626-0

Sharanappa AK, Khan HH, Sahu R (2017) Efficacy of certain insecticide against rice stem borer, *Scirpophaga incertulas* (Walker) on rice, *Oryza sativa L.* J Entomol Zool Stud 5(5):719–721

Sharma U, Srivastava A (2019) Field efficacy of some insecticides and biopesticides against rice hispa *Dicladispa armigera* (Olivier). Indian J Entomol 81(1):166–169. https://doi.org/10.5958/0974-8172.2019.00008.7

Sharma R, Ram L, Devi R (2017) Efficacy of white muscardine fungus (*Beauveria bassiana*) on rice hispa (*Dicladispa armigera*). Indian J Agric Res 51(3):296–298. https://doi.org/10.18805/ijare.v51i03.7922

Shi W, Li C, Li M, Zong X, Han D, Chen Y (2016) Antimicrobial peptide melittin against *Xanthomonas oryzae pv. oryzae*, the bacterial leaf blight pathogen in rice. Appl Microbiol Biotechnol 100(11):5059–5067. https://doi.org/10.1007/s00253-016-7400-4

Sholl A, Pearson R, Ruth K, Way MO (2020) Efficacy of selected botanical water extracts for rice water weevil control, 2017. Arthropod Manage Tests 45(1):1–1. https://doi.org/10.1093/amt/tsaa005

Shrestha BK, Karki HS, Groth DE, Jungkhun N, Ham JH (2016) Biological control activities of rice-associated *Bacillus sp.* strains against sheath blight and bacterial panicle blight of rice. PLoS One 11(1):1–18. https://doi.org/10.1371/journal.pone.0146764

Shtienberg D, Zilberstaine M, Oppenheim D, Herzog Z, Manulis S, Shwartz H, Kritzman G (2001) Efficacy of oxolinic acid and other bactericides in suppression of *Erwinia amylovora* in pear orchards in Israel. Phytoparasitica 29:143–154. https://doi.org/10.1007/BF02983958

Singh RP, Jha PN (2016) Alleviation of salinity-induced damage on wheat plant by an ACC deaminase-producing halophilic bacterium *Serratia sp.* SL-12 isolated from a salt lake. Symbiosis 69:101–111. https://doi.org/10.1007/s13199-016-0387-x

Singh RA, Das B, Ahmed KM, Pal V (1980) Chemical control of bacterial leaf blight of rice. Int J Pest Manag 26(1):21–25. https://doi.org/10.1080/09670878009414278

Singh S, Sidhu JS, Huang N, Vikal Y, Li Z, Brar DS, Dhaliwal HS, Khush GS (2001) Pyramiding three bacterial blight resistance genes (*xa5, xa13* and *Xa21*) using marker-assisted selection into Indica rice cultivar PR106. Theor Appl Genet 102(6–7):1011–1015. https://doi.org/10.1007/s001220000495

Singh S, Prasad CS, Nath L, Tiwari GN (2008) Eco-friendly management of *Scirpophaga incertulas* (Walk.) and *Cnaphalocrocis medinalis* (Guen.) in basmati rice. Ann Plant Prot Sci 16(1):11–16

Singh R, Yadav RS, Javeria S (2015a) Management of bacterial leaf blight of basmati rice caused by *Xanthomonas oryzae pv. oryzae* with some available antibiotics and plant products. Int J Innov Appl Res 3(11):1–6. https://doi.org/10.1128/AEM.02768-14

Singh P, Singh R, Dhaka SS, Kumar D, Kumar H, Kumari N (2015b) Bioefficacy of insecticides and bio-pesticides against yellow stem borer, Scirpophaga incertulus (walk.) and their effect on spiders in rice crop. South Asian J Food Technol Environ 1(2):179–183

Singh UB, Malviya D, Singh S, Pradhan JK, Singh BP, Roy M, Imram M, Pathak N, Baisyal BM, Rai JP, Sarma BK (2016) Bio-protective microbial agents from rhizosphere eco-systems trigger plant defense responses provide protection against sheath blight disease in rice (*Oryzasativa L.*). Microbiol Res 192:300–312. https://doi.org/10.1016/j.micres.2016.08.007

Skamnioti P, Gurr SJ (2009) Against the grain: safeguarding rice from rice blast disease. Trends Biotechnol 27(3):141–150. https://doi.org/10.1016/j.tibtech.2008.12.002

Sontakke BK, Dash AN (2000) Field efficacy of some new granular insecticides against major pests of rice. Indian J Entomol 62(4):353–357

Soomro AS, Mazari SN, Hulio MH, Soomro JA, Junejo GQ (2020) Efficacy of different insecticides against rice leaf folder (*Cnaphalorosis medinalis*) under field conditions. Int J Appl Sci Biotechnol 8(2):211–215. https://doi.org/10.3126/ijasbt.v8i2.29582

Subramaniam S, Loganathan K, Thiruvengadam R, Ramasamy S (2013) Ecofriendly approaches of potential microbial bioagents in management of sheath rot disease in rice caused by *Sarocladium oryzae* (Sawada). Plant Pathol J 12(2):98–103

Sujithra M, Chander S (2013) Simulation of rice brown plant-hopper, *Nilaparvata lugens* (*Stal.*) population and crop-pest interactions to assess climate change impact. Clim Chang 121(2):331–347. https://doi.org/10.1007/s10584-013-0878-1

Sunder S, Singh R, Dodan DS (2005) Management of bacterial blight of rice with botanical extracts and non-conventional chemicals. Plant Dis Res 20(1):12–17

Sunder S, Singh RAM, Agarwal R (2014) Brown spot of rice: an overview. Indian Phytopathol 67(3):201–215. https://doi.org/10.22271/tpr.2019.v6.i1.021

Sundin GW, Castiblanco LF, Yuan X, Zeng Q, Yang CH (2016) Bacterial disease management: challenges, experience, innovation and future prospects: challenges in bacterial molecular plant pathology. Mol Plant Pathol 17(9):1506–1518. https://doi.org/10.1111/mpp.12436

Syed-Ab-Rahman SF, Carvalhais LC, Omar D (2020) Development of plant-based emulsion formulations to control bacterial leaf blight and sheath brown rot of rice. Heliyon 6(1):1–8. https://doi.org/10.1016/j.heliyon.2019.e03151

Tandon V, Srivastava A (2018) Efficacy of some insecticides and biopesticides for the management of white stem borer, *Scirpophaga fusciflua* Hampson in paddy. Agric Sci Digest- Res J 38(1):55–57. https://doi.org/10.18805/ag.D-4693

Tang R, Babendreier D, Zhang F, Kang M, Song K, Hou ML (2017) Assessment of *Trichogramma japonicum* and *T. chilonis* as potential biological control agents of yellow stem borer in rice. Insects 8(1):1–12. https://doi.org/10.3390/insects8010019

Tanwar RK, Prakash A, Panda SK, Swain NC, Garg DK, Singh SP, Sathyakumar S, Bambawale OM (2010) Rice swarming caterpillar (*Spodoptera mauritia*) and its management strategies. Tech Bull 24:1–20

Thimmegowda PR, Arun S, Ambika DS, Prasad PS, Chandrashekar M (2012) Efficacy of antibiotics and biorational pesticides against bacterial blight of paddy. Madras Agric J 99(7/9):592–596

Tirnaz S, Batley J (2019) DNA methylation: toward crop disease resistance improvement. Trends Plant Sci 24(12):1137–1150. https://doi.org/10.1016/j.tplants.2019.08.007

Torres MA, Lumansoc J, Demayo CG (2010) Variability in head shapes in three populations of the rice bug *Leptocorisa oratorius* (*Fabricius*) (Hemiptera: *Alydidae*). Egypt Acad J Biol Sci A Entomol 3(1):173–184. https://doi.org/10.21608/eajbsa.2010.15252

Tran TT, Nguyen MN, Phan MN, Nguyen MH, Szurek B, Koebnik R, Le-Huy H, Viet CH, Cunnac S (2008) Confirmation of bacterial leaf streak of rice caused by *Xanthomonas oryzae pv. oryzicola* in Vietnam. J Appl Microbiol 31:366. https://doi.org/10.1094/PDIS-03-15-0289-PDN

Tyagi H, Rajasubramaniam S, Rajam MV (2008) RNA-interference in rice against rice tungro bacilliform virus results in its decreased accumulation in inoculated rice plants. Transgenic Res 17:897–904. https://doi.org/10.1007/s11248-008-9174-7

Van Hop D, Hoa PT, Quang ND, Ton PH, Ha TH, Van Hung N, Van NT, Van Hai T, QUY NT, Dao NT, Thom VT (2014) Biological control of *Xanthomonas oryzae pv. oryzae* causing rice bacterial blight disease by *Streptomyces toxytricini* VN08-A-12, isolated from soil and leaf-litter samples in Vietnam. Biocontrol Sci 19(3):103–111. https://doi.org/10.4265/bio.19.103

Van-Nguyen N, Ferrero A (2006) Meeting the challenges of global rice production. Paddy Water Environ 4(1):1–9. https://doi.org/10.1007/s10333-005-0031-5

Velusamy P, Immanuel JE, Gnanamanickam SS, Thomashow L (2006) Biological control of rice bacterial blight by plant-associated bacteria producing 2, 4-diacetylphloroglucinol. Can J Microbiol 52(1):56–65. https://doi.org/10.1139/w05-106

Velusamy P, Immanuel JE, Gnanamanickam SS (2013) Rhizosphere bacteria for biocontrol of bacterial blight and growth promotion of rice. Rice Sci 20(5):356–362. https://doi.org/10.1016/S1672-6308(13)60143-2

Vidhyasekaran P, Ling DH, Borromeo ES, Zapata FJ, Mew TW (1990) Selection of brown spot-resistant rice plants from *Helminthosporium oryzae* toxin-resistant calluses. Ann Appl Biol 7(3):515–523. https://doi.org/10.1111/j.1744-7348.1990.tb04818.x

Vidhyasekaran P, Kamala N, Ramanathan R, Rajappan K, Paranidharan V, Velazhahan R (2001) Induction of systemic resistance by Pseudomonas fluorescens Pf1 against Xanthomonas oryzae pv. oryzae in rice leaves. Phytoparasitica 29:155–166. https://doi.org/10.1007/BF02983959

Vromant N, Rothuis AJ, Cuc NT, Ollevier F (1998) The effect of fish on the abundance of the rice caseworm *Nymphula depunctalis* (*Guenee*) (Lepidoptera: *Pyralidae*) in direct seeded, concurrent rice-fish fields. Biocontrol Sci Tech 8(4):539–546. https://doi.org/10.1080/09583159830054

Wang PY, Zhou L, Zhou J, Wu ZB, Xue W, Song BA, Yang S (2016) Synthesis and antibacterial activity of pyridinium-tailored 2, 5-substituted-1, 3, 4-oxadiazole thioether/sulfoxide/sulfone derivatives. Bioorg Med Chem Lett 26(4):1214–1217. https://doi.org/10.1016/j.bmcl.2016.01.029

Wang K, Peng Y, Chen J, Peng Y, Wang X, Shen Z, Han Z (2019a) Comparison of efficacy of RNAi mediated by various nanoparticles in the rice striped stem borer (*Chilo suppressalis*). Pestic Biochem Physiol 165:1–38. https://doi.org/10.1016/j.pestbp.2019.10.005

Wang S, Gan X, Wang Y, Li S, Yi C, Chen J, He F, Yang Y, Hu D, Song B (2019b) Novel 1, 3, 4-oxadiazole derivatives containing a cinnamic acid moiety as potential bactericide for rice bacterial diseases. Int J Mol Sci 20(5):10–20. https://doi.org/10.3390/ijms20051020

Wang ZZ, Liu YQ, Min SH, Huang JH, Chen XX (2019c) Parasitoid wasps as effective biological control agents. J Integr Agric 18(4):705–715. https://doi.org/10.1016/S2095-3119(18)62078-7

Wei Y, Li L, Hu W, Ju H, Zhang M, Qin Q, Zhang S, Li G (2020) Suppression of rice blast by bacterial strains isolated from cultivated soda saline-sodic soils. Int J Environ Res Public Health 17(14):1–12. https://doi.org/10.3390/ijerph17145248

Wilson RA, Talbot NJ (2009) Under pressure: investigating the biology of plant infection by *Magnaporthe oryzae*. Nat Rev Microbiol 7(3):185–195. https://doi.org/10.1038/nrmicro2032

Xu W, He J, He M, Han F, Chen X, Pan Z, Wang J, Tong M (2011) Synthesis and antifungal activity of novel sulfone derivatives containing 1, 3, 4-oxadiazole moieties. Molecules 16(11):9129–9141. https://doi.org/10.3390/molecules16119129

Xu S, Pan X, Luo J, Wu J, Zhou Z, Liang X, He Y, Zhou M (2015a) Effects of phenazine-1-carboxylic acid on the biology of the plant-pathogenic bacterium *Xanthomonas oryzae pv. oryzae*. Pestic Biochem Physiol 117:39–46. https://doi.org/10.1016/j.pestbp.2014.10.006

Xu HX, Zheng XS, Lu ZX (2015b) No. 34 Striped stem borer, Chilo suppressalis. vol 34. Institute of Plant Protection, Chinese Academy of Agricultural Sciences & China Society of Plant Protection. Crop Diseases and Insect Pests in China, China Agriculture Press, Beijing, pp 124–130

Yamaguchi I (2004) Overview on the chemical control of rice blast disease. In: Kawasaki S (ed) Rice blast: interaction with rice and control. Springer, Dordrecht, pp 1–13. https://doi.org/10.1007/978-0-306-48582-4_1

Yang L, Wang S, Ma Q, Song Z, Hou R, Huang S, Cheng D, Zhang Z (2020) Fabrication of sulfoxaflor-loaded natural polysaccharide floating hydrogel microspheres against *Nilaparvata lugens* (Stal) in rice fields. Pest Manag Sci 76:1–10. https://doi.org/10.1002/ps.5855

Yasin SI, Shah JA, Ali SS, Kanwal F (2007) Effect of different chemicals and antibiotics on bacterial leaf blight (*Xanthomonas oryzae pv. oryzae*) of rice. Pak J Phytopathol 19(2):211–213

Yasmin S, Zaka A, Imran A, Zahid MA, Yousaf S, Rasul G, Arif M, Mirza MS (2016) Plant growth promotion and suppression of bacterial leaf blight in rice by inoculated bacteria. PLoS One 11(8):1–19. https://doi.org/10.1371/journal.pone.0160688

Yasmin S, Hafeez FY, Mirza MS, Rasul M, Arshad HM, Zubair M, Iqbal M (2017) Biocontrol of bacterial leaf blight of rice and profiling of secondary metabolites produced by rhizospheric *Pseudomonas aeruginosa* BRp3. Front Microbiol 8:1–23. https://doi.org/10.3389/fmicb.2017.01895

Yoshida S (1981) Fundamentals of rice sciences. The International Rice Research Institute, Manila

Yuan XH, Song LW, Zhang JJ, Zang LS, Zhu L, Ruan CC, Sun GZ (2012) Performance of four Chinese *Trichogramma* species as biocontrol agents of the rice striped stem borer, *Chilo suppressalis*, under various temperature and humidity regimes. J Pest Sci 85(4):497–504. https://doi.org/10.1007/s10340-012-0456-8

Yugander A, Ladhalakshmi D, Srinivas PM, Laha GS (2015) Evaluation of selected plant extracts against *Xanthomonas oryzae pv. oryzae* causing bacterial blight of rice. World J Pharm Res 4(10):1555–1561

Zhang H, Wang S (2013) Rice versus *Xanthomonas oryzae pv. oryzae*: a unique pathosystem. Curr Opin Plant Biol 16:188–195. https://doi.org/10.1016/j.pbi.2013.02.008

Zhang Q, Li J, Xue Y, Han B, Deng XW (2008a) Rice 2020: a call for an international coordinated effort in rice functional genomics. Mol Plant 1(5):715–719. https://doi.org/10.1093/mp/ssn043

Zhang S, Luo HG, Zhang QD, Xu YY, Zou CH, Guo ML, He SM, Fang XW, Zhang JX, Chen QZ (2008b) Effects of nitrogen and potassium fertilizer applications on yield and occurrence of major diseases and insect pests of rice. J Huazhong Agric Univ 27:732–735

Zhang RS, Liu YF, Luo CP, Wang XY, Liu YZ, Qiao JQ, Yu JJ, Chen ZY (2012) *Bacillus amyloliquefaciens* lx-11, a potential biocontrol agent against rice bacterial leaf streak. J Plant Pathol 94(3):609–619. http://www.jstor.org/stable/45156288

Zhang MZ, Mulholland N, Beattie D, Irwin D, Gu YC, Chen Q, Yang GF, Clough J (2013) Synthesis and antifungal activity of 3-(1, 3, 4-oxadiazol-5-yl)-indoles and 3-(1, 3, 4-oxadiazol-5-yl) methyl-indoles. Eur J Med Chem 63:22–32. https://doi.org/10.1016/j.ejmech.2013.01.038

Zhang J, Hou W, Ren T, Cong R, Zhao Z, Lu J, Li X (2020a) Applying potassium fertilizer improves sheath rot disease tolerance and decreases grain yield loss in rice (*Oryza sativa L.*). Crop Prot 20:1–36. https://doi.org/10.1016/j.cropro.2020.105392

Zhang YC, Feng ZR, Zhang S, Pei XG, Zeng B, Zheng C, Gao CF, Yu XY (2020b) Baseline determination, susceptibility monitoring and risk assessment to triflumezopyrim in *Nilaparvata lugens* (Stal). Pestic Biochem Physiol 20:1–27. https://doi.org/10.1016/j.pestbp.2020.104608

Zheng YT, Zhang TT, Wang PY, Wu ZB, Zhou L, Ye YQ, Zhou X, He M, Yang S (2017) Synthesis and bioactivities of novel 2-(thioether/sulfone)-5-pyrazolyl-1, 3, 4-oxadiazole derivatives. Chin Chem Lett 28(2):253–256. https://doi.org/10.1016/j.cclet.2016.06.055

Zhou Y, Bao J, Zhang D, Li Y, Li H, He H (2020) Effect of heterocystous nitrogen-fixing cyanobacteria against rice sheath blight and the underlying mechanism. Appl Soil Ecol 153:1–9. https://doi.org/10.1016/j.apsoil.2020.103580

Zhu J, Zhu Z, Feng J, Cai X, Zhong X, Cheng J (2011) Effect of sowing/transplanting time on occurrence of main locally-overwintering insect pests and diseases. Acta Agric Zhejiang 23(2):329–334

Zibaee A, Bandani AR, Kafil M, Ramzi S (2008) Characterization of a-amylase in the midgut and the salivary glands of rice striped stem borer, *Chilo suppressalis* Walker (Lepidoptera: *Pyralidae*). J Asia Pac Entomol 11(4):201–205. https://doi.org/10.1016/j.aspen.2008.09.003

Chapter 6
Comet Assay: Is it a Sensitive Tool in Ecogenotoxicology?

Meenakshi Sundari Rajendran, Rajkumar Prabhakaran, Sivanandam Vignesh, and Baskaran Nagarathinam

Abstract The soil is the resident of the biotic components of the terrestrial ecosystem. Due to industrialization and other human activities, organic pollutants accumulated in the soil; these accumulated soil pollutants were interacted with the biotic components' genome and cause genotoxicity in the genetic material and led to various diseases. Hence, there is an urgent necessity to monitor the genetic toxicity in the terrestrial ecosystem or soil-based organism. The genotoxicity was assessed by multiple methods, viz. Ames assay and micronucleus assay; however, these methods have some pitfalls. Recently, the simple, sensitive, and rapid technique called single-cell gel electrophoresis or comet assay for detecting DNA damage caused by the pollutants in the terrestrial ecosystem models such as earthworms, plants and mammals were developed. Therefore, in this chapter, we have discussed the types of ecosystem, genotoxicity versus cytotoxicity, genotoxicity assay types, the importance of genotoxicity assay, comet assay protocol for the identification of genotoxicity, applications of comet assay as a tool in ecogenotoxicology and its application in translational research.

Keywords Ecosystem · Genotoxicity · Organic pollutants · Genotoxicants · Comet assay

M. S. Rajendran
Department of Biochemistry, Karpagam University, Coimbatore, Tamil Nadu, India

R. Prabhakaran
Research and Development, DSIR approved lab, VVD and Sons Private limited, Thoothukudi, Tamil Nadu, India

S. Vignesh · B. Nagarathinam (✉)
Department of Technology Dissemination, Indian Institute of Food Processing Technology (IIFPT), Thanjavur, Tamil Nadu, India
e-mail: vignesh@iifpt.edu.in; baskaran@iifpt.edu.in

K. M. Gothandam et al. (eds.), *Environmental Biotechnology Volume 4*, Environmental Chemistry for a Sustainable World 68,
https://doi.org/10.1007/978-3-030-77795-1_6

6.1 Introduction

In 1935, Tansley (Botanist) coined the term "Ecosystem". An ecosystem can be defined as interactions between living and nonliving components in their environment systemically. An ecosystem can be divided into two types: 1. Terrestrial ecosystem and 2. aquatic ecosystem. The terrestrial ecosystem is land-based, whereas; an aquatic ecosystem is water-based interactions between the biotic and abiotic components. The biotic and abiotic components interactions occur via nutrient cycling and energy flow. The terrestrial ecosystem's major component is soil, a complex and dynamic medium of the earth's biosphere. The properties of soil are mainly based on the mineralogical composition of the parent rock, and that it also depends on biological activity, climate effects and topography (Weber et al. 2012). Mineral of the soil ranges from extremely fine clay particles to coarse sand (Table 6.1). The soil's organic materials come by complete or partial decomposition of plant and/or animal tissues and are achieved by microbial biota. Even though 2–5% is the volumetric basis of organic matter in the soil, the values can range from 1% to the highest range of 80%; and this highly organic composition of the soil is called peat (Winegardner 1995; Pollumaa et al. 2000).

Nowadays, the ecosystem was contaminated by organic pollutants due to anthropogenic waste accumulation and industrial waste accumulation (Schnaak et al. 1997; El-Shahawi et al. 2010; Rajkumar et al. 2017; Prabhakaran et al. 2018). When the organic materials undergo pyrolysis by the industrial chemical release, sewage treatments, chemical fertilizers in agricultural lands result in a higher concentration of polyaromatic hydrocarbon. The chemical structure of different types of polyaromatic hydrocarbons is shown in Fig. 6.1. Ingestion and dermal contact of the polyaromatic hydrocarbon contaminated soil develop cancers. In India, it has been estimated that exposure of polyaromatic hydrocarbons about 74% through the dermal pathway and 26% of exposure occurred through soil ingestion in children; in adults, 84% exposure of polyaromatic hydrocarbon took the dermal pathway and 16% exposure occurred via soil ingestion (Wang et al. 2011; Ferreira-Baptista and De Miguel 2005). Chlorinated hydrocarbons are used as fumigants, pesticides, industrial solvents, glues & nail polish remover etc. and cause chromosomal

Table 6.1 Volumetric basis of soil

Volumetric basis of soil	**In %**
Mineral	45
Organic material	2–5
Air	20–30
Water	20–30
Types of soil based on particle size	
Soil type	**Particle size in diameter (mm)**
Sand	0.5–2
Silt	0.002–0.5
Clay	<0.002

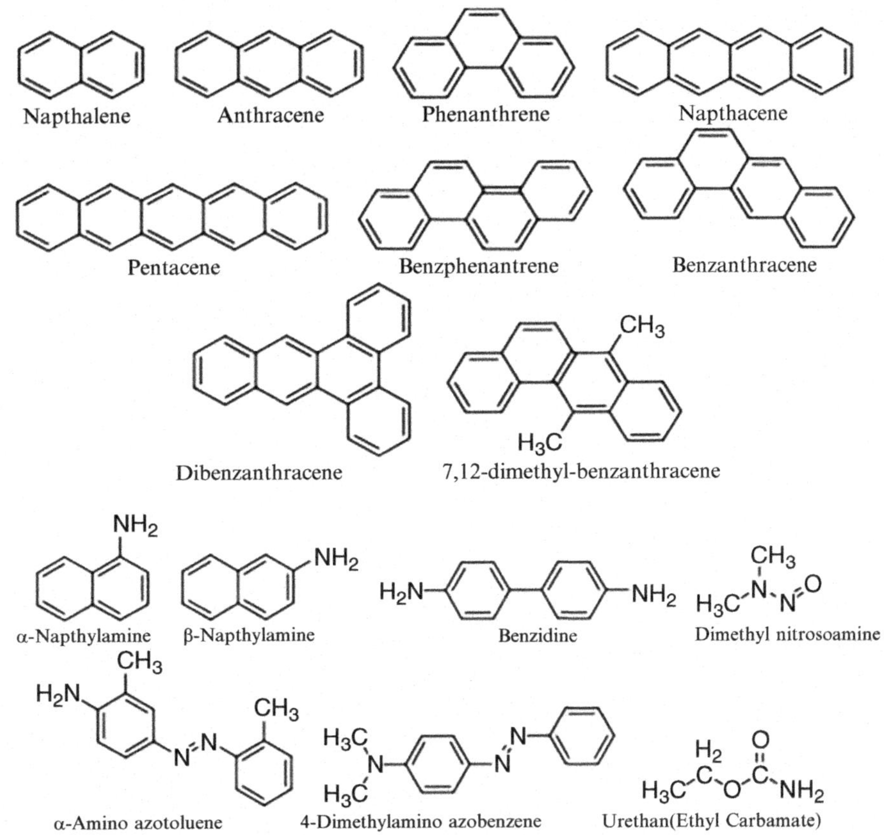

Fig. 6.1 Chemical structure of carcinogenic chemicals

aberration. α- and β napthylamine, Benzidine are examples of aromatic amines suspected to be a carcinogenic compound. An inorganic substance such as nickel, arsenic and chromium are well-known carcinogens. These accumulated soil pollutants caused the genotoxicity of the biotic components in the terrestrial ecosystem.

Genotoxicity is a chemical/molecule that causes damage in the genetic material and structural modification of the genetic material or chromosomal aberrations and a mutation in the genes leading to various diseases. In general, the substance responsible for damaging the genetic material is called genotoxins or genotoxicants. Various methods assessed the genotoxicity in the soil organism it includes the Ames test, micronucleus and comet assays. Nevertheless, the simple, sensitive, and rapid technique called single-cell gel electrophoresis or simply as comet assay, to detect DNA damage caused by the pollutants in the terrestrial ecosystem models earthworm, plants and mammals are enormously useful (Sarrif et al. 2006). Therefore, in this chapter, we brief about the ecosystem, genotoxicity, importance of genotoxicity assay, comet assay protocol for identifying genotoxicity and its applications.

6.2 Ecosystem

Several demands of human life completely rely on the services provided by the earth, and the components of the ecosystem provide sustainability and fulfill humans' needs. The ecosystem provides pollination, biological pest control, soil structure and fertility, nutrient cycling, and hydrological services for humans' well-being. The land-based ecosystem provides food, shelter etc., human civilization and industrialization, and developmental activities resulted in a daunting challenge to ecosystem maintenance. The ecosystem gains energy from the sun and the materials from the rocks or air and transfers the energy to the living organisms within the ecosystem; then, they release the energy and materials to the environment again. This energy flow is cycling between biotic and abiotic components (Cotelle and Férard 1999).

6.2.1 Biotic Components

Plants, animals and decomposers are the essential biotic components of the terrestrial ecosystem. Plants can produce their food by capturing solar energy and synthesizing carbohydrates are called photosynthesis; thereby, bringing carbon into the ecosystem. Thus, they are the primary producers of energy in the ecosystem. Animals are the consumers of the energy produced by plants, and they are the critical components of ecosystems. The animals transfer energy and materials to the ecosystem by influencing plants and soil microbes' quantity and activities. Decomposer microorganisms are the crucial biotic component of the ecosystem. They are involved in breaking down dead organic material, thereby releasing CO_2 to the atmosphere and nutrients in forms available to other microbes and plants. Decomposition of dead organic matter is essential for the energy flow and nutrient cycling to the plants and other biotic components (Hättenschwiler and Vitousek 2000).

6.2.2 Abiotic Components

Abiotic components are the non-living components essential for the living components to survive/live on the earth. Water, atmosphere and soil are the crucial abiotic factors.

Water: Essential component for all living things in the ecosystem. They provide hydrogen and oxygen.

Atmosphere or Air: Atmosphere is the supplier of carbon and nitrogen to the ecosystem.

Soil: It supplies all the essential minerals such as potassium, magnesium, sodium, calcium etc., and other nutrients required by organisms.

6.2.3 Terrestrial Ecosystem

The major part of the terrestrial ecosystem is soil, a complex and dynamic medium of the earth's biosphere. The properties of soil are mainly based on the mineralogical composition of the parent rock, and that it also depends on biological activity, climate effects and topography.

Complete or partial decomposition of plant and animal tissues forms the soil's organic material, achieved by microbial biota. Even though 2–5% is the volumetric basis of organic matter in the soil, the values can range from 1% to the highest range of 80%; and this highly organic composition of the soil is called peat (Winegardner 1995). The plant that grows in the soil depends on the quality and the texture of the soil. Soil should have the following properties to support the plant growth: (1) soil must have an adequate supply of nutrients and an excellent nutrient-holding capacity; (2) it should allow infiltration, have an exceptional water-holding ability, and resist evaporative water loss; (3) it must have a porous structure that permits good aeration; (4) pH should be near neutral; and (5) they must have a low salt content. Humus is the residue of partly decomposed organic matter, and it is highly concentrated in the bottom of the "O- horizon". Composting is defined as processing and recycling various wastes by fostering organic waste decay under more or less controlled conditions. The resulting compost is similar to humus. Soil structure is maintained by the animals that feed the detritus present on or in the soil. For instance, earthworms said to be the farmer's friend feed on detritus, and they ingest mineral soil particles. The minerals in the guts of earthworms are glued, and their excretions are relatively stable clumps of inorganic particles plus humus. The burrowing activity of earthworms loses the clumps, which will soften the soil structure that makes the soil structure ideal for infiltration, aeration, and workability.

6.3 Human Activities Influence the Ecosystem

The ecosystem is always being interconnected functions, and they can't function in isolated forms. Because of this interconnection, we humans of the ecosystem have a major influence on it. Human activities influence the ecosystem either directly or indirectly and make drastic changes to the environment such as global warming, nitrogen deposition, erosion, diversions- soil and water resource; fire control and land-use change in disturbance regime and species introduction and extinctions of organisms for functional types.

6.3.1 Organic Pollutants Due to Industrialization

Novel chemicals are being introduced into the environment due to industrialization. For example, synthetic organic chemicals include Dichlorodiphenyltrichloroethane and polychlorinated biphenyls. Polyaromatic hydrogens are the industrial compounds that have major impacts on the land-based ecosystem as they are mixing with the soil and changing the soil texture quality and fertility. When these organic materials underwent pyrolysis by the industrial chemical release, sewage treatments, chemical fertilizers in agricultural land result in a higher concentration of polyaromatic hydrogen. The chemical structure of different types of polyaromatic hydrogens is given below. Ingestion and dermal contact of the polyaromatic hydrogen contaminated soil develop cancers. In India, it has been estimated that exposure of polyaromatic hydrogens via dermal pathway is 74%, and the rest of the 26% of exposure occurred via soil ingestion in children; 84% exposure took the dermal pathway, and 16% exposure occurred through soil ingestion in adults (Wang et al. 2011; Ferreira-Baptista and De Miguel 2005; Zhang 2015).

6.4 Genotoxicity Versus Cytotoxicity

Genotoxicity is defined as a chemical/molecule that causes damage in the genetic material and structural modification of the genetic material or chromosomal aberrations and a mutation in the genes leading to various diseases. Cytotoxicity is the term defined as cellular level damage caused by toxins. Toxins are the substance which is toxic to biotic components. It may be chemical/molecule causing environmental pollution in abiotic components such as soil, air, water. This toxic pollutant of the abiotic system will interact with the biotic system causing toxicity. The study that deals with toxicity in the environment are called ecotoxicology.

6.4.1 Mutagens Versus Carcinogens

A chemical that causes mutations is called "mutagens". The mutagens that cause cancer are called oncogenes. Teratogens are the toxins that cause birth defects or toxins that cause physical and functional defects in an embryo. In general, the substance responsible for damaging the genetic material is called genotoxins (Fig. 6.2). α- and β napthylamine, Benzidine are examples of aromatic amines suspected to be a carcinogenic compound. Inorganic substances such as nickel, arsenic and chromium are well-known carcinogens.

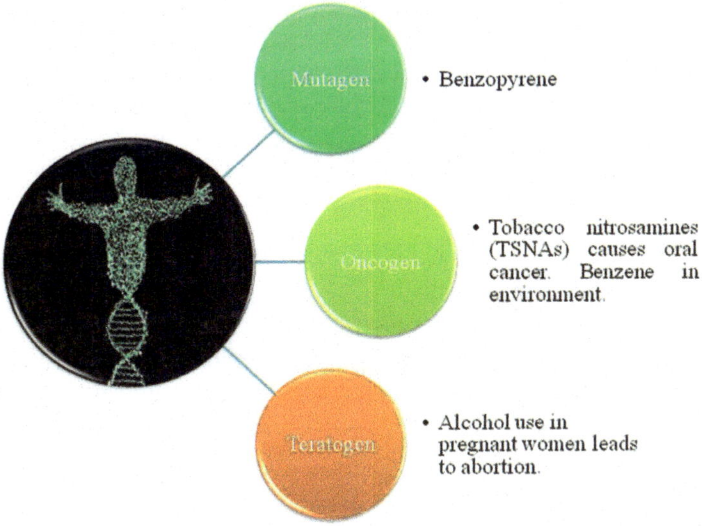

Fig. 6.2 Diagrammatic representation of classification of genotoxicity concepts with examples

6.4.2 Genotoxicity Assay and Its Types

To know whether the novel chemical impacts the ecosystem's organisms, we need to measure the level of damage in an organism's genome. The method which measures the level of toxicity in the genome is called genotoxicity assay.

Genotoxicity assay carried out by different methods starting from the screening of compounds to testing the level of damage in the cell division and DNA (Ames 1983).

Includes the following different types of methods:

1. Ames test (microbial screening for mutation)
2. Micronucleus assay (Chromosomal aberrations at the cellular level)
3. Comet assay (measures DNA damage).

6.4.3 Importance of Genotoxicity Assay

Genotoxicity assay is important for testing new chemical entities of pre-clinical studies. It is useful to ensure drug safety. Data of new chemical entities tested for genotoxicity will assess the drug application and predict that it is risky or safe to use in humans. New chemical entities hazardous can be detected by studying chromosomal aberrations, recombination, structural modifications of gene and mutations. The test can also say the chemical's potentiality and characterize the chemical as oncogenic/mutagenic (Baskaran et al. 2011).

6.4.3.1 Ames Test

Strains of *Salmonella typhimurium* and *Escherichia coli* Strains have defined mutations, and they are used for making require specific amino acids *S. typhimurium* – histidine. *E. coli* –tryptophan.

Back mutation of their genes allows growth in medium without histidine or tryptophan.

6.4.3.2 Micronucleus Assay

Micronucleus can be defined as a small nucleus that forms whenever a chromosome or a chromosome fragment is not incorporated into the daughter nuclei during cell division.

Micronucleus Induction

Micronuclei are formed from acentric fragments or whole chromosomes, which fail to attach to the mitotic spindle. The acentric fragments, associated with two hit aberrations or terminal or interstitial deletions, are not attached to the mitotic spindle. During anaphase separation of the chromatids, these fragments are excluded from the daughter nuclei. They may appear as small membrane-enclosed chromatin bodies outside the nucleus; these are known as micronuclei. They can be identified in the daughter cells' cytoplasm as small round or oval bodies with the same staining properties as the main nucleus. Savage, in his chromosome 'clock' or 'web,' beautifully illustrates the relationship between chromosomal aberrations and micronuclei. Micronucleus test is considered a simple alternative to chromosomal aberration analysis for the genotoxicity testing of radiation and other chromosome and spindle toxins.

The earlier methods used to study micronuclei could not distinguish between cells that have divided only once or have completed more than one division after irradiation. Fenech (1993, 2007) developed a method to score micronuclei in cultured human lymphocytes, which have divide only once after treatment, by using a cytokinesis blocking agent, Cytochalasin B. They defined the following criteria for identifying micronuclei in Cytochalasin treated binucleated peripheral lymphocytes: Morphologically identical but smaller than the main nuclei, diameter between 1/16 and 1/3 of the main nuclei, non-refractile. Not linked to the main nuclei via a nucleoplasmic bridge may sometimes overlap the main nuclei's boundaries. However, some types of cells undergo nuclear budding, which could be mistaken for micro nucleation. It may be prudent to exclude micronuclei, which contact the main nuclei from scoring in such cases.

Merits of Micronuclei Test

Micronuclei technique has several advantages over the chromosomal aberration scoring, like, detection of a eugens, easiness of scoring, the possibility of automation, statistical power and low chance of false-positive results, because micronuclei can occur only in dividing cells. The comet and micronuclei tests' positives are due to different mechanisms; the comet assay identifies reparable injuries or alkali-label sites while the micronuclei test detects injuries that survive at least one mitotic cycle (Van Goethem et al. 1997).

6.4.3.3 Comet Assay

During the last few decades, there has been a great interest in developing a rapid and simple and economical method to identify exposure to environmental agents that can affect individual health due to DNA damage. The comet assay is a rapid and sensitive technique to measure sites sensitive to basic pH and DNA breaks in individual cells. This method was described for the first time by Ostling and Johanson (1984) and Singh et al. (1988) by introducing alkaline conditions to this technique.

The assay technique consists of evaluating cells kept in agarose gel on a microscope slide, submitted to electrophoresis and dyed with ethidium bromide or acridine orange. Cells with damaged DNA form a comet, consisting of a head, and a tail, formed by DNA fragments (Fig. 6.3). The amount of DNA that has migrated is correlated with the damage (Singh et al. 1988; Morillas et al. 2002). Comet assay is used extensively in Medicine and Toxicology due to its capacity and sensitivity in demonstrating DNA breaks, both single and double breaks, and alkali-label sites; thus, it also has wide application in biology (Fairbain et al. 1995a; Stavreva et al. 1998). The alkaline conditions cause the separation of the paired bases, enabling the detection of simple chain ruptures (McKelvey-Martin et al. 1993; Albertini and Kirsch-Volders 1997).

Calculation of DNA Comets

For calculating the DNA damage in cells by measuring the tail length of comet slides, there has been a number of commentary through research has been published. Collins et al. (2004), in his "The comet assay for DNA damage and repair: principles, applications, and limitations," given that it is important to avoid bias in the identification of cells to measure comets. Cells with large tails may overlap and thus may not be selected for measurement that could violate the assumptions underlying statistical tests that the cells represent a random sample. Three DNA migration measures are commonly used: tail length, tail moment and % of the DNA in the tail. Metrics on the tail or head length and moment are measured in arbitrary units and may vary from study to study or from laboratory to laboratory.

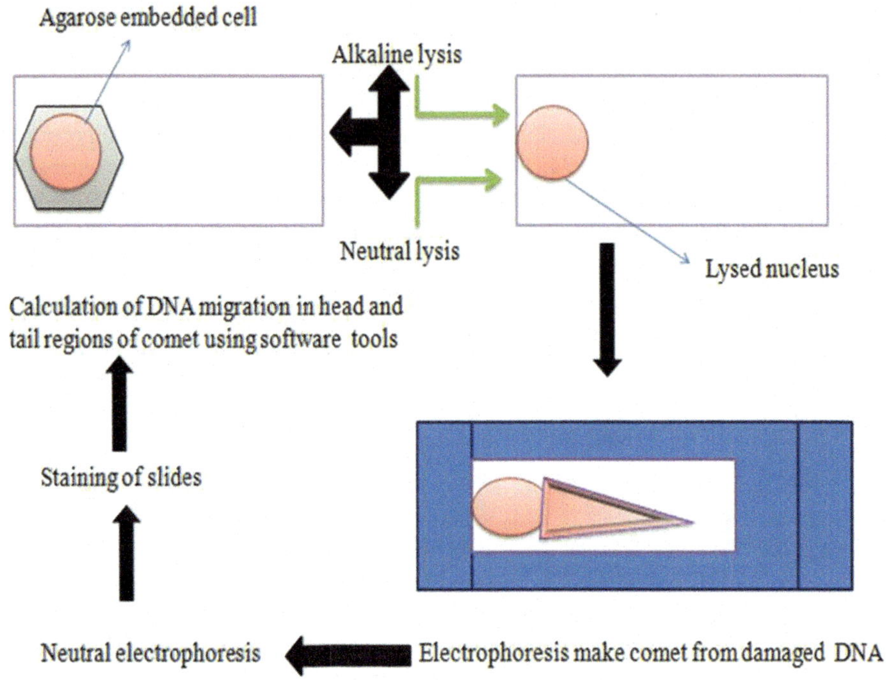

Fig. 6.3 Schematic summary of the comet assay protocol

Tail length is considered unsatisfactory as a measure because the length only increases at relatively low damage levels and is sensitive to the background intensity of the image analysis system, which affects the criteria for determining the end of the tail. Tail moment, an index taking account both the migration of the genetic material and the relative amount of DNA in the tail, can be calculated in a number of ways. For instance, tail length and the % tail DNA are called an Olive tail moment. The percentage of tail DNA is a measure of the relative fluorescent intensity in the head and tail. The percentage of DNA tail values are constrained to a maximum of 100 and a minimum of 0 with no variability at the extremes and maximum variability at intermediate values such as 50%. Hartmann et al. (2003), for instance, describes that 'there is much to recommend the use of tail DNA percentage'. Relative tail intensity (the % tail DNA) was linearly related to DNA damage over a wide range of damage and DNA break frequency. % tail DNA covered a wide range of damage (from 0 to100%), and Collins viewed it as the most useful measure. The comet assay tail DNA % was independent of the image analysis program's threshold settings and gave the 'feel' about how the comet looks like? Bowden et al. (2003) developed a 'tail profile', which identifies more DNA damage than measured by the tail moment. The group of scientist derived a 'profile plot', which represented a visual image of a series of comets on a slide, and enabled to identify features in the data that were not otherwise apparent.

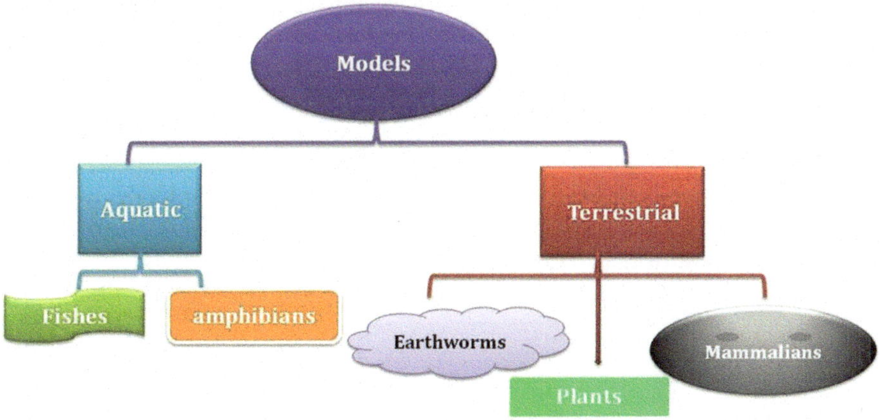

Fig. 6.4 Classification of models for genotoxicity monitoring in ecotoxicology

Applications of Comet Assay

The single-cell gel electrophoresis or comet assay is one such state-of-the-art technique for quantitating DNA damage and repair *in vivo* and *in vitro* in any eukaryotic cell and some prokaryotic cells. This technique is rapid, non-invasive, sensitive, visual, and inexpensive compared to the conventional techniques and is a powerful tool to study factors modifying mutagenicity and carcinogenicity. It has rapidly gained importance in the fields of genetic toxicology and human biomonitoring. Collins et al. (2004) gave an overview of the comet assay's uses. The Comet workshop showed the range of applications from traditional genotoxicity *in vivo* and *in vitro* studies and had an application in ecotoxicology. Other examples of its use include investigations on industrial chemicals, pharmaceuticals, biocides, agrochemicals and food chemicals such as additives and biomarker study in cancer and nutrition studies. Aquatic and terrestrial approaches could assess environmental health. Figure 6.4 represents the models used for ecotoxicology research.

6.4.4 Terrestrial Ecotoxicology and Ecogenotoxicology

As the terrestrial ecosystem is land-based, the soil is the main abiotic component to monitor the pollutants that cause toxicity to biotic components. Persistent organic pollutants such as dichlorodiphenyltrichloroethane, polycyclic aromatic hydrocarbons, polychlorinated biphenyls, and hexachlorocyclohexane in soil cause pollution and toxicity. This pollution is caused due to the use of pesticides, motor vehicles, industrial activities, incomplete combustion in heating plants and so on (Manz et al. 2001; Voldner and Li 1995). Attention to be paid in risk assessment of soil mutagens and carcinogens has potentially detrimental effects on microbial, animal population and human health (Chapman et al. 2002). Screening and identifying chemicals that cause toxicity in the soil provide biomarker information, a valuable ecotoxicological tool (Wells et al. 2001; Kloepper-Sams and Owens 1993). Single-cell gel electrophoresis or comet assay, a simple, sensitive, and rapid technique for detecting DNA damage (strand breaks, modified bases, and DNA-DNA and DNA-protein cross-links), is a widely used technique. In vivo testing of the cells exposed to toxins in the environment, their genetic materials are tested for measuring the toxicological effects of the toxicants (Fairbairn et al. 1995a), and the study is called ecogenotoxicology (Fairbairn et al. 1995b).

6.4.4.1 Earthworm Studies

The role of earthworms in agriculture and in maintaining soil fertility is inevitable. Today the farmers are using more pesticides and herbicides to protect the crops. The usage of pesticides and insecticides causes soil infertility, and toxicity to the organism depends on it. Many studies have reported using earthworms as a model to study the genotoxicity of the pesticides and herbicides, and a comparison of different earthworms for sensitivity has also been studied. Chromium (VI) is highly toxic in vertebrates. Hexavalent chromium or chromium (VI) is a soil pollutant, and the major sources are electroplating industries and tanneries. Elevation of hexavalent chromium in soil caused by byproducts of coal combustion used in thermal power plants, sewage application in agriculture process and metal finishing industries (Molnár et al. 1989).

The earthworm is well studied as a model for heavy metal toxicity (Homa et al. 2003; Lukkari et al. 2004). Earthworms are suitable for eco-toxicity testing because they can be maintained easily in labs and have a short lifecycle, and most importantly, they are placed in the terrestrial ecosystem. Earthworms have been used to determine the toxicity caused by hexavalent chromium. As the earthworm population has a major role in terrestrial food chains, they could determine soil contamination and good indicator toxicity testing. Chromium metabolism study was well defined in the species *Eisenia fetida* (Arillo and Melodia 1991). Earthworm coelomocytes are the circulating leukocytes, and they play a major role in cellular defense mechanisms, antimicrobial defense, wound healing, encapsulation reactions, phagocytosis (Valembois et al. 1992, 1994; Cooper 1996; Cossarizza et al. 1996). Figure 6.5 Explain the protocol for coelomocytes extrusion from an earthworm for research.

Pinocytosis and plastic adherence immune systems of earthworms get affected due to different heavy metal exposures even at low levels (Homa et al. 2003). In *Pheretima posthuma,* the adverse effect of Cr(VI) on growth and reproduction was identified (Abbasi and Soni 1983). *In vitro* and *in vivo* exposures *of* Cr (VI) in *Dichogaster curgensis* were assessed using single-cell gel electrophoresis, i.e., comet assay, and the observation of the results revealed that in vitro exposure caused DNA damage in a dose-dependent manner whereas in vivo exposure showed unexpected results of decrease arbitrary units at higher concentrations 70 and 100 ppm

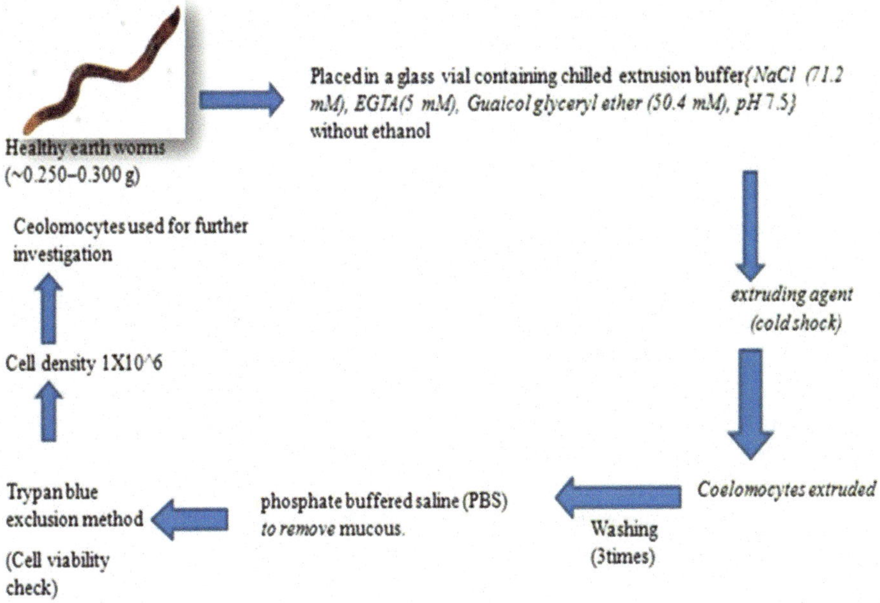

Fig. 6.5 Schematic explanation of the protocol for coelomocyte extrusion from an earthworm for research

of Cr (VI) while increased DNA damage up to 30 ppm arbitrary units. This speculation of the result may be due to different types of cross-links and lower DNA mobility (Vasseur and Bonnard 2014) performed comet assay for several earthworm comparative studies. An equivalent sensitivity has been observed in both *Eisenia fetida* and *Aporrectodea caliginosa*, as assessed by Comet Assay (Klobučar et al. 2011). Comparative assessment for the sensitivity to Cd genotoxicity after a 48 h-exposure of five earthworm species (*Amynthas diffringens, A. caliginosa, E. fetida, Dendrodrilus rubidus* and *Microchaetus benhami*) and the results indicated *D. rubidus* showed the highest increase in DNA breaks and *E. fetida* the second most sensitive species presented the highest percent of DNA in the tail, compared with the control (Fourie et al. 2007).

6.4.4.2 Comet Assay in Construction of Food Web

Using comet assay toxicity caused by radiation in the ecosystem of Japan's Fukushima nuclear power plant accident in 2011 was assessed, and the impact of radiation as studied at multi-trophic level approach. Comparison study between wild boars (*Sus scrofa leucomystax*) and earthworms (*Megascolecidae*) and level of atmospheric radiation with low (0.28μSv/h) and high (2.85μSv/h) using comet assay made a model food web construction. In that model food web, wild boar as the top predator because there was no difference in the genome of the low and high radiation-exposed animals, but in the case of earthworms had a significantly greater extent of DNA damage in high-level radiation regions than that of low-level radiation regions species (Fujita et al. 2014).

6.4.4.3 Plant Studies

Based on the DNA integrity, Comet assay has been used to study the effect of age of the plant (Koppen et al. 1999) as well as DNA repair kinetics (Gichner and Plewa 1998) in isolated nuclei from leaves of tobacco plants. A small but significant increase in DNA damage compared to controls was noted in heterozygous tobacco and potato plants grown on soil contaminated with heavy metal (Gichner et al. 2006). Plants are a very good model for assessing ecotoxicological effects in soil by performing comet assay in different parts of the plants such as leaf cell, root's nucleus and can be combined for genotoxicity studies. The protocol has to be specialized and challenging for various plan species and remains to be challenging. Despite the challenging task of performing comet assay in the plant, Table 6.2 will give details of research works in ecotoxicology using plants.

6.4.4.4 Avians and Rodents

Avian and rodents are sentinel species to study genotoxicants of the terrestrial ecosystem. Avians and rodents are good bioindicators to study the genotoxicants, particularly in disaster effect in genomes. One example of such a study was carried out to track the genomic damage in *Ciconia ciconia and Milvus migrans due to* a massive toxic spill of acid waste containing metals that occurred in the year 1998 threatened the wildlife in a National park in Spain. It was observed that exposed birds had an increased level of genotoxic damage significantly compared with controls from non-contaminated locations, and still, the toxic spill appears to be affecting the wildlife even after 4 years of the mining disaster; and based on the DNA damage detection by comet assay, it has been proved to be ineffective to clean up the spill waste (Pastor et al. 2001; Pastor et al. 2004; Baos et al. 2006). Different rodent

Table 6.2 Ecotoxicological research in plant by performing Comet assay

Plant species	Tested for	Research team
Pisum sativum	Cr(VI) effects	Rodriguez et al. (2011)
Allium cepa	Chlorfenvinphos and fenbuconazole effects	Türkoğlu (2012)
Tobacco	Cadmium-zinc (Cd-Zn) interactions	Tkalec et al. (2014)
Crepis capillaris	The DNA damage that is induced by the chemical mutagen maleic hydrazide (MH) demonstrates the correlation between the occurrence of B chromosomes and the mutagen.	Kwasniewska and Mikolajczyk (2014)

Table 6.3 Examples of rodent species studied in ecotoxicology using Comet Assay

Species name	Used to	Advantage of studying comet assay
Apodemus sylvaticus	1. Assess the effects of dumping sites 2. Urban or traffic pollution assessment (Borras and Nadal 2004)	Sensitive and reliable tool for the detection of exposure to environmental genotoxicants
Mus spretus	Comparison study between mice lives in the city and industrial area (Mateos et al. 2008)	Wild mice could be used as a valuable tool in pollution monitoring
Microtus pennsylvanicus	effects of pesticide exposure (Knopper et al. 2005)	Measure the genotoxicant levels
Ctenomys torquatus	Biomonitor the coal region (Erdtmann 2000; da Silva et al. 2000a, b)	Coal and byproducts not only induce DNA damage in blood cells but also in other tissues, mainly the liver, kidney, and lung.

species have been widely used to detect the eco-genotoxicity of various scenarios. Table 6.3 Summarize application of comet assay on different mice species in ecotoxicology.

6.4.4.5 Translational Ecotoxicology Research from Rodents to Humans: A Landmark Achievement

International Agency for Research on Cancer monographs and the U.S. National Toxicology Program carcinogenicity database selected 208 chemicals and studied their exposures in multiple mouse organs by using DNA strand breaks in the late 1990s (Azqueta and Collins 2013). After this research work, it made a landmark to research and laid the study to be validated by the Japanese Center for the Validation of Alternative Methods (JaCVAM) and recently an Organization for Economic Co-operation and Development (OECD) genotoxicity guidelines in 2016 adopted comet assay guidelines for the alkaline version of the in vivo comet assay [single cell gel (pH. 13)] in animals (Collins 2017).

6.5 Conclusion

Comet assay is a powerful tool for measuring the potentiality of genotoxic chemicals. Here we made some examples in the terrestrial ecosystem for studying genotoxicants using comet assay protocols. Undoubtedly, the comet assay is a cost-effective, rapid, and significant technique, and its applications are wider to the component of the ecosystem to study ecogenotoxicology from microbes to humans. Herein, we have discussed the application of comet assay in the genotoxicology of the terrestrial ecosystem. Comet assay is one of the most useful techniques to further research new chemical entries and toxins. However, further research has to be done about the pollutants of the ecosystem and their effects at genome levels.

Acknowledgments The author, Meenakshi Sundari R, thank Dr. Sunilkumar (Scientist), who taught the comet assay technique, Division of reproductive and Cytotoxicology, National Institute of Occupational Health, Ahmedabad, India.

Conflict of Interest The authors declare no potential conflict of interest.

References

Abbasi SA, Soni R (1983) Stress-induced enhancement of reproduction in earthworm *Octochaetus pattoni* exposed to chromium (vi) and mercury (ii)—implications in environmental management. Int J Environ Stud 22(1):43–47. https://doi.org/10.1080/00207238308710100

Albertini S, Kirsch-Volders M (1997) Summary and conclusions on the MNT in vitro and implication on testing strategies. Mutat Res/Genet Toxicol Environ Mutagen 392(1–2):183–185. https://doi.org/10.1016/S0165-1218%2897%2900056-6

Ames BN (1983) Dietary carcinogens and anticarcinogens. Oxygen radicals and degenerative diseases. Science 221:1256–1264. https://doi.org/10.1007/978-1-4613-2465-2_24

Arillo A, Melodia F (1991) Reduction of hexavalent chromium by the earthworm *Eisenia foetida* (Savigny). Ecotoxicol Environ Saf 21(1):92–100. https://doi.org/10.1016/0147-6513(91)90011-D

Azqueta A, Collins AR (2013) The essential comet assay: a comprehensive guide to measuring DNA damage and repair. Arch Toxicol 87(6):949–968. https://doi.org/10.1007/s00204-013-1070-0

Baos R, Jovani R, Pastor N, Tella JL, Jiménez B, Gómez G, Hiraldo F (2006) Evaluation of genotoxic effects of heavy metals and arsenic in wild nestling white storks (Ciconia ciconia) and black kites (Milvus migrans) from southwestern Spain after a mining accident. Environ Toxicol Chem Int J 25(10):2794–2803. https://doi.org/10.1897/05-570R.1

Baskaran N, Rajasekaran D, Manoharan S (2011) Coumarin protects 7,12-dimethylbenz(a) anthracene-induced genotoxicity in the bone marrow cells of golden Syrian hamsters. Int J Nutr Pharmacol Neurol Dis 1:167–173. https://doi.org/10.4103/2231-0738.84209

Borras M, Nadal J (2004) Biomarkers of genotoxicity and other end-points in an integrated approach to environmental risk assessment. Mutagenesis 19(3):165–168. https://doi.org/10.1093/mutage/geh023

Bowden CL, Joseph R Calabrese, Gary Sachs, Lakshmi N Yatham, Shaheen Akthar Asghar, Magne Hompland, Paul Montgomery, Nancy Earl, Tonya M Smoot, Joseph DeVeaugh-Geiss, Lamictal 606 Study Group (2003) A placebo-controlled 18-month trial of lamotrigine and lithium maintenance treatment in recently manic or hypomanic patients with bipolar I disorder. Arch Gen Psychiatry 60(4):392–400. https://doi.org/10.1001/archpsyc.60.4.392

Chapman PM, Ho KT, Munns WR, Solomon K, Weinstein MP (2002) Issues in sediment toxicity and ecological risk assessment. Mar Pollut Bull 44(4):271–278. https://doi.org/10.1016/S0025-326X(01)00329-0

Collins AR (2017) The use of bacterial repair endonucleases in the comet assay. In: Drug safety evaluation. Humana Press, New York, pp 173–184. https://doi.org/10.1007/978-1-4939-7172-5_9

Collins AR, Cadet J, Mőller L, Poulsen HE, Vina J (2004) Are we sure we know how to measure 8-oxo-7, 8-dihydroguanine in DNA from human cells? Arch Biochem Biophys 423(1):57–65. https://doi.org/10.1016/j.abb.2003.12.022

Cooper EL (1996) Earthworm immunity. In: Invertebrate immunology. Springer, Berlin/Heidelberg, pp 10–45. https://doi.org/10.1007/978-3-642-79735-4_2

Cossarizza A, Cooper EL, Suzuki MM, Salvioli S, Capri M, Gri G, Franceschi C (1996) Earthworm leukocytes that are not phagocytic and cross-react with several human epitopes can kill human tumor cell lines. Exp Cell Res 224(1):174–182. https://doi.org/10.1006/excr.1996.0125

Cotelle S, Férard JF (1999) Comet assay in genetic ecotoxicology: a review. Environ Mol Mutagen 34(4):246–255. https://doi.org/10.1002/(SICI)1098-2280(1999)34:4<246::AID-EM4>3.0.CO;2-V

da Silva J, de Freitas TR, Heuser V, Marinho JR, Bittencourt F, Cerski CTS, Delgado E, Borràs M, Nadal J (2000a) Genotoxic assessment of urban dumping sites: comet test in wood mouse circulating lymphocytes. Toxicol Lett 116:91. https://doi.org/10.3389/fgene.2015.00180

da Silva J, de Freitas TR, Marinho JR, Speit G, Erdtmann B (2000b) An alkaline single-cell gel electrophoresis (comet) assay for environmental biomonitoring with native rodents. Genet Mol Biol 23(1):241–245. https://doi.org/10.1590/S1415-47572000000100042

El-Shahawi MA, Hamza A, Bashammakh W, Al-Saggaf (2010) An overview on the accumulation, distribution, transformations, toxicity and analytical methods for the monitoring of persistent organic pollutants. Talanta 80(5):1587–1597. https://doi.org/10.1016/j.talanta.2009.09.055

Erdtmann B (2000) Effects of chronic exposure to coal in wild rodents (Ctenomys torquatus) evaluated by multiple methods and tissues. Mutat Res/Genet Toxicol Environ Mutagen 470(1):39–51. https://doi.org/10.1016/S1383-5718(00)00094-2

Fairbairn JJ, Khan MW, Ward KJ, Loveridge BW, Fairbairn DW, O'Neill KL (1995a) Induction of apoptotic cell DNA fragmentation in human cells after treatment with hyperthermia. Cancer Lett 89(2):183–188. https://doi.org/10.1016/0304-3835(94)03668-9

Fairbairn DW, Olive PL, O'Neill KL (1995b) The comet assay: a comprehensive review. Mutat Res/Rev Genet Toxicol 339(1):37–59. https://doi.org/10.1016/0165-1110(94)00013-3

Fenech M (1993) The cytokinesis-block micronucleus technique and its application to genotoxicity studies in human populations. Environ Health Perspect 101(suppl 3):101–107. https://doi.org/10.1289/ehp.93101s3101

Fenech M (2007) Cytokinesis-block micronucleus cytome assay. Nat Protoc 2(5):1084. https://doi.org/10.1038/nprot.2007.77

Ferreira-Baptista L, De Miguel E (2005) Geochemistry and risk assessment of street dust in Luanda, Angola: a tropical urban environment. Atmos Environ 39(25):4501–4512. https://doi.org/10.1016/j.atmosenv.2005.03.026

Fourie F, Reinecke SA, Reinecke AJ (2007) The determination of earthworm species sensitivity differences to cadmium genotoxicity using the comet assay. Ecotoxicol Environ Saf 67(3):361–368. https://doi.org/10.1016/j.ecoenv.2006.10.005

Fujita Y, Yoshihara Y, Sato I, Sato S (2014) Environmental radioactivity damages the DNA of earthworms of Fukushima Prefecture, Japan. Eur J Wildl Res 60(1):145–148. https://doi.org/10.1007/s10344-013-0767-y

Gichner T, Plewa MJ (1998) Induction of somatic DNA damage as measured by single cell gel electrophoresis and point mutation in leaves of tobacco plants. Mutat Res/Fundam Mol Mech Mutagen 401(1–2):143–152. https://doi.org/10.1016/S0027-5107(98)00003-7

Gichner T, Patková Z, Száková J, Demnerová K (2006) Toxicity and DNA damage in tobacco and potato plants growing on soil polluted with heavy metals. Ecotoxicol Environ Saf 65(3):420–426. https://doi.org/10.1016/j.ecoenv.2005.08.006

Hartmann A, Agurell E, Beevers C, Brendler-Schwaab S, Burlinson B, Clay P, Tice RR (2003) Recommendations for conducting the in vivo alkaline Comet assay. Mutagenesis 18(1):45–51. https://doi.org/10.1093/mutage/18.1.45

Hättenschwiler S, Vitousek PM (2000) The role of polyphenols in terrestrial ecosystem nutrient cycling. Trends Ecol Evol 15(6):238–243. https://doi.org/10.1016/s0169-5347(00)01861-9

Homa J, Niklinska M, Plytycz B (2003) Effect of heavy metals on coelomocytes of the earthworm Allolobophora chlorotica: the 7th international symposium on earthworm ecology Cardiff Wales 2002. Pedobiologia 47(5–6):640–645. https://doi.org/10.1078/0031-4056-00239

Klobučar GI, Stambuk A, Srut M, Husnjak I, Merkaš M, Traven L, Cvetković Z (2011) Aporrectodea caliginosa, a suitable earthworm species for field based genotoxicity assessment? Environ Pollut. 159(4):841–9. https://doi.org/10.1016/j.envpol.2011.01.009

Kloepper-Sams PJ, Owens JW (1993) Environmental biomarkers as indicators of chemical exposure. J Hazard Mater 35(2):283–294. https://doi.org/10.1016/0304-3894(93)80012-5

Knopper LD, Mineau P, Mcnamee JP, Lean DRS (2005) Use of comet and micronucleus assays to measure genotoxicity in meadow voles (Microtus pennsylvanicus) living in golf course ecosystems exposed to pesticides. Ecotoxicology 14(3):323–335

Koppen G, Toncelli LM, Triest L, Verschaeve L (1999) The comet assay: a tool to study alteration of DNA integrity in developing plant leaves. Mech Ageing Dev 110(1–2):13–24. https://doi.org/10.1016/s0047-6374(99)00038-x

Kwasniewska J, Mikolajczyk A (2014) Influence of the presence of B chromosomes on DNA damage in Crepis capillaris. PLoS One 9(1):e87337. https://doi.org/10.1371/journal.pone.0087337

Lukkari T, Taavitsainen M, Väisänen A, Haimi J (2004) Effects of heavy metals on earthworms along contamination gradients in organic rich soils. Ecotoxicol Environ Saf 59(3):340–8. https://doi.org/10.1016/j.ecoenv.2003.09.011

Manz M, Wenzel KD, Dietze U, Schüürmann G (2001) Persistent organic pollutants in agricultural soils of Central Germany. Sci Total Environ 277(1–3):187–198. https://doi.org/10.1016/S0048-9697(00)00877-9

Mateos S, Daza P, Domínguez I, Cárdenas JA, Cortés F (2008) Genotoxicity detected in wild mice living in a highly polluted wetland area in south western Spain. Environ Pollut 153(3):590–593. https://doi.org/10.1016/j.envpol.2007.09.008

McKelvey-Martin VJ, Green MHL, Schmezer P, Pool-Zobel BL, De Meo MP, Collins A (1993) The single cell gel electrophoresis assay (comet assay): a European review. Mutat Res/Fundam Mol Mech Mutagen 288(1):47–63. https://doi.org/10.1016/0027-5107(93)90207-V

Molnár L, Fischer E, Kallay M (1989) Laboratory studies on the effect, uptake and distribution of chromium in *Eisenia foetida* (Annelida, Oligochaeta). Zool Anz 223(1–2):57–66. http://pascal-francis.inist.fr/vibad/index.php?action=getRecordDetail&idt=19711733

Morillas MJ, Guillamet E, Surralles J, Creus A, Marcos R (2002) Spontaneous and induced genetic damage in T lymphocyte subsets evaluated by the Comet assay. Mutat Res/Genet Toxicol Environ Mutagen 514(1–2):39–48. https://doi.org/10.1016/S1383-5718(01)00319-9

Ostling O, Johanson KJ (1984) Microelectrophoretic study of radiation-induced DNA damages in individual mammalian cells. Biochem Biophys Res Commun 123(1):291–298. https://doi.org/10.1016/0006-291X(84)90411-X

Pastor N, López-Lázaro M, Tella JL, Baos R, Hiraldo F, Cortés F (2001) Assessment of genotoxic damage by the comet assay in white storks (*Ciconia ciconia*) after the Donana ecological disaster. Mutagenesis 16(3):219–223. https://doi.org/10.1093/mutage/16.3.219

Pastor N, Baos R, López-Lázaro M, Jovani R, Tella JL, Hajji N, Cortés F (2004) A 4 year follow-up analysis of genotoxic damage in birds of the Doñana area (south west Spain) in the wake of the 1998 mining waste spill. Mutagenesis 19(1):61–65. https://doi.org/10.1093/mutage/geg035

Pollumaa L, Kahru A, Eisenträger A, Reiman R, Maloveryan A, Rätsep A (2000) Toxicological investigation of soils with the solid-phase Flash assay: comparison with other ecotoxicological tests. Altern Lab Anim 28(3):461–472. https://doi.org/10.1177/026119290002800314

Prabhakaran R, Rajkumar SN, Ramprasath T, Selvam GS (2018) Identification of promoter P cadR, in silico characterization of cadmium resistant gene cadR and molecular cloning of promoter P cadR from Pseudomonas aeruginosa BC15. Toxicol Ind Health 34(12):819–833. https://doi.org/10.1177/0748233718795934

Rajkumar P, Ramprasath T, Selvam GS (2017) A simple whole cell microbial biosensors to monitor soil pollution. New pesticides and soil sensors. Academic, pp 437–481. https://doi.org/10.1016/B978-0-12-804299-1.00013-8

Rodriguez E, Azevedo R, Fernandes P, Santos CA (2011) Cr (VI) induces DNA damage, cell cycle arrest and polyploidization: a flow cytometric and comet assay study in Pisum sativum. Chem Res Toxicol 24(7):1040–1047. https://doi.org/10.1021/tx2001465

Sarrif A, Kyrtopoulos SA, Autrup H, Farmer P, Kirsch-Volders M, Kleinjans JC, Knudsen LE, Mathar U, Wiegand HJ, Vrijhof H (2006) Environmental genotoxins in children and adults: introduction and overview. Mutat Res 608(2):97–99

Schnaak W, Küchler T, Kujawa M, Henschel KP, Süßenbach D, Donau R (1997) Organic contaminants in sewage sludge and their ecotoxicological significance in the agricultural utilization of sewage sludge. Chemosphere 35(1–2):5–11. https://doi.org/10.1016/S0045-6535(97)88285-9

Singh NP, McCoy MT, Tice RR, Schneider EL (1988) A simple technique for quantitation of low levels of DNA damage in individual cells. Exp Cell Res 175(1):184–191. https://doi.org/10.1016/0014-4827(88)90265-0

Stavreva DA, Ptáček O, Plewa MJ, Gichner T (1998) Single cell gel electrophoresis analysis of genomic damage induced by ethyl methanesulfonate in cultured tobacco cells. Mutat Res/Fundam Mol Mech Mutagen 422(2):323–330. https://doi.org/10.1016/S0027-5107(98)00213-9

Tkalec M, Štefanić PP, Cvjetko P, Šikić S, Pavlica M, Balen B (2014) The effects of cadmium-zinc interactions on biochemical responses in tobacco seedlings and adult plants. PLoS One 9(1):e87582. https://doi.org/10.1371/journal.pone.0087582

Türkoğlu Ş (2012) Determination of genotoxic effects of chlorfenvinphos and fenbuconazole in Allium cepa root cells by mitotic activity, chromosome aberration, DNA content, and comet assay. Pestic Biochem Physiol 103(3):224–230. https://doi.org/10.1016/j.pestbp.2012.06.001

Valembois P, Lassegues M, Roch P (1992) Formation of brown bodies in the coelomic cavity of the earthworm Eisenia fetida andrei and attendant changes in shape and adhesive capacity of constitutive cells. Dev Comp Immunol 16(2–3):95–101. https://doi.org/10.1016/0145-305X(92)90010-A

Valembois P, Seymour J, Lassègues M (1994) Evidence of lipofuscin and melanin in the brown body of the earthworm Eisenia fetida andrei. Cell Tissue Res 277(1):183–188. https://doi.org/10.1007/BF00303095

Van Goethem F, Lison D, Kirsch-Volders M (1997) Comparative evaluation of the in vitro micronucleus test and the alkaline single cell gel electrophoresis assay for the detection of DNA damaging agents: genotoxic effects of cobalt powder, tungsten carbide and cobalt–tungsten carbide. Mutat Res/Genet Toxicol Environ Mutagen 392(1–2):31–43. https://doi.org/10.1016/s0165-1218(97)00043-8

Vasseur P and Bonnard M (2014) Ecogenotoxicology in earthworms: A review. Current Zoology 60(2). https://doi.org/10.1093/czoolo/60.2.255

Voldner EC, Li YF (1995) Global usage of selected persistent organochlorines. Sci Total Environ 160:201–210. https://doi.org/10.1016/0048-9697(95)04357-7

Wang W, Huang MJ, Kang Y, Wang HS, Leung AO, Cheung KC, Wong MH (2011) Polycyclic aromatic hydrocarbons (PAHs) in urban surface dust of Guangzhou, China: status, sources and human health risk assessment. Sci Total Environ 409(21):4519–4527. https://doi.org/10.1016/j.scitotenv.2011.07.030

Weber J, Tyszka R, Kocowicz A, Szadorski J, Debicka M, Jamroz E (2012) Mineralogical composition of the clay fraction of soils derived from granitoids of the Sudetes and fore-Sudetic block, Southwest Poland. Eur J Soil Sci 63(5):762–772. https://doi.org/10.1111/j.1365-2389.2012.01482.x

Wells PG, Wells PG, Depledge MH, Butler JN, Manock JJ, Knap AH (2001) Rapid toxicity assessment and biomonitoring of marine contaminants—exploiting the potential of rapid biomarker assays and microscale toxicity tests. Mar Pollut Bull 42(10):799–804. https://doi.org/10.1016/S0025-326X(01)00054-6

Winegardner DL (1995) An introduction to soils for environmental professionals. CRC Press. https://doi.org/10.1002/ep.670170205

Chapter 7
Drosophila melanogaster as a Model to Study Acrylamide Induced Toxicity and the Effects of Phytochemicals

Pallavi Dan, Swetha Senthilkumar, V. P. Narayanan Nampoothri, Abhinaya Swaminathan, and Sahabudeen Sheik Mohideen

Abstract Acrylamide, a water-soluble vinyl polymer, is a common environmental toxin present in several industrial effluents and most importantly in high-temperature cooked foods like potato chips, meat, cookies, etc. Studies conducted over a decade have revealed acrylamide's potency in eliciting detrimental effects in different physiological systems. Researchers have also noticed its presence in the biological samples collected from employees working in mining and ore industries and deemed it to be classified as an occupational hazardous chemical. In this case, we have reviewed acrylamide induced neurotoxicity in *D. melanogaster* model and prospective phytochemicals for its treatment. This study highlights acrylamide's role in modulating cellular redox chains and divulges the probable mechanism for induced toxicity. The effect of acrylamide on its major target sites like the neuronal synapse and kinesin proteins are explored as the basis for acrylamide induced toxicity. We also focused our study at the cellular level and other model organisms to better understand acrylamide-induced neuronal apoptosis and mitochondrial depolarisation events. Furthermore, the neuroprotective effects of common phytoconstituents like curcumin and thymoquinone against acrylamide-induced neurotoxicity are discussed.

Keywords Acrylamide · *D. melanogaster* · Phytochemicals · Toxicology · Neurotoxicity

---

P. Dan · S. Senthilkumar · V. P. Narayanan Nampoothri · A. Swaminathan · S. S. Mohideen (✉)
Department of Biotechnology, School of Bio-engineering, SRM Institute of Science and Technology, Chennai, Tamil Nadu, India
e-mail: sahabuds@srmist.edu.in

K. M. Gothandam et al. (eds.), *Environmental Biotechnology Volume 4*, Environmental Chemistry for a Sustainable World 68,
https://doi.org/10.1007/978-3-030-77795-1_7

7.1 Introduction

Acrylamide with the chemical formula C_3H_5NO has become one of the most researched chemicals globally due to its polar nature that accounts for its easy solubility in water, alcohol and ether. Its existence as monomeric or polymeric form enables it to be used in various industrial processes such as paper, textile, cosmetics, synthesis of dyes, sewage, and drinking water treatment plants (Acrylamide National Pollutant Inventory 2020).

The extensive usage of acrylamide in treated water and a broad range of essential merchandise like textiles and paper industries have led to environmental contamination. These industrial effluents primarily contain acrylamide in its monomeric form, which is predicted to contaminate the groundwater sources and other aquatic environments. Acrylamide can thus be classified as an environmental toxin; moreover, acrylamide being easily absorbed through the mucous membrane, lungs, skin possesses greater threat as an occupational hazard to people working in ore industries, etc. (Acrylamide|National Pollutant Inventory 2020; Moorman et al. 2012).

In 2002, Swedish scientists reported acrylamide in thermally processed foods; following this, many health organisations, including the US food and drug administration, World Health Organisation, the European Food Safety Authority, have acknowledged the need for further research. Acrylamide has been classified as a probable carcinogen to humans by the International Agency for Research on Cancer (Acrylamide and Cancer Risk 2020; Biedermann et al. 2002).

Further studies have confirmed the formation of acrylamide in thermally processed foods such as meat, potato fries, baked products, etc., as a result of the Maillard reaction. It is mainly observed that acrylamide is mostly produced in foods containing amino acids such as asparagine such as coffee cereals, and starchy foods like potatoes (Gökmen and Şenyuva 2006; Liu et al. 2018; Xu et al. 2016). Physiologically based toxicokinetic models were developed to study the tolerable daily intake levels for acrylamide. Significantly a tolerable daily intake of 40μg/kg bw/day of acrylamide induced neurotoxicity and 2.6μg/kg bw/day was estimated to induce cancer (Tardiff et al. 2010).

Acrylamide-induced toxicity studies became essential; its presence in food items deemed it to be researched further to fully comprehend its effect on all living organisms, especially humans. Hence, scientists investigated acrylamide induced toxicity using various models, and this led us to observe that acrylamide induced models show developmental delay, neurotoxicity, genotoxicity, reproductive toxicity and carcinogenicity (Ranjini and Manonmani 2019; Senthilkumar et al. 2020; Zhao et al. 2017a, b).

Knowledge of the effects and the mechanisms has enabled the researchers to focus on compounds that can ameliorate the impact of acrylamide induced oxidative stress, which can then result in the prevention of further damage. Phytochemicals are very well known for their anti-oxidant properties, which became the primary focus for scientists. Several acrylamide-induced models were subjected to various phytochemicals such as curcumin, thymoquinone, geraniol, etc., to name a few.

Subsequently, these studies have shown a significant improvement in reducing the behavioural and developmental defects induced by acrylamide (Prasad 2014b; Senthilkumar et al. 2020).

In this chapter, we have reviewed the use of *D. melanogaster* as a model to understand the different effects of acrylamide induced neurotoxicity and the use of phytochemicals as a potential treatment for the above-said effects.

7.2 Acrylamide Induced Toxicity – *D. melanogaster* Model

D. melanogaster has been one of the extensively researched models by scientists around the world. It is believed that this model contains nearly 77% functional homologs of the human disease-causing genes and is, therefore, a very tractable genetic model to study neurotoxicology and neurodegenerative diseases (Kumar and Harish 2019; Pandey and Nichols 2011).

D. melanogaster, with its well characterised genetic systems and relatively short generation time, is used to research toxicology studies (Sundararajan et al. 2019). Researchers have also well studied the role of P-450 and Glutathione-S-Transferase family genes in the metabolism of xenobiotics/drugs. This knowledge further enables us to understand the drug's effects and its mechanism of action (Yang et al. 2007).

The underlying mechanism of acrylamide induced toxicity is believed to be the generation of reactive oxygen species in the models. It is noted that the oxidised form of acrylamide, commonly known as glycidamide plays a major role in altering the cellular redox chain, thereby inducing oxidative stress (Prasad et al. 2012; Zamani et al. 2017a, b). It is most commonly noted that oxidative stress triggers the activation of mitogen-activated protein kinases, which play a major role in modulating key cellular responses like apoptosis (Tabeshpour et al. 2020; Zamani et al. 2017a, b).

Assuming oxidative stress as the basis of the toxic effects caused by acrylamide, many studies have shown changes in the behavioural, developmental and reproductive characteristics (Senthilkumar et al. 2020.). It is also seen that with increasing concertation of acrylamide significantly affected the lifespan of the flies. Basing the behavioural patterns concerning the locomotor functions in larvae and flies, it was noted that acrylamide adversely affects the motor neurons at both larval and adult stages (Prasad et al. 2014a; Senthilkumar et al. 2020). An exact mechanism of action for acrylamide is not yet well understood, however with all the evidence from previously conducted studies; we can derive that acrylamide is a well-known neurotoxin (Kumar and Harish 2019; Prasad et al. 2019). In this chapter, we will further focus on the acrylamide induced neurotoxicity studies in *D. melanogaster* models.

7.3 Acrylamide Induced Neurotoxicity in *D. melanogaster* Models

Acrylamide induced neurotoxicity has been widely studied for over a decade to speculate the mechanism of action. Different studies indicated the nerve endings, also called the synaptic endings of the sensory and motor neurons, as the site of action (Lopachin 2004; Yao et al. 2014). This leads to neural damage due to oxidative stress and inhibition of motor protein kinases (Kumar and Harish 2019). Extensive studies have shown that acrylamide being electrophilic, can add a sulfhydryl group to membrane fusion proteins, thus forming protein adducts that inhibit the neurotransmission (Shinomol et al. 2013; Lopachin 2004).

Comparatively, *D. melanogaster* serves to be an easy model to study neurotoxicity due to its level of complexity and ease of rearing. The major attributes that lead to neurotoxicity are oxidative damage on locomotor neurons, dopamine levels, acetylcholinesterase, butyl choline esterase and kinesin motor proteins.

Locomotor functions concerning larvae and adult flies involve complex neural networks. The central complex substructures, primarily the ellipsoid body ring neurons along with the mushroom body, have been extensively studied to determine their strong role in *D. melanogaster* locomotor ability. It is widely noted that thermogenetic or optogenetic stimulation of the central complex opens up the possibility of neural pathways concerning drug locomotor effects (Scaplen et al. 2019). Several studies have shown that acrylamide has induced severe locomotor deficits in a dose-dependent manner (Senthilkumar et al. 2020). The locomotor functions of the third instar larvae and the adult flies have been studied through crawling and negative geotaxis assays, respectively, and it was noted that acrylamide had affected the flies severely at different stages (Shinomol et al. 2013; Prasad 2014a).

The dopamine levels play a major role in all organisms; in *D. melanogaster*, also it plays a vital role in modulating the dopaminergic neurotransmission essential for motor functions, behaviour, etc. (Neckameyer 1996). Tyrosine hydroxylase is the key player in dopamine biosynthesis, and any fluctuation in this affects the neural dopamine levels significantly (Riemensperger et al. 2011). Acrylamide models were noted to show a significant decrease in dopamine levels, thereby affecting (decreased) the dopaminergic neurotransmission, which can be correlated to the locomotor deficit in *D. melanogaster* models (Kumar and Harish 2019; Prasad and Muralidhara 2012, 2019).

The cholinergic functions can be studied by assessing the levels of acetylcholine esterase and butyrylcholinesterase. Studies have shown that acrylamide-exposed workers sensed ataxia and muscular weakness correlated to impaired cholinergic transmission in the neuromuscular junctions (Konings et al. 2003; Lionetto et al. 2013). Increased acetylcholine esterase activity and butyrylcholine esterase were noted in acrylamide treated *D. melanogaster* models; this is understood to decrease cholinergic neurotransmitters acetylcholine/butyrylcholine in the synaptic cleft. These changes lead to the downregulation of the corresponding receptors and subsequent cholinergic neurotransmission and its related functions, thus triggering apoptosis (Kumar and Harish 2019; Schmatz et al. 2009). This neurotransmitter's altered levels are the most established parameter for establishing acrylamide-induced neurotoxicity (Prasad and Muralidhara 2019; Senthilkumar et al. 2020).

Microtubule motor protein kinesins are observed to be a major site of the acrylamide effect; it is earlier noted that acrylamide plays a major role in the inhibition of the kinesin molecule, thereby affecting the fast anterograde axonal transport (Kumar and Harish 2019; Sickles et al. 1996). Kinesins play a major role in transporting neurotransmitters or enzymes responsible for neurotransmitter generation or regeneration along the microtubules destined for axolemma, mitochondria and synaptic vessels (Sickles et al. 2002). The first kinesin protein, KIF5B, belonging to the kinesin-1 family, was studied in *D. melanogaster*-Oregon-K models recently. It was noted that the kinesin levels were significantly affected by acrylamide exposure. At a sub-toxic level of 5 mM acrylamide, depleted levels of kinesin levels can be related to their decreased locomotor and neurological functions (Kumar and Harish 2019).

7.4 Effects of Acrylamide Induced Toxicity in the Cellular Level

The most prevalent way acrylamide affects the cell is by increasing reactive oygen species production, thereby oxidative stress. Different apoptotic pathways get activated, leading to cell death. DNA adducts are formed on key genes like p53 causing cancer (Fig. 7.1).

Current research indicates that acrylamide toxicity influences three major systems in our body, namely the immune system, reproductive system and neurological system. The Immune system responses to acrylamide treatment of splenocytes of mice revealed dose-dependent apoptosis and relied on the Caspase 9 pathway (Zamani et al. 2017a, b). When macrophages are treated with acrylamide, a change in the senescence production gene and increased, ROS production was observed (Kim et al. 2015). 30-day exposure of acrylamide to female BALB/c mice obstructed immunity levels by decreased lymphocyte, Natural Killer cells, Concanavalin A and Interleukin-6 production (Fang et al. 2014). Acrylamide poses a danger to mouse oocyte development as studies revealed increased reactive oygen species production, which confirms oxidative stress.

Further experimentation showed acrylamide-induced damage of spindle organization directly linked to ɣ-tubulin and MAP-K (Duan et al. 2015). The cumulus cells surrounding the oocytes of mouse origin express increased Caspase - 3 productions in the presence of acrylamide, leading to high incidences of apoptosis (Liu et al. 2015). Reproductive toxicity caused by 10 mg/kg-d acrylamide in male rats resulted in Leydig cells' high production, increased serum testosterone levels and irregular expression of sGCα1 blood pressure regulator. Thus, acrylamide has detrimental effects on both the male and female reproductive cells (Wang et al. 2010).

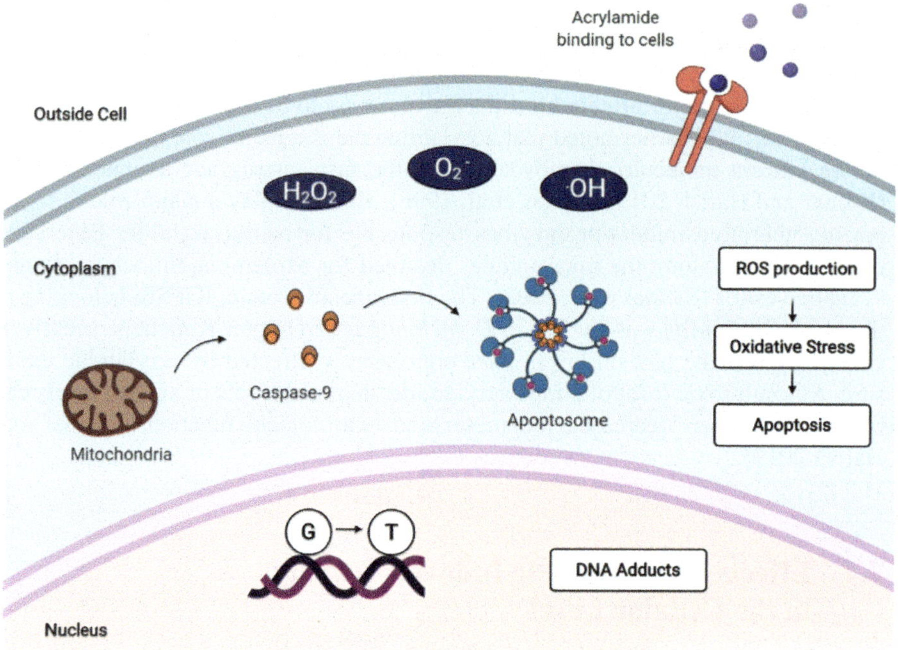

Fig. 7.1 Acrylamide toxicity ultimately leads to cell death and cancer. ROS – Reactive Oxygen Species; DNA – Deoxyribonucleic acid

Acrylamide can cause cytotoxicity and morphological changes on the lung epithelial cells BEAS-2B cells (Kacar et al. 2019). The viability of the cells was reduced after 24 h. Various studies have reported that acrylamide induces oxidative stress by raising ROS, Malondialdehyde and H_2O_2, corresponding to its concentration (Yousef and El-Demerdash 2006; Mehri et al. 2012; Yilmaz et al. 2017). Also, an upregulation of the Nrf-2 protein can be observed, which further confirms acrylamide's effect on oxidative stress. The number of apoptotic cells in BEAS-2B was comparatively higher when treated with acrylamide than the control. The cells showed classic apoptotic symptoms of blebbing, nuclear condensation, vacuolization and shrinkage.

In the study conducted by Atoff et al. (2015) on the SH-SY5Y cell line (neuroblastoma) and C17.2 cell line (neural progenitor), decreased proliferation urged a reduction in the number of viable cells. Also, a decrease in the count of neurons could be observed in C17.2 cells. In SH-SY5Y cells, changes in axonal growth and reduction in the number of neurites per cell could be observed, thus indicating less differentiated neuronal cell culture. The conjugation of acrylamide in the neurofilaments and microtubular system components caused inhibition or delay in the axonal growth (Atoff et al. 2015). Acrylamide decreased the viability of primary astrocytes, U-251 MG, U-1240 MG and U-87 MG cell lines in a time and dose-dependent cytotoxicity. When these cells were exposed to 2 mM acrylamide for 24–72 h, a reduction in the cell viability was observed. Also, a reduction in the viability of the primary astrocytes' cells was observed after treatment with 0.1 mM acrylamide for 24 h. In the primary astrocytes, a dose-dependent increase in the apoptotic proteins' levels correspond to acrylamide levels. Other apoptotic regulators such as Bax increased while pro-caspase 3, 8, 9 and Bcl-2 decreased. Acrylamide concentration of 0.1 and 0.5 mM resulted in elevated expression of Bcl-2; meanwhile, higher dosages of acrylamide at 1 and 2 mM was required for a substantial increase in Bax/Bcl-2 ratio. This further indicates that apoptotic responses increased in the primary astrocytes. Mitochondrial depolarization was also observed in these cell lines after 48 h of treatment with 2 mM acrylamide (Lee et al. 2014).

7.5 Neurotoxic Studies on Animals

By far, acrylamide poses the most threat to the nervous system as a highly potent neurotoxin. It can cause severe toxicity in humans as it can be absorbed by ingestion or inhalation, as it is highly permeable and can cross the cell monolayer through passive diffusion. Through the placenta and blood-brain barrier, it can be passed on to the fetus as well (Friedman and Levin 2008; Alkarim et al. 2015). The No Observable Adverse Effect Level in animals is 0.2–0.5μg/kg/day and the lowest observable adverse effect level is 2μg/kg/day, respectively. In rats, the oral lethal dose (LD_{50}) is 107–251 mg/kg, for guinea pigs 150–180 mg/kg and in mouse, it was 107–170 mg/kg. Studies conducted on various laboratory animals like cats, rats, mice, monkeys etc., have shown that continuous exposure to acrylamide can cause ataxia, hindlimb foot splay, and skeletal muscle weakness, respectively (Exon 2006). A recent study conducted on male rats with 21 days of continuous intraperitoneal injection of acrylamide induced neurotoxicity, and a decrease in body weight and distal axonopathy could be observed. In the central and peripheral nervous systems, histopathological alterations like neuronal degeneration and necrosis of the glial cell's activation are associated with degeneration of the axons and abnormalities in the myelin sheath can be observed. In the hippocampal region, disorganization in the hippocampus's dentate gyrus and vacuolation of granule cells was observed (Elblehi et al. 2020). When the lumbar spinal cord of male rats was exposed to acrylamide, it caused degeneration in the pyknotic neurons, the focal area of gliosis and vacuolation of the neuropil (Oda 2017).

Acrylamide can trigger the development of micronucleated erythrocytes in male B6CF31 mice (Zeiger et al. 2009). Also, chronic administration of acrylamide in rats for 5 weeks decreased erythrocyte deformability (Arıhan et al. 2011). Studies have shown that the toxic effects of acrylamide can cause a delay the development of hippocampal neurons in juvenile rats. It can also reduce the locomotor activity and grip strength of the hindlimb (He et al. 2017; Lai et al. 2017). The abnormal posture of the hindlimb, ataxia, dragging the hindlimb and weakness in the muscle was observed in acrylamide intoxicated rats (Zhang et al. 2017). Earlier studies have stated the neurotoxic effects of acrylamide in adult animals. Axonal lesions in the central and peripheral nervous systems, weight loss, skeletal muscle weakness and ataxia were observed (LoPachin and Gavin 2012; Tian et al. 2015). Also, pathological features such as peripheral nerve axonal swelling and degeneration can be observed (Shi et al. 2012; He et al. 2017).

Zebrafish has been widely used as a neurotoxic model for various toxicological studies. In the study conducted by Prats et al. (2017), 5-day post-fertilization old zebrafish larvae were exposed to 1 mM acrylamide for 3 days. It was found that there were alterations in the motor functions, and presynaptic nerve terminals at the neuromuscular junction were affected. Also, transcriptional proteins involved in the synaptic vesicle cycle were changed (Prats et al. 2017). When adult zebrafish (5-day post-fertilization) old were exposed to the same concentration of acrylamide (1 mM), they showed depression with anxiety behaviour. Expression of regeneration-associated genes, growth-associated proteins and alpha-tubulin at the axonal level was significantly downregulated. On the other hand, myelin basic protein expression and glial fibrillary basic protein that is expressed in oligodendrocytes were significantly upregulated (Faria et al. 2019). Acrylamide has been shown to cause structural disruptions in the brain of 6- day post-fertilization larvae when exposed to 2.5 mM concentration for 24 h (Komoike and Matusoka 2016). In another study conducted by Prang et al. 2006, when 4 -day post-fertilization zebrafish were exposed to 6.25 mM acrylamide for 8 h, loss of myelin in the forebrain and midbrain and neuronal apoptosis was observed (Prang et al. 2006).

7.6 The Role of Phytochemicals in Reducing the Acrylamide Toxicity

There are several studies that reveal the action of various phytochemicals against the toxicity induced by acrylamide. The acrylamide toxicity and the protective action of various chemicals are studied in several models. Briefly, rat models, *D. melanogaster* models and various cell lines are used for the studies. Table 7.1 shows some important studies and significant findings. Recent studies found that Kaempferol can be used against acrylamide toxicity. They have studied the effect of the Kaempferol in a rat model, showing that acetylcholine activity in the brain was significantly reduced. The biochemical and histopathological results show the curative effect of the Kaempferol. Initially, acrylamide administration caused a significant increase in transaminase activities, lipid profile, creatinine and urea, and a decline in hemoglobin. A significant increase in lipid peroxidation levels was observed, whereas the levels of superoxide dismutase, catalase and glutathione were decreased. In the kaempferol treated group, a significant increase in the levels of superoxide dismutase, glutathione, and catalase was observed.

Quercetin is a known flavonoid that is normally seen in vegetables and fruits; the estimated intake should be up to 25 mg/day in the USA. It has many protective effects on the health of humans, which includes cardiovascular protection,

Table 7.1 Phytochemicals in reducing the Acrylamide toxicity

Sl: No	Phytochemicals	Model of study	Dosage/Concentration	Key parameters/ Observations	Reference
1	Kaempferol (3,5,7-trihydroxy-2-(4-hydroxyphenyl)-4H-1-benzopyran-4-one)	In vivo- Rat	Acrylamide dose-38.27 mg/kg p.o. (1/third of LD50) for 10 days Kaempferol- 5,10,20,40 mg/kg for 3 days	Kaempferol can be used as a dietary agent against toxicity induced by acrylamide. Promising drug against acrylamide toxicity by improving the activity of antioxidant enzymes and reducing oxidative stress.	
2	Quercetin (3,3′,4′,5,7-pentahydroxyflavone)	In vivo- Rat	Acrylamide dose-38.27 mg/kg p.o. (1/3rdof LD50) for 10 days Quercetin- 5,10,20,40 mg/kg for 3 days	Acrylamide exposure caused a decline in acetylcholinesterase activity, a decrease in antioxidant enzymes such as superoxide dismutase & catalase and a significant increase in lipid peroxidation. Quercitin has a protective ability against acrylamide induced cellular damage and treatment helps to restore the antioxidant enzymes level and acetylcholinesterase activity.	Uthra et al. (2017)
		In vivo-Rat	Acrylamide-50 mg/kg Quercitin-10 mg/kg for 5 days	An increase in guanosine level decrease in glutathione-s- transferase was evident with acrylamide administration. Quercitin is evident for protecting the liver tissue from acrylamide induced damages. Quercitin treatment showed a significant reduction in tissue damage when histopathological examination was done using H&E Staining.	Ansar et al. (2016)

(continued)

Table 7.1 (continued)

Sl: No	Phytochemicals	Model of study	Dosage/Concentration	Key parameters/ Observations	Reference
3	Thymoquinone (2-Isopropyl-5-methylbenzo-1,4-quinone)	In vivo- Rat	Acrylamide-50 mg/kg ip for 11 days. Thymoquinone- 2.5,5,10 mg/kg ip for 11 days	Acrylamide exposure caused gait inabilities, malondialdehyde content increased and reduced glutathione contents. Thymoquinone treatment prevented gait abnormalities and elevated the glutathione and reduced malondialdehyde contents	Tabeshpour et al. (2019)
4	Thymoquinone and curcumin	In vivo- D. *melanogaster*	Acrylamide-2 mM Thymoquinone-10μM Curcumin-10μM	Treatment with thymoquinone and curcumin increased the survivability, behavioural improvement, level of reactive oxygen species reduced, restored acetylcholinesterase activity. IHC & m-RNA studies shows that toxicity induced by acrylamide can be reduced with treatment.	Senthilkumar et al. (2020)
5	Geraniol ((2*E*)-3,7-Dimethyl-2,6-octadien-1-ol) & Curcumin	In vivo-Rat	Acrylamide-50 mg/kg i.p 3 times /week for 4 weeks Geraniol:100 mg/kg bw/d; Curcumin:50 mg/kg bw/d, 4 weeks	Acrylamide administration resulted in oxidative stress, elevation in cytosolic calcium level, increased acetyl cholinesterase activity Both phytochemicals show therapeutic effect. Restored the dopamine level.	Sathya et al. (2014)
		In vivo- D. *melanogaster*	Acrylamide-5 mm for 7 days. Geraniol:5–10μM Curcumin:5–10μM	Acrylamide administration resulted in oxidative stress, elevation in cytosolic calcium level, increased acetyl cholinesterase activity Both phytochemicals show therapeutic effect. Restored the dopamine level.	Sathya et al. (2014)

Sl: No	Phytochemicals	Model of study	Dosage/Concentration	Key parameters/ Observations	Reference
6	Extra virgin oil and its fraction	In vivo- Rat	AlCl3–50 mg/kg & Acrylamide-20 mg/kg Extra virgin oil - 300μl for 21 days Hydrophilic fraction - 1 ml Lipophilic fraction - 300μl for 21 days	Combined exposure of aluminium and acrylamide induced cardiac toxicity and reduced levels of vitamin C, high-density lipoprotein cholesterol, Total cholesterol, glutathione and NPSH level. Catalase, superoxide dismutase activities reduced. Treatment with extra virgin oil and its fraction ameliorated the toxicity and histopathology results shows the restoration of cardiac muscles.	Ghorbel et al. (2015)
7	Turmeric	In vivo-Rat	Acrylamide-0.34 g/kg for 11 days Turmeric- 0.5% of total food for 11 days	Glutathione, superoxide dismutase, malondialdehyde, creatinine, urea, uric acid, Total protein analysis conclude that turmeric ameliorated the toxicity induced by acrylamide.	Gehan et al. (2008)
8	Catechin &neem leaves extract	In vivo -Rat	Acrylamide-2.8 mM for 6 weeks Catechin-1.5% Neem leaves extract (200 mg/kg)	Lipid peroxidase, glutathione & catalase level was regulated with treatment. Histopathological examination of liver, lungs, heart and spleen tissues shows the anti-inflammatory antioxidant effect of both phytochemicals.	Mansour et al. (2008)

(continued)

Table 7.1 (continued)

Sl: No	Phytochemicals	Model of study	Dosage/Concentration	Key parameters/ Observations	Reference
9	Linoleic acid (cis, cis-9,12-Octadecadienoic acid)	In vivo-Rat	Acrylamide-38.27 mg/kg for 10 days Linoleic acid-5,10,20,40 mg/kg for 10–13 days	Acrylamide administration caused a significant reduction in glutathione and haemoglobin levels and a high increase in the level of lipid peroxidase. Linoleic acid administration regulated levels of enzymes and their activity The histopathological examination concludes that linoleic acid can reduce the acrylamide toxicity	Shrivastava et al. (2019)
10	*Bacoppamonnieri* leaf powder extract	In vitro- N27 cell line In-vivo- *D. melanogaster* In vivo- Rat	Acrylamide- 5 mM for 24 h (for cells), 5 mM for 72 h (for *D. melanogaster*), i.p 40 mg/kg (for rat) BM extact- 6µg/ml (for cells), 0.05% w/v (for *D. melanogaster*), 2% w/w for 30 days (for rat)	BM extract showed protective action against acrylamide toxicity in the cell line, *D. melanogaster* and rat models. Cell viability increased with pre-treatment of BM extract also, total glutathione content was restored. Locomotor dysfunction due to acrylamide was recovered, and total glutathione, reactive oxygen species and lipid peroxidase level was restored. In the rat model, the catalase and *Glutathione S-transferase* level in the brain cortex and liver was restored with treatment.	Shinomol et al. (2013)

Sl: No	Phytochemicals	Model of study	Dosage/Concentration	Key parameters/ Observations	Reference
11	Morin	In vivo- Rat	Acrylamide- 38.27 mg/kg b.w., p.o. Morin- 50 and 100 mg/kg b.w., p.o.	Superoxide dismutase, GPx, catalase, lipid peroxidase and glutathione levels were restored with Morin treatment. Immunohistochemical and histopathological changes and expression of EGFR, nephrin, p53 and AQP2 show the protective effect. Morin helps to reverse the levels of apoptotic, autophagic and inflammatory parameters.	Kandemir et al. (2020)
12	Cerium oxide (CeO_2) nanoparticles	In vitro- Hep G2 cell line	Acrylamide 200 mM (IC50) Nano-ceria (50, 100, and 200 mM) for 30 min pre-treatment	Cell viability, which is decreased due to acrylamide was restored with treatment. Reactive oxygen species, superoxide dismutase and glutathione level regulated with treatment	Azari et al. (2018)
13	Myricitrin (3′,4′, 5′, 7-five hydroxy flavone-3-O-α-Lrhamnoside)	In vitro- Caco-2 cell line	Acrylamide-5 mM Myricitrin (2.5–10 μg/mL)	Myricitin shows free radical scavenging property and helps to reduce the reactive oxygen species *generation* caused *due to* acrylamide.	Chen et al. (2013)
14	Curcumin	In vitro-Hep G2 cell line	Curcumin 2.5μg/ml Acrylamide 2.5 mM	Curcumin was effective in protecting the cells from radicals by antioxidant mechanism. Curcumin can reduce the genotoxicity & cytotoxicity induced by acrylamide	Cao et al. (2008)

(continued)

Table 7.1 (continued)

Sl: No	Phytochemicals	Model of study	Dosage/Concentration	Key parameters/ Observations	Reference
15	Cyanidin-3-glucoside	In vitro- human malondialdehyde -MB-231 cells	Cyanidin-3-glucoside- (10–100 micro molar) for 4 h Acrylamide-5 mM	Cyanidin treatment reduced cytotoxicity, reactive oxygen species level, increased glutathione. Inhibited the expression of cytochrome P450 2E1.	Song et al. (2013)
16	Procyanidin B2 and a cocoa polyphenolic extract	In vitro- human Caco-2 cells	Acrylamide-(1-20 mM) Procyanidin B2-10μM for 20 h Cocoa phenolic extract-10micro gram per ml. for 20 h	Apoptotic pathway was blocked by the action of phytochemicals and protected the cells from acrylamide induced toxicity. Erk & Jnk pathways were regulated with treatment and also reduced the reactive oxygen species level.	Rodríguez-Ramiro et al. (2011)

anti-ischemic (Liu et al. 2014; Larson et al. 2012), anti-inflammatory, anti-allergy, antitumor, antithrombotic (Aguirre et al. 2014) and neuroprotection (Costa et al. 2013). It is also reported that Quercetin has a protective effect on neurodegenerative disorders such as Alzheimer's disease by reducing the cellular stress, and oxidative injury can alter intracellular signalling and thereby protect the cells and tissue from damage (Lakhanpal and Rai 2007).

The oral administration of Quercetin can improve learning and memory ability. It is also reported that it can pass through the blood-brain barrier, thereby helping to slow down the development of degenerative diseases (Wattanathorn et al. 2007). Quercetin also shows a protective effect against toxicity induced by acrylamide administration. Acrylamide exposure caused a decline in acetylcholinesterase activity, a significant increase in lipid peroxidation. Also, the levels of reduced glutathione and anti-oxidant enzymes superoxide dismutase and catalase were decreased in the kidney, liver and brain (Uthra et al. 2017). Another study revealed that Quercitin is hepatoprotective against acrylamide toxicity. It increased the GST activity, reduced DNA damage, and decline in guanosine level. Quercitin treatment also showed a significant reduction in tissue damage when Histopathological examination was done using H&E staining (Sabah Ansar et al. 2016).

Thymoquinone is a constituent in *Nigella sativa*, is known for its anti-oxidant properties and is nowadays used as a phytochemical for various studies, and several mechanisms are also suggested (Tavakkoli et al. 2017). Nigella sativa oil shows that the pre-treatment of rats reduces glutathione levels, catalase, and a significant increase in nitric oxide levels. Also, a decrease in the Na^+, K – ATPase activity and an increase in the acetylcholine esterase activity was observed. These neuroprotective effects of thymoquinone may be due to its properties like anti-inflammation and free radical quenching capabilities (Ezz et al. 2011). Direct interaction of thymoquinone with free radicals helps to increase glutathione level (Farkhondeh et al. 2018).

Jamshid et al. studied that thymoquinone has a protective effect against acrylamide toxicity. The study reveals that the acrylamide administration of 50 mg/kg IP for 11 days in rats caused reduced gait abilities, increased Malondialdehyde content, and reduced glutathione contents in the sciatic nerve. Also, from western blot studies, they found that P-ERK/ERK ratio reduced with acrylamide administration; also, myelin basic protein declined and increased caspase 9&3 levels, increase in the ratios of P-JNK/JNK, Bax/Bcl-2, P-P38/P38. The administration of thymoquinone, along with acrylamide helps to prevent the toxic effect of acrylamide. Gait abnormalities were prevented, a significant increase in glutathione, decline in Malondialdehyde levels, apoptotic markers level reduced and also elevated the myelin basic protein content and P-ERK/ERK ratio. Thymoquinone thus showed neuroprotection against acrylamide toxicity by anti-oxidant and antiapoptotic mechanism (Jamshid et al. 2019).

Recent studies reveal that thymoquinone, curcumin, and co-treatment can reverse acrylamide-induced toxicity in the *D. melanogaster* model. The study revealed an increase in survival capacity, behavioural improvement followed by reduced reactive oxygen species levels and restored acetylcholinesterase activity.

Immunohistochemistry & m-RNA studies showed that toxicity induced by acrylamide could be reduced with phytochemical treatments (Senthilkumar et al. 2020).

Geraniol is a monoterpene commonly seen in spices, herbs, essential oils, coriander, and lavender. It is used as an agent in beverages and food, perfumes, cleaning agents, and cosmetics due to its aroma (Lapczynski et al. 2008). It has certain pharmaceutical properties like anti-inflammatory & anti-oxidant (Ahmad et al. 2011). Recent studies in *in-vivo* models showed that the supplementation of Geraniol could suppress the toxicity induced by ferric nitrilotriacetate and shows protective action against tumour and liver cancers (Cardozo et al. 2011).

Curcumin is a main active constituent of turmeric, a widely used spice for various medicinal needs. The neuroprotective action of curcumin is widely studied than geraniol, and it is accepted as an anti-oxidant, also effective against inflammation (Mythri and Bharath 2012). It can also act against oxidative stress, thereby helping against ageing and other diseases that occur due to oxidative stress (Aggarwal and Hari Kumar 2009).

Several studies in various models show that neurodegenerative and neurological disorders can be ameliorated with curcumin (Jagatha et al. 2008). Both pharmaceuticals effectively modulate the levels of oxidative stress markers and retain the functions in neurotransmission when studied in the rat model. Mainly the study reveals that the treatment is effective in the cortex and cerebellum region, which suggests that further clinical studies can be done with these phytochemicals or a combination of the same can be done.

Researchers had studied the effect of both phytochemicals in the *D. melanogaster* model also. Interestingly the mortality rate decreased, acetylcholinesterase activity and dopamine level restored with the treatment, and from this study, they suggested that geraniol is as potential as curcumin in neuroprotective action, and both are useful against acrylamide induced toxicity (Prasad et al. 2014b).

In 2015 researchers investigated Extra Virgin Olive Oil's pharmaceutical effect, a lipophilic fraction and hydrophilic fraction against the cardiac toxicity induced by acrylamide and aluminium with co-exposure. *In vivo* rat models were chosen for the study, and treatment shows a significant reduction in toxicity. Co-exposure of the toxicants resulted in cardiac toxicity; the levels of malondialdehyde and advanced oxidation protein product were increased. The activities of superoxide dismutase, catalase, glutathione peroxidase and non-protein thiol decreased significantly. Also, the levels of Vitamin C are reduced. Also, it affected the Low-density lipoprotein *cholesterol*, High-density lipoprotein *cholesterol* & Total cholesterol level and their ratios. An increase in total cholesterol and Low-density lipoprotein *cholesterol* levels was observed, whereas the levels of High-density lipoprotein *cholesterol* and *triglycerides* were decreased. The anti-oxidant effects of olive oil and its fraction helped to overcome the toxicity. DNA damage was reduced, and histopathology results evidence the protective effect of Extra Virgin Olive Oil and its fraction. It helps restore the cardiac muscle cells and lipids to normal values. This study emphasizes that the Extra Virgin Olive Oil can be used to decrease the cardiac toxicity and genotoxicity induced by acrylamide and aluminium (Ghorbel et al. 2015).

Turmeric is known for its anti-oxidant capacity and is used for many uses because of its medicinal property. A study conducted on in vivo rat model shows that turmeric, which is given as 0.5% of total food for 11 days, can ameliorate the toxicity due to acrylamide exposure. They conducted various assay for glutathione, superoxide dismutase, Malondialdehyde, Creatinine, Urea, Uric acid, Total protein and analysed that turmeric can be used as an ingredient in the diet, which can help to reduce the disturbances that occurred to various organs such as the brain, lungs and kidney (Gehan et al. 2008).

Catechin is known for its anti-oxidant properties and is mainly found in green tea and also known for its anti-carcinogenic, anti-inflammatory and antimicrobial properties. Mogda et al. demonstrated that catechin and neem leaves extract can protect against the adverse effects due to intake of acrylamide. Lipid peroxidase, glutathione & catalase level was regulated with treatment, and the histopathological examination of liver, lungs, heart and spleen tissues shows the ant inflammatory& anti-oxidant effect of both phytochemicals. Also, blood analysis for RBCs, WBCs, Lymph, MID, Gran, and Hb values suggests the same protective effect against leukopenia and anaemia. Also, various vitamins such as Vitamin C, Vitamin A, Vitamin E & β-Carotene decreased in serum with acrylamide intake and restored with the treatment (Mogda et al. 2008).

Vitamin F, which is Linoleic acid, was studied against toxicity due to acrylamide and found out that linoleic acid can be used against acrylamide toxicity in rat models. They induced toxicity by providing acrylamide for 10 days at a dose of 38.27 mg/kg, which in turn caused a reduction in body size, body weight, induced skin irritation decline in various enzymatic activity and levels and thereby caused toxicity in various tissues like kidney, liver and brain (Uthra et al. 2017). Treatment with linoleic acid helps to reduce the toxicity caused due to exposure to acrylamide. Haemoglobin % increased, kidney function markers were restored, acetylcholinesterase activity in the brain was restored, lipid peroxidation and glutathione were restored. Levels of superoxide dismutase and catalase came back to normal. Histopathological results clearly stated the effect of Vitamin F, and thus, Linoleic acid can be considered a potent phytochemical against acrylamide toxicity (Mansour et.al 2008, Shrivastava et al. 2019).

Cerium oxide (CeO2) nanoparticles (nano-ceria), due to their catalytic anti-oxidant properties, can be developed as a therapeutic for acrylamide-induced cytotoxicity and oxidative damage. HepG2 human cancer cell lines were initially pre-treated with nano-ceria (50, 100, and 200 mM) for 30 min, followed by the stimulation with acrylamide at a concentration of 200 mM (IC50). After 24 h of incubation, it is analysed that the cell viability was increased in the pre-treated cells along with alleviation in acrylamide induced reactive oygen species generation and lipid peroxidase and dejected glutathione level. It is also evident that the reactive oygen species scavenging property of the compound is arbitrated via reversible reactive oygen species binding and elimination of free radicals with switching between the Ce3þ (reduced) and Ce4þ (oxidized) forms at the particle surface (Azari et al. 2018).

Myricitrin (3′,4′, 5′, 7-five hydroxy flavone-3-O-α-L rhamnoside) bioflavonoid extracted mainly from Chinese bayberry bark. The potent antioxidative and reactive oygen species scavenging ability is due to its polyhydroxy structure. The cell viability can also be increased by the intake of myricitrin, which can be used to investigate as a remedy for acrylamide mediated cytotoxicity in the Caco-2 cells. The study preferred Caco-2 cells because acrylamide is mainly absorbed via the gastrointestinal tract to the body. Thus, acrylamide-mediated cytotoxicity can be studied at its primary level (Chen et al. 2013).

Various other in-vitro studies were conducted worldwide with different phytochemicals against the toxicity induced by acrylamide. Some of them are discussed in this chapter. Table 7.1 shows the effect of various phytochemicals against acrylamide induced toxicity.

Curcumin was studied in HEP G2 cell lines, and due to the anti-oxidant properties, it helped reduce the free radical generation and showed scavenging activity. Also, the cytotoxicity induced by acrylamide was significantly reduced with the treatment of curcumin. Also, the breakage of DNA strands reduced with treatment (Cao et al. 2008). In another study, anthocyanidin, namely, cyaniding-2-glucoside, was studied. The research reveals that Cyanidin treatment reduced Cytotoxicity, reactive oygen species level, increased glutathione, inhibited the expression of Cytochrome P450 2E1. GPx1, GSTP1, c-GCS and CYP2E1 levels of expression were studied using western blot analysis, and treatment showed increased expression of these proteins, revealing the cytoprotective action of the anthocyanin (Song et al. 2013). A group of researchers in Spain studied another anthocyanidin procyanidin b2 and cocoa phenolic extract. They studied human Caco2 cells and showed that these drugs are capable of modulating Erk and JNK pathways. Also, capable of reducing reactive oygen species levels, thereby inhibiting the cytotoxicity that occurred due to acrylamide (Rodríguez-Ramiro et al. 2011).

In 2013, researchers showed that 30 days of *Bacopa monnieri* leaf powder's dietary intake shows a significant effect against acrylamide toxicity. They also found that pre-treatment with N27 cells shows significant protection against oxidative damage and acrylamide toxicity. They have also studied the effect of *Bacopa monnieri* in the *D. melanogaster* model and found out pre-treatment and co-treatment with *Bacopa monnieri* extract show a protective effect against glutathione depletion locomotor dysfunction (Shinomol et al. 2013).

The effect of Morin against acrylamide toxicity was studied by researchers in Turkey and found out the Hepato and nephroprotective effect of Morin. They have conducted the study in rat models and found out that superoxide dismutase, glutathione peroxidase, catalase, lipid peroxidase and glutathione levels were restored with Morin treatment. Immunohistochemical and Histopathological changes and expression of EGFR, nephrin, p53 and AQP2 show the protective effect of Morin. Also, Morin helps to reverse the levels of apoptotic, autophagic and inflammatory parameters (Kandemir et al. 2020).

7.7 Conclusion

Acrylamide induced toxicity has become prevalent for many years and requires further inspection. Acrylamide has proven to instigate toxicity in *D. melanogaster* primarily through oxidative stress. Compared to other models, the flies reproduce fast and in high volumes, which makes them available for many different types of studies. In particular, *D. melanogaster* is extremely useful to observe changes in neurotoxicity too. Through various behavioural studies and individually examining the brain by biochemical studies, it can be correlated to the entire system's effects. Exposure to acrylamide in *D. melanogaster* decreased dopamine, acetylcholine/butyrylcholine production, examination in cells showed the formation of DNA adducts linking it to cancer. Symptoms of neuronal degeneration and underdeveloped brain were observed in other animal models, rats and zebrafish. Apart from the neural system, the immune system and the reproductive system were also affected. Phytochemicals have gained an attraction to combat this problem. Kaempferol and Quercetin increased the production of anti-oxidant enzymes in an acrylamide affected model. Quercitin and thymoquinone have shown added neuroprotective functions. Similarly, the various properties of other phytochemicals are tapped to manage acrylamide toxicity. Thus, phytochemicals can hinder acrylamide toxicity in *D. melanogaster* and other in vitro and in vivo models.

Acknowledgement The authors are thankful to Dr. S. Sahabudeen for providing guidance and support for preparing the manuscript. The authors also thank the reviewers and the editorial team members for providing their valuable suggestions to improve the manuscript.

References

Acrylamide and Cancer Risk. Retrieved April 3, 2020., from https://www.cancer.org/cancer/cancer-causes/Acrylamide.html

Acrylamide|National Pollutant Inventory. Retrieved April 2, 2020., from http://www.npi.gov.au/resource/Acrylamide

Aggarwal BB, Hari Kumar KB (2009) Potential therapeutic effects of curcumin, the anti-inflammatory agent, against neurodegenerative, cardiovascular, pulmonary, metabolic, autoimmune and neoplastic diseases. Int J Biochem Cell Biol 41(1):40–59. https://doi.org/10.1016/j.biocel.2008.06.010

Aguirre L, Portillo MP, Hijona E, Bujanda L (2014) Effects of resveratrol and other polyphenols in hepatic steatosis. World J Gastroenterol 20(23):7366–7380. https://doi.org/10.3748/wjg.v20.i23.7366

Ahmad ST, Arjumnd W, Seth A, Nafees S, Rashid S, Ali N, Sultana S (2011) Preclinical renal cancer chemopreventive efficacy of geraniol by modulation of multiple molecular pathways. Toxicology 290(1):69–81. https://doi.org/10.1016/j.tox.2011.08.020

Alkarim S, Elassouli S, Ali S, Ayuob N, Elassouli Z (2015) Effects of low dose Acrylamide on rat reproductive organs structure, fertility and gene integrity. Asian Pac J Reprod 4(3):179–187. https://doi.org/10.1016/j.apjr.2015.05.001

Ansar S, Siddiqi NJ, Zargar S, Ganaie MA, Abudawood M (2016) Hepatoprotective effect of Quercetin supplementation against Acrylamide induced DNA damage in wistar rats. BMC Complement Altern Med 16(1):327. https://doi.org/10.1186/s12906-016-1322-7

Arıhan O, Seringec NB, Gürel EI, Dikmenoglu NH (2011) Effects of oral Acrylamide intake on blood viscosity parameters in rats. Clin Hemorheol Microcirc 47(1):45–52. https://doi.org/10.3233/CH-2010-1364

Atoff K, Kertika D, Lundqvist J, Oredsson S, Forsby A (2015) Acrylamide affects proliferation and differentiation of the neural progenitor cell line C17.2 and the neuroblastoma cell line SH-SY5Y. Toxicol In Vitro 35:100–111. https://doi.org/10.1016/j.tiv.2016.05.014

Azari A, Shokrzadeh M, Zamani E, Amani N, Shaki F (2018) Cerium oxide nanoparticles protects against acrylamide induced toxicity in HepG2 cells through modulation of oxidative stress. Drug Chem Toxicol 42(1):54–59. https://doi.org/10.1080/01480545.2018.1477793

Biedermann M, Biedermann-Brem S, Noti A, Grob K, Egli P, Mandli H (2002) Two GC-MS methods for the analysis of Acrylamide in foods. Mitt Lebensmittelunters Hyg 93(6):638–652

Cao J, Liu Y, Jia L, Jiang LP, Geng CY et al (2008) Curcumin attenuates acrylamide-induced cytotoxicity and Genotoxicity in HepG2 cells by ROS scavenging. J Agric Food Chem 56(24):12059–12063. https://doi.org/10.1021/jf8026827

Cardozo MT, de Conti S, Ong TP, Scolastici C, Purgatto E, Horst MA, Bassoli BK, Moreno FS (2011) Chemopreventive effects of β-ionone and geraniol during rat hepatocarcinogenesis promotion: distinct actions on cell proliferation, apoptosis, HMG CoA reductase and RhoA. J Nutr Biochem 22(2):130–135. https://doi.org/10.1016/j.jnutbio.2009.12.007

Chen W, Feng L, Shen Y, Su H, Li Y, Zhuang J et al (2013) Myricitrin inhibits acrylamide-mediated cytotoxicity in human Caco-2 cells by preventing oxidative stress. Biomed Res Int 2013:724183. https://doi.org/10.1155/2013/724183

Costa LG, Tait L, de Laat R, Giordano G, Pellacani C, Cole TB, Furlong CE (2013) Modulation of paraoxonase 2 (PON2) in mouse brain by the polyphenol quercetin: a mechanism of neuroprotection. Neurochem Res 38(9):1809–1818. https://doi.org/10.1007/s11064-013-1085-1

Duan X, Wang Q, Chen K, Zhu C, Liu J, Sun S (2015) Acrylamide toxic effects on mouse oocyte quality and fertility in vivo. Sci Rep 5(1):11562. https://doi.org/10.1038/srep11562

Elblehi SS, El Euony OI, El-Sayed YS (2020) Apoptosis and astrogliosis perturbations and expression of regulatory inflammatory factors and neurotransmitters in acrylamide-induced neurotoxicity under ω3 fatty acids protection in rats. Neurotoxicol 76:44–57. https://doi.org/10.1016/j.neuro.2019.10.004

Exon JH (2006) A review of the toxicology of acrylamide. J Toxicol Environ Health B Crit Rev 9(5):397–412. https://doi.org/10.1080/10937400600681430

Ezz HS, Khadrawy YA, Noor NA (2011) The neuroprotective effect of curcumin and Nigella sativa oil against oxidative stress in the pilocarpine model of epilepsy: a comparison with valproate. Neurochem Res 36(11):2195–2204. https://doi.org/10.1007/s11064-011-0544-9

Fang J, Liang CL, Jia XD, Li N (2014) Immunotoxicity of acrylamide in female BALB/c mice. Biomed Environ Sci 27(6):401–409. https://doi.org/10.3967/bes2014.069

Faria M, Valls A, Prats E, Bedrossiantz J et al (2019) Further characterization of the zebrafish model of acrylamide acute neurotoxicity: gait abnormalities and oxidative stress. Sci Rep 9:7075. https://doi.org/10.1038/s41598-019-43647-z

Farkhondeh T, Samarghandian S, Shahri AMP, Samini F (2018) The neuroprotective effects of thymoquinone: a review. Dose-Response 16(2):1559325818761455. https://doi.org/10.1177/1559325818761455

Friedman M, Levin CE (2008) Review of methods for the reduction of dietary content and toxicity of acrylamide. J Agric Food Chem 56(15):6113–6140. https://doi.org/10.1021/jf0730486

Gehan MM, Hanaa H El-S, Emily TH, Mohammed K Abd E R (2008) Turmeric may protect cells from oxidative stress by acrylamide in vivo. J Egypt Toxicol 38:123–129

Ghorbel I, Khemakhem M, Boudawara O, Marrekchi R, Jamoussi K (2015) Effects of dietary extra virgin olive oil and its fractions on anti-oxidant status and DNA damage in the heart of rats co-exposed to aluminium and acrylamide. Food Funct 6(9):3098–3108. https://doi.org/10.1039/c5fo00342c

Gökmen V, Şenyuva HZ (2006) A simplified approach for the kinetic characterization of acrylamide formation in fructose-asparagine model system. Food Addit Contam 23(4):348–354. https://doi.org/10.1080/02652030500482355

He Y, Tan D, Bai B, Wu Z, Ji S (2017) Epigallocatechin-3-gallate attenuates Acrylamideinduced apoptosis and astrogliosis in rat cerebral cortex. Toxicol Mech Method. https://doi.org/10.1080/15376516.2017.1279251

Jagatha B, Mythri RB, Vali S, Bharath MM (2008) Curcumin treatment alleviates the effects of depletion of glutathione in vitro and in vivo: therapeutic implications for Parkinson's disease explained via in silico studies. Free Radic Biol Med 44(5):907–917. https://doi.org/10.1016/j.freeradbiomed.2007.11.011

Jamshid T, Soghra M, Khalil A, Hossein H (2019) Neuroprotective effects of thymoquinone in acrylamide-induced peripheral nervous system toxicity through MAPKinase and apoptosis pathways in rat. Neuro Chem Res 44(5):1101–1112. https://doi.org/10.1007/s11064-019-02741-4

Kacar S, Sahinturk V, Kutlu MH (2019) Effect of acrylamide on BEAS-2B normal human lung cells: cytotoxic oxidative, apoptotic and morphometric analysis. Acta Histochem 121(5):595–603. https://doi.org/10.1016/j.acthis.2019.05.005

Kandemir FM, Yildırım S, Kucukler S, Caglayan C, Darendelioğlu E, Dortbudak MB (2020) Protective effects of Morin against acrylamide-induced hepatotoxicity and nephrotoxicity: a multi-biomarker approach. Food Chem Toxicol 138:111190. https://doi.org/10.1016/j.fct.2020.111190

Kim K, Park B, Rhee D, Pyo S (2015) Acrylamide induces senescence in macrophages through a process involving ATF3, ROS, p38/JNK, and a telomerase-independent pathway. Chem Res Toxicol 31(12):1426. https://doi.org/10.1021/acs.chemrestox.8b00323

Komoike Y, Matusoka M (2016) Endoplasmic reticulum stress-mediated neuronal apoptosis by acrylamide exposure. Toxicol Appl Pharm 310:68–77. https://doi.org/10.1016/j.taap.2016.09.005

Konings EJ, Baars AJ, Van Klaveren JD, Spanjer MC, Rensen PM, Hiemstra M, van Kooji JA, Peters PWJ (2003) Acrylamide exposure from foods of the Dutch population and an assessment of the consequent risks. Food Chem Toxicol 41(11):1569–1579. https://doi.org/10.1016/s0278-6915(03)00187-x

Kumar PP, Harish HP (2019) Low molecular weight chitosan (~20 kDa) protects acrylamide induced oxidative stress in D. melanogaster by restoring dopamine and KIF5B levels. Carbohydr Polym 222:115005. https://doi.org/10.1016/j.carbpol.2019.115005

Lai SM, Gu ZT, Zhao MM, Li XX, Ma YX, Luo L, Liu J (2017) Toxic effect of acrylamide on the development of hippocampal neurons of weaning rats. Neural Regen Res 12(10):1648–1654. https://doi.org/10.4103/1673-5374.217345

Lakhanpal P, Rai DK (2007) Quercetin: a versatile flavonoid. Internet J Med Update 2(2):22–37

Lapczynski A, Bhatia SP, Foxenberg RJ, Letizia CS, Api AM (2008) Fragrance material review on geraniol. Food Chem Toxicol 46(11):160–170. https://doi.org/10.1016/j.fct.2008.06.048

Larson AJ, Symons JD, Jalili T (2012) Therapeutic potential of Quercetin to decrease blood pressure: review of efficacy and mechanisms. Adv Nutr 3(1):39–46. https://doi.org/10.3945/an.111.001271

Lee GJ, Wang YS, Chou CC (2014) Acrylamide-induced apoptosis in rat primary astrocytes and human astrocytoma cell lines. Toxicol In Vitro 28(4):562–570. https://doi.org/10.1016/j.tiv.2014.01.005

Lionetto MG, Caricato R, Calisi A, Giordano ME, Schettino T (2013) Acetylcholinesterase as a biomarker in environmental and occupational medicine: new insights and future perspectives. Biomed Res Int 2013:321213. https://doi.org/10.1155/2013/321213

Liu H, Guo X, Chu Y, Lu S (2014) Heart protective effects and mechanism of Quercetin preconditioning on anti-myocardial ischemia reperfusion (IR) injuries in rats. Gene 545(1):149–155. https://doi.org/10.1016/j.gene.2014.04.043

Liu S, Jiang L, Zhong T, Kong S, Zheng R, Kong F et al (2015) Effect of acrylamide on oocyte nuclear maturation and cumulus cells apoptosis in mouse in vitro. PLoS One 10(8):e0135818. https://doi.org/10.1371/journal.pone.0135818

Liu J, Liu X, Man Y, Liu Y (2018) Reduction of acrylamide content in bread crust by starch coating. J Sci Food Agric 98(1):336–345. https://doi.org/10.1002/jsfa.8476

Lopachin RM (2004) The changing view of acrylamide neurotoxicity. Neurotoxicology 25(4):617–630. https://doi.org/10.1016/j.neuro.2004.01.004

LoPachin RM, Gavin T (2012) Molecular mechanism of acrylamide neurotoxicity: lessons learned from organic chemistry. Environ Health Perspect 120(12):1650–1657. https://doi.org/10.1289/ehp.1205432

Mansour M, EM I, Kholy -EL MM, Madawy-EL AS (2008) Antioxidant and histopathological effect of catechin and neem leaves extract in acrylamide toxicity of rats. Egypt J Comp Path Clinic Path 21(1):290–313

Mehri S, Abnous K, Mousavi SH, Shariaty VM, Hosseinzadeh H (2012) Neuroprotective effect of crocin on acrylamide-induced cytotoxicity in PC12 cells. Cell Mol Neurobiol 32(2):227–235. https://doi.org/10.1007/s10571-011-9752-8

Mogda K, Mansour, Ibrahim EM, El-Kholy MM, El-Madawy SA (2008) Anti-oxidant and histopathological effect of catechin and neem leaves extract in acrylamide toxicity of rats. Egypt J Comp Path Clinic Path 21(1):290–313

Moorman WJ, Reutman SS, Shaw PB, Blade LM, Marlow D et al (2012) Occupational exposure to acrylamide in closed system production plants: air levels and biomonitoring. J Toxicol Environ Health A 75(2):100–111. https://doi.org/10.1080/15287394.2011.615109

Mythri RB, Bharath MMS (2012) Curcumin: a potential neuroprotective agent in Parkinson's disease. Curr Pharm Des 18(1):91–99. https://doi.org/10.2174/138161212798918995

Neckameyer WS (1996) Multiple roles for dopamine in *Drosophila melanogaster* development. Dev Biol 176(2):209–219. https://doi.org/10.1006/dbio.1996.0128

Oda SS (2017) Metformin protects against experimental Acrylamide neuropathy in rats. Drug Dev Res 78(7):349–359. https://doi.org/10.1002/ddr.21400

Pandey UB, Nichols CD (2011) Human disease models in *Drosophila melanogaster* and the role of the fly in therapeutic drug discovery. Pharmacol Rev 63(2):411–436. https://doi.org/10.1124/pr.110.003293

Prang C, Roy NM, Ton C, Lin Y, McGrath P (2006) Neurotoxicity assessment using zebrafish. J Pharmacol Tox Met 55(1):103–112. https://doi.org/10.1016/j.vascn.2006.04.004

Prasad SN (2014a) Mitigation of acrylamide-induced behavioural deficits, oxidative impairments and neurotoxicity by oral supplements of geraniol (a monoterpene) in a rat model. Chem Biol Interact 223:27–37. https://doi.org/10.1016/j.cbi.2014.08.016

Prasad SN (2014b) Neuroprotective effect of geraniol and curcumin in an acrylamide model of neurotoxicity in *Drosophila melanogaster* : relevance to neuropathy. J Insect Physiol 60:7–16. https://doi.org/10.1016/j.jinsphys.2013.10.003

Prasad SN, Muralidhara (2012) Evidence of acrylamide induced oxidative stress and neurotoxicity in *Drosophila melanogaster* - its amelioration with spice active enrichment: relevance to neuropathy. Neurotoxicology 33(5):1254–1264. https://doi.org/10.1016/j.neuro.2012.07.006

Prasad SN, Muralidhara (2019) Neuromodulatory effects of aqueous extract of Coriandrum sativum seeds against acrylamide induced toxicity in *Drosophila melanogaster*. Int J Pharm Sci Res 10(2):1127–1135. https://doi.org/10.26452/ijrps.v10i2.393

Prats E, Gomez-Canela C, Ben-Lulu S, Zev T et al (2017) Modelling acrylamide acute neurotoxicity in zebrafish larvae. Sci Rep 7(1):13952. https://doi.org/10.1038/s41598-017-14460-3

Ranjini A, Manonmani HK (2019) Resveratrol protects acrylamide/glycidamide induced genotoxicity in hepatic cell line. Pharm Innov 8(1):56–64

Riemensperger T, Isabel G, Coulom H, Neuser K, Seugnet L, Kume K (2011) Behavioral consequences of dopamine deficiency in the *Drosophila* central nervous system. Proc Natl Acad Sci U S A 108(2):834–839. https://doi.org/10.1073/pnas.1010930108

Rodríguez-Ramiro I, Ramos S, Bravo L, Goya L, Martín MA (2011) Procyanidin B2 and a cocoa polyphenolic extract inhibit acrylamide-induced apoptosis in human Caco-2 cells by preventing oxidative stress and activation of JNK pathway. J Nutr Biochem 22(12):1186–1194. https://doi.org/10.1016/j.jnutbio.2010.10.005

Sathya NP, Muralidhara (2014) Mitigation of acrylamide-induced behavioral deficits, oxidative impairments and neurotoxicity by oral supplements of geraniol (a monoterpene) in a rat model. Chem Biol Interact 223:27–37. https://doi.org/10.1016/j.cbi.2014.08.016

Scaplen KM, Mei NJ, Bounds HA, Song SL, Azanchi R, Kaun KR (2019) Automated real-time quantification of group locomotor activity in *Drosophila melanogaster*. Sci Rep 9(1):4427. https://doi.org/10.1038/s41598-019-40952-5

Schmatz R, Mazzanti CM, Spanevello R, Stefanello N, Gutierres J et al (2009) Resveratrol prevents memory deficits and the increase in acetylcholinesterase activity in streptozotocin-induced diabetic rats. Eur J Pharmacol 610(1–3):42–48. https://doi.org/10.1016/j.ejphar.2009.03.032

Senthilkumar S, Raveendran R, Madhusoodhan S, Sundar M, Shankar S et al (2020) Developmental and behavioral toxicity induced by acrylamide exposure and amelioration using phytochemicals in Drosophila melanogaster. J Hazard Mater 394:122533. https://doi.org/10.1016/j.jhazmat.2020.122533

Shi J, Ma Y, Zheng M, Ruan Z, Liu J, Tian S, Zhang D, He X, Li G (2012) Effect of sub-acute exposure to acrylamide on GABAergic neurons and astrocytes in weaning rat cerebellum. Toxicol Ind Health 28(1):10–20. https://doi.org/10.1177/0748233711401264

Shinomol GK, Raghunath N, Bharath MM, Muralidhara (2013) Prophylaxis with Bacopa monnieri attenuates acrylamide induced neurotoxicity and oxidative damage via elevated antioxidant function. Cent Nerv Syst Agents Med Chem 13(1):3–12. https://doi.org/10.2174/1871524911313010003

Shrivastava S, Nirala KS, Mohamad SR, Shukla S, Sharma A, Uthara C (2019) Protective efficacy of vitamin F against acrylamide induced toxicity: studies on oxidative stress biomarkers. Open Biomarkers J 9:62-69. https://doi.org/10.2174/1875318301909010062

Sickles DW, Brady ST, Testino A, Friedman MA, Wrenn RW (1996) Direct effect of the neurotoxicant acrylamide on kinesin based microtubule motility. J Neurosci Res 46(1):7–17. https://doi.org/10.1002/(SICI)1097-4547(19961001)46:1<7::AID-JNR2>3.0.CO;2-P

Sickles DW, Stone JD, Friedman MA (2002) Fast axonal transport : a site of acrylamide neurotoxicity ? Neurotoxicology 23(2):223–251. https://doi.org/10.1016/s0161-813x(02)00025-6

Song J, Zhao M, Liu X, Zhu Y, Hu X, Chen F (2013) Protection of cyanidin-3-glucoside against oxidative stress induced by acrylamide in human MDA-MB-231 cells. Food Chem Toxicol 58:306–310. https://doi.org/10.1016/j.fct.2013.05.003

Sundararajan V, Dan P, Kumar A, Venkatasubbu GD (2019) Applied surface science *Drosophila melanogaster* as an in vivo model to study the potential toxicity of cerium oxide nanoparticles. Appl Surf Sci 490:70–80. https://doi.org/10.1016/j.apsusc.2019.06.017

Tabeshpour J, Mehri S, Abnous K, Hosseinzadeh H (2019) Neuroprotective effects of Thymoquinone in acrylamide-induced peripheral nervous system toxicity through MAP kinase and apoptosis pathways in rat. Neurochem Res 44(5):1101–1112. https://doi.org/10.1007/s11064-019-02741-4

Tabeshpour J, Mehri S, Abnous K, Hosseinzadeh H (2020) Role of oxidative stress, MAPKinase and apoptosis pathways in the protective effects of Thymoquinone against acrylamide-induced central nervous system toxicity in rat. Neurochem Res 45(2):254–267. https://doi.org/10.1007/s11064-019-02908-z

Tardiff RG, Gargas ML, Kirman CR, Carson ML, Sweeney LM (2010) Estimation of safe dietary intake levels of acrylamide for humans. Food Chem Toxicol 48(2):658–667. https://doi.org/10.1016/j.fct.2009.11.048

Tavakkoli A, Ahmadi A, Razavi BM, Hosseinzadeh H (2017) Black seed (*Nigella sativa*) and its constituent thymoquinone as an antidote or a protective agent against natural or chemical toxicities. Iran J Pharm Res 16:2–23

Tian SM, Ma YX, Shi J, Lou TY, Liu SS, Li GY (2015) Acrylamide neurotoxicity on the cerebrum of weaning rats. Neural Regen Res 10(6):938–943. https://doi.org/10.4103/1673-5374.158357
Uthra C, Shrivastava S, Jaswal A, Sinha N, Reshi MS, Shukla S (2017) Therapeutic potential of Quercetin against acrylamide induced toxicity in rats. Biomed Pharmacother 86:705–714. https://doi.org/10.1016/j.biopha.2016.12.065
Wang H, Huang P, Lie T, Li J, Hutz RJ, Li K, Shi F (2010) Reproductive toxicity of acrylamide-treated male rats. Reprod Toxicol 29(2):225–230
Wattanathorn J, Phachonpai W, Priprem A, Suthiparinyanont S (2007) Intranasal administration of quercetin liposome decreases anxiety-like behaviour and increases spatial memory. Am J Agric Biol Sci 2:31–35. https://doi.org/10.3844/ajabssp.2007.31.35
Xu F, Oruna-Concha MJ, Elmore JS (2016) The use of asparaginase to reduce acrylamide levels in cooked food. Food Chem 210:163–171. https://doi.org/10.1016/j.foodchem.2016.04.105
Yang J, McCart C, Woods DJ, Terhzaz S, Greenwood KG, ffrench-Constant RH, Dow JA (2007) A Drosophila systems approach to xenobiotic metabolism. Physiol Genomics 30(3):223–231. https://doi.org/10.1152/physiolgenomics.00018.2007
Yao X, Zhang Y, Yan L, Yao L, Guan W, Zeng F, Cao F, Zhang Y (2014) Acrylamide exposure impairs blood-cerebrospinal fluid barrier function. Neural Regen Res 9(5):555–560. https://doi.org/10.4103/1673-5374.130080
Yilmaz BO, Yildizbayrak N, Aydin Y, Erkan M (2017) Evidence of Acrylamide-and glycidamide-induced oxidative stress and apoptosis in Leydig and Sertoli cells. Hum Exp Toxicol 36(12):1225–1235. https://doi.org/10.1177/0960327116686818
Yousef MI, El-Demerdash FM (2006) Acrylamide-induced oxidative stress and biochemical perturbations in rats. Toxicology 219(1–3):133–141. https://doi.org/10.1016/j.tox.2005.11.008
Zamani E, Shaki F, Abediankenari S, Shokrzadeh M (2017a) Acrylamide induces immunotoxicity through reactive oxygen species production and caspase-dependent apoptosis in mice splenocytes via the mitochondria-dependent signaling pathways. Biomed Pharmacother 94:523–530. https://doi.org/10.1016/j.biopha.2017.07.033
Zamani E, Shokrzadeh M, Fallah M, Shaki F (2017b) A review of acrylamide toxicity and its mechanism. Pharm Biomed Res 3(1):1–7. https://doi.org/10.18869/acadpub.pbr.3.1.1
Zeiger E, Recio L, Fennell TR, Haseman JK, Snyder RW, Friedman M (2009) Investigation of low-dose response in the *in vivo* induction of micronuclei and adducts by acrylamide. Toxicol Sci 107(1):247–257. https://doi.org/10.1093/toxsci/kfn214
Zhang B, Shao H, Wang H, Xiu H, Chen X, LI ZS, CAO P, Zhu D, Yang Yi Guang, Xiao JW, Li B (2017) Acrylamide-induced Subacute neurotoxic effects on the cerebral cortex and cerebellum at the synapse level in rats. Biomed Environ Sci 30(6):432-443. https://doi.org/10.3967/bes2017.057
Zhao M, Lewis Wang FS, Hu X, Chen F, Chan HM (2017a) Acrylamide induced neurotoxicity in primary astrocytes and microglia: roles of the Nrf2-ARE and NF-κB pathways. Food Chem Toxicol 106(PtA):25–35. https://doi.org/10.1016/j.fct.2017.05.007
Zhao M, Wang F, Hu XS, Chen F, Chan HM (2017b) Effect of acrylamide-induced neurotoxicity in a primary astrocytes/microglial co-culture model. Toxicol In Vitro 39:119–125. https://doi.org/10.1016/j.tiv.2016.11.007

Chapter 8
Mesoporous Silica Nanoparticles Are Nanocarrier for Drug Loading and Induces Cell Death in Breast Cancer

Lakshminarasimhan Harini, Karthikeyan Bose, T. Mohan Viswanathan, Nachimuthu Senthil Kumar, Krishnan Sundar, and Thandavarayan Kathiresan

Abstract Breast cancer is a frequent malignancy in women globally, and it is curable at an early stage. However, the later stage is non-curable with currently available therapy. Because of its high invasiveness and heterogeneity in nature is limiting drug activity. Nowadays, chemotherapeutic molecules are created more side effects; therefore, they need an alternative strategy for overcoming this problem. Natural compounds are biologically active molecules that are prevention of disease with lesser toxicity. Curcumin is an active polyphenolic compound and hydrophobic. In this regard, mesoporous silica nanoparticles are effectively utilized as a nanocarrier for loading various therapeutic molecules, including curcumin and other natural products. The curcumin-loaded nanocarriers are targeted delivery and sustained release at intracellular level then induced cell death effectively. Depending on the conjugate, the natural compounds effectively loading enhances the sustained release from the carrier. Therefore, mesoporous silica nanocarriers significantly play a major role in targeted delivery, sustained release and induce breast cancer cell death.

Keywords Apoptosis · Curcumin · Chemotherapy · Comet assay · Drug loading · Drug delivery

L. Harini · T. M. Viswanathan · K. Sundar · T. Kathiresan (✉)
Department of Biotechnology, School of Bio and Chemical Engineering, Kalasaligam Academy of Research and Education, Krishnankoil, India
e-mail: sundarkr@klu.ac.in; t.kathiresan@klu.ac.in

K. Bose
Oregon Health and Science University, Knight Cardiovascular Institute (KCVI), Portland, OR, USA

N. S. Kumar
Department of Biotechnology, Mizoram University, Aizawl, Mizoram, India

K. M. Gothandam et al. (eds.), *Environmental Biotechnology Volume 4*, Environmental Chemistry for a Sustainable World 68,
https://doi.org/10.1007/978-3-030-77795-1_8

8.1 Introduction

In India, breast cancer incidence is predicted to be 90,723 in the year 2015, which is less when compared with world breast cancer occurrence. Of the breast cancer patients, India holds 17% of cases globally; however, the mortality of breast cancer patients in India is around 54%, which is much higher than in other countries (Bawa et al. 2016). As long as 12% of patients were detected in a later stage of breast cancer in developed countries, 50% of them were diagnosed in India (Saraswathy and Gong 2013). Because of poor cancer awareness and delayed diagnosis is the main reason for high mortality in India. The metastasis of breast cancer is a high capacity to spread to other parts of the body. The breast cancer cells are migrated through both axillary lymph nodes and blood vessels during metastasis (Cardoso 2018). The metastatic breast cancer incidence is more commonly in bone (67%), axillary lymph node (30–50%), liver (40%), lung (37%), and brain (13%) (Golse and Adam 2017). Metastatic breast cancer is currently difficult to curable, and most patients with median survive approximately 2–3 years (Shien and Doihara 2014).

Metastasis is the later stage of tumorigenesis and establishes depending on the size of the primary tumor, lymph node, and undifferentiated histopathology. Initially, minimized tumor cells circulate from the original site to form micrometastasis, then rapidly developing metastasis through a set of proteins (Kweia et al. 2010). The protein biomarker for cancer metastasis is epidermal growth factor receptor 2, urokinase plasminogen activator and plasminogen activator inhibitor (Wang et al. 2011). The nucleoside diphosphate kinase 1, kisspeptins, *KAI1,* carbamoyl-phosphate synthetase 2, aspartate transcarbamylase, and dihydroorotase, breast cancer metastasis suppressor 1, and mitogen-activated protein kinase 4 were used as gene biomarkers to suppress metastasis (Weigelt et al. 2005).

Till now, chemotherapy, anti-angiogenesis therapy, hormonal therapy, and antibody-based therapy were implemented against metastatic breast cancer (Weigelt et al. 2005). However, metastasis is non-curable and induced more side effects. To overcome this problem, using nutraceuticals with anti-cancer activity is less toxic and able to chemotherapeutic challenges. Among the nutraceuticals, curcumin is an efficient molecule with pleiotropic effects, including anti-cancer activity against many cancerous cell lines (Harini et al. 2017, 2019). However, the major limitation of curcumin *in vivo* conditions is more hydrophobicity and less bioavailability. The solubility and bioavailability of curcumin were enhanced by conjugating with nanocarriers such as mesoporous silica nanoparticles. These nanocarriers are porous in nature with modifiable pore sizes ranging from 2–50 nm for drug loading. Surface modifications of these nanocarriers can target and sustain release and then avoid side effects of the drug. Therefore, the present chapter focused on how drug-loaded mesoporous silica nanoparticles are induced apoptosis in cancerous cells.

8.2 Breast Cancer

Globally, breast cancer is the leading cause of mortality in the female population. Nearly 1.7 million people were diagnosed with breast cancer in 2012. Change in lifestyle and family history was considered as a significant risk factor in breast cancer development. Besides, breast cancers have many subtypes, and genetic causes play a critical role in developing cancers.

8.2.1 Types of Breast Cancer

Breast cancer is the most frequently occurring women cancer in the world. Breast cancer is a heterogeneous disease and consists mainly of divided into three major subtypes they are 1 histological, 2 intrinsic, and 3, surrogate intrinsic subtype. The histological subtype is classified into preinvasive and invasive subtype among them, preinvasive is further classified into ductal carcinoma *in situ* and lobular carcinoma in situ. The invasive subtype includes ductal carcinoma of no special type and lobular carcinoma of minimal fibrous response. The second intrinsic subtypes are five main categories: luminal A, luminal B, human epidermal growth factor receptor 2 (HER2)-enriched, normal-like and basal-like breast cancers (Parker 2009). Third, surrogate intrinsic subtype consists of luminal A-like, luminal B-like HER2–, luminal B-like HER2+, HER2-enriched and, triple-negative breast cancer. Among them, 70% of breast cancers are articulate through estrogen receptor α (ER). The estrogen receptor positive breast cancer is slowly activated in luminal subtype A and aggressively activate in the endocrine-resistant luminal B subtype (Elizabeth et al. 2012). The estrogen receptor positive is encoded by the estrogen receptor 1 gene belonging to the nuclear receptor superfamily (Olefsky 2001). It consists of two transcriptional activation domains: AF-1 (N-terminal ligand-independent) and AF-2 (C-terminal ligand-dependent) domain. These ER-ligand domains directly bind with specific DNA sequences of estrogen-responsive elements, regulating tumor progression (Sun et al. 2007). The activator protein-1 can tether with another transcriptional factor-like nuclear κB factor and activation of tumor progression.

8.2.2 Genetics of Breast Cancer

Many factors are involved in predisposing women's breast cancer development, such as early menarche, late pregnancy, late menopause, obesity, dietary factors, and high radiations. The further interplay between inherited genes and environmental interactions is a modified genotypic manifestation of cancer development (Kortenkamp 2008). The hallmarks of selective signatures are useful for cancer's diagnostic and therapeutic purposes (Berger and Mardis 2018). The most frequently muted genes in early breast cancer is tumor protein p53 (41%), phosphatidylinositol-4,5-bisphosphate 3-kinase, catalytic subunit alpha (30%), myelocytomatosis (20%), phosphatase and tensin homolog (16%), cyclin D1 (16%), receptor tyrosine-protein kinase erbB-2 (13%), fibroblast growth factor receptor 1 (11%) and GATA3 (10%) (Nik-Zainal 2016). These genes conceal cell cycle alteration through either activation of cell proliferation or cell death suppression, later inducing an oncogenic pathway. The luminal A and B subtypes are expressed their mutations in PIK3CA, GATA3, mitogen activated protein kinase (Brambilla et al. 2015; Bulusu et al. 2016). The basal-like breast cancers are typically mutated in tumor protein p53, retinoblastoma protein, MYC, and breast cancer gene (Bourdeanu and Luu 2014). More than 75% of BRCA1 is mutated in triple negative breast cancer and basal-like subtypes (Breckenridge et al. 2003).

Genomic instability, like telomere dysfunction, is an increased copy number of gene variations are also induce breast cancer (Kimbung et al. 2015; Kweia et al. 2010). Another striving factor is developing drug resistance against cancer cells by altering therapeutic targets, and drug efflux elicits tumor cells against drugs. The heterogeneity of breast cancer enhances the development of drug-resistant and complication of drug targeting towards cancerous cells (Housman et al. 2014).

8.3 Modern Breast Cancer Therapy and Limitation

Depending on the tumor size and metastasis, therapy is mainly divided into two types; they are adjuvant and neoadjuvant therapy. The tumors are primarily found in a single location, and the rest of them are negligible size initially undergoing surgery followed by either chemotherapy or radiation therapy is adjuvant type. The primary tumors like are found in many locations are pursued by neoadjuvant therapy. Initially undergoing either chemotherapy or radiation therapy to reduce the tumor size then performing surgery. After surgery, the elimination of the remaining tumor is through chemotherapy or radiation therapy.

8.3.1 Surgery

Beginning of cancer treatment, when the primary tumor is around 2.5 cm if pursued chemotherapy, the tumors are dissolved and mixed with blood, which leads to tumor lysis syndrome. To avoid the consequences, surgery is preferred when the breast cancer size is more than 3 cm diameter. Surgical removal is known as lumpectomy or mastectomy after that following either chemotherapy or radiation therapy. The tumor tissues and the small portion of normal tissues are surgically removed through wide excision. A quarter of the breast tissues are removed along with the chest muscles, and 2–3 cm radius of the primary tumor is quadrantectomy. Breast-conserving therapy refers to surgery followed by a moderate dose of radiation therapy to eliminate minimal residues of cancer (Sharma et al. 2010). Further, lymphatic mapping and sentinel lymph node biopsy is also performed to reduce the morbidity of cancers.

8.3.2 Chemotherapy

The chemotherapeutic molecules are treating cancer cells to reduce the size or eliminate tumors effectively. The effective neoadjuvant/adjuvant chemotherapy regimens in breast cancer are generally following in the combination of (i) doxorubicin with cyclophosphamide, (ii) 5-fluorouracil, with doxorubicin, and cyclophosphamide, (iii) taxanes, with vinorelbine and other alkaloids, (iv) gemcitabine, capecitabine, ixabepilone, and eribulin to reduce or eliminate the tumor. Apart from this general therapy, the specified therapy is also suggesting types of breast cancer. A luminal-like subtype is recommended for combining endocrine therapy (anastrozole, exemestane, fulvestrant, goserelin, and letrozole) with cyclin-dependent kinase 4 and 6 (CDK4/6) inhibitors (abemaciclib, palbociclib, and ribociclib). Breast cancer gene (BRCA1) mutated, and triple negative breast cancer mainly targets poly-ADP ribose polymerase inhibitors (rucaparib, veliparib, and niraparib) with platinum derivatives (cisplatin, carboplatin, and oxaliplatin). The triple negative breast cancer suggested that paclitaxel plus atezolizumab is a superior option and other recommendations are taxane or capecitabine with bevacizumab (anti-vascular endothelial growth factor antibody) and capecitabine, with vinorelbine. The human epidermal growth factor receptor 2 positive (HER2+) subtype first line of combination is trastuzumab and pertuzumab, then taxane or vinorelbine, which leads to increases overall survival.

Further, chemotherapies are targeting receptor proteins such as tyrosine kinase inhibitor (lapatinib, pertuzumab, gefitinib, erlotinib, canertinib and, neratinib) (Sinn and Kreipe 2013), and epidermal growth factor receptor inhibitors (Gefitinib, anilinoquinazoline Erlotiniband, quinazolinamine). The vascular endothelial growth factor and platelet-derived growth factor inhibitors (sunitinib malate, Motesanib, Vandetanib, and, Vatalanib) are mainly antiangiogenic therapy. The downstream of ras is tipifarnib and lorafarnib. Epigenetic modifiers such as (i) histone deacetylases or DNA methyltransferase inhibitors are nucleoside analogs 5-azacytidine (5-aza) and 5-aza-2′-deoxycytidine (decitabine), zebularine and 5-fluoro-2′-deoxycytidine, (ii) HDACi include vorinostat (suberoylanilidehydroxamic acid) (Lustberg and Ramaswamy 2011).

Many drugs are available regarding breast cancer even though the exact mechanism behind the drugs with anti-cancer activity and their protein interactions remained unclear. The major drawback of the anti-cancer drugs was severe side effects such as nausea, diarrhea, fatigue, neutropenia, and elevation of liver enzymes, epistaxis, and conjunctival disorder (Bourdeanu and Luu 2014). Therefore, there is a need to develop a drug that overcomes the above side effects.

8.3.3 *Radiation Therapy*

Radiation therapy is a wide range of primary management of breast cancer, it is a highly efficient treatment, with 40% of cancer cures worldwide. Radiation therapy leads to cancer cure in many patients, along with surgery and radiation therapy. It is mainly using bone cancer and other types of metastasis. The major serious side effects of radiation therapy include ischemic heart disease and pneumonitis (Rutqvist 1993).

8.4 Role of Natural Compounds in Breast Cancer Therapy

Nutraceuticals are biologically active molecules and derived from natural sources, they are mainly involved in the prevention or treatment of diseases with lesser toxicity (Katiyar et al. 2012). Natural products are classified based on their functional groups; (1) carotenoids include β-carotene, lutein, lycopene, tocotrienol, and saponins.(2) phenolic acids containing resveratrol, curcumin, and glucosinolates. (3) phytoestrogens include genistein, daidzein, lignans, and (4) dietary fiber varieties, including soluble and insoluble fiber, dithiolthiones omega 3 fatty acids, and monounsaturated fatty acids (Talekar et al. 2015). Anticarcinogenic natural products such as curcumin, diindolylmethane, genistein, indole-3-carbinol, lutein, quercetin, and sulforaphane are in a clinical trial. Some of the natural compounds with anti-cancer activity are tabulated as shown in Table 8.1.

Polyphenol is a double-edged sword having both antioxidant and pro-oxidant activity. Catechol is carbonyl and hydroxyl groups (3 and 5 positions) of polyphenols are determining the free radical scavenging activity. The mechanism of anticancer activities is included metal chelation, inhibition of oxidases, and stimulates enzymes with antioxidant properties. On the other hand, oxidative induction of polyphenols is the generation of reactive oxygen species (ROS), which consequently affects the mitochondrial-mediated apoptosis, oxidative DNA damage, and regulating cell cycle (Mocanu et al. 2015; Fresco et al. 2006; Talekar et al. 2015).

8.4.1 *Curcumin Structure and Function*

Curcumin is an active polyphenolic compound present in rhizome of *Curcuma longa* and chemically known as 1,7-bis(4-hydroxy-3-methoxyphenyl)-1,6-heptadiene-3,5-dione or diferuloylmethane. The curcumin is consists of two aromatic rings connecting with seven carbon bonds, which are interacting with biomolecules depending on its nature. Curcumin is metabolized and reduced to form curcumin glucuronide, curcumin sulfate, and hexahydrocurcuminol in the gut and liver (Moorthi et al. 2011). Biological properties of curcumin are interacting

Table 8.1 Natural compounds with anti-cancer activity and their source

Anti-cancer compounds	Source	References
Curcumin	Turmeric	Kocaadam and Şanlier (2017) and Baldi et al. (2020)
Cyanidins	Grapes	Ananga et al. (2017) and Hudlikar et al. (2020)
Diindolylmethane/ Indole-3-carbinol	Brassica v	Deb et al. (2018)
Epigallocatechin gallate (EGCG)	Green tea	Gan et al. (2018) and Shafie et al. (2019)
Fisetin	Strawberries, Apples	Min et al. (2017) and Takahashi et al. (2020)
Genistein	Soybean	Chae et al. (2019) and Hafidh (2017)
Gingerol	Gingers	Zhang et al. (2017) and Furlan and Bren (2020)
Kaempferol	Tea, Broccoli, grapefruit	Bonta (2020)
Lycopene	Tomato	Koul et al. (2019)
Phenethylisothiocyanate	Cruciferous vegetable	Hsia et al. (2018)
Resveratrol	Grapes	Rauf et al. (2018)
Rosmarinic acid	Rosemary	Moore et al. (2016)
Sulforaphane	Cruciferous vegetable	Liang et al. (2019)
Triterpenoids	Wax-like coating of fruits and herbs	Kohno et al. (2017)
Vitamin D	Mushroom	Pandya et al. (2019)
Vitamin E	Vegetable oils	Abraham et al. (2019)

with other molecules through covalent or non-covalent interactions. Among them, carbon bonds having 2 keto-enol forms and end of benzene ring containing two hydrogen bond acceptors are active functional groups of curcumin. The antioxidant property of curcumin depends on the activation of functional groups. The curcumin is non-covalently binding with the minor groove of DNA, hydrophobic pocket of proteins, and choline of lipids. The covalent bonding of curcumin is neutralizing reactive oxygen species (ROS) activity by The O-methoxy phenolic OH group (Priyadarsini 2013, 2014). Curcumin is associating with key signaling proteins such as protein kinase A, protein kinase B, Ca^{2+} – dependent protein kinase, nuclear factor kappa-light-chain-enhancer of activated beta (NFK-β), and cytochrome p450 (Rahimi et al. 2016).

The therapeutics of curcumin is extended to anti-cancer activity such as colorectal, pancreatic, breast, lung, prostate cancers, multiple myeloma, head and neck squamous cell carcinoma. Besides, curcumin is effectively curing inflammation and infection-related ailments, Alzheimer's disease, atherosclerosis, diabetes, and chronic bacterial prostatitis (Gupta et al. 2013). The major therapeutic limitation of curcumin is low aqueous solubility and bioavailability. Curcumin administered at 2 g/kg in rats resulted in a serum concentration is $1.35 \pm 0.23 \mu g/mL$ (Cheng et al. 2001). Though the bioactive concentration of curcumin is micromolar, the water solubility is at nanomolar concentration. Another drawback of curcumin is instability at pH 7.4 and a half-life time of 20 min. The curcumin up-taking cells are limited and lesser concentration of cytoplasm. Adverse effects of curcumin on clinical trials include diarrhea, headaches, rashes, yellow stool, and nausea. Elevation of serum alkaline phosphatase and lactate dehydrogenase are unmanageable abdominal pain observed in curcumin administered patients at 48 h (Moran et al. 2016; Nagahama et al. 2016; Gupta et al. 2013).

8.5 Cancer Drug Targeting and Delivery

The tumor micro environments contain abnormal vasculature leads to increase tumor progression, inhibits drug targets and drug delivery. The lesser drug availability is found in the tumor microenvironment because of the absence of lymph vessels and reduces blood vessels. The stromal cells and hypoxic regions are produced impaired blood flow is limited drug exposure to tumor sites (Talekar et al. 2015). The tumor-targeting drug is inhibited by anatomical, physiological, chemical, and clinical barriers (Jain et al. 2015). In addition, vascular endothelium, perivascular space, cell membranes, nucleus, blood-brain barrier, hepatic degradation, renal filtration, drug efflux, and density of the tumor microenvironment are preventing drug targets. To overcome this problem, the therapeutic molecules are loaded with nanocarrier are using targeted delivery and sustained release.

8.5.1 Active Drug Targeting

The drugs are conjugating with the surface of target molecules like nanoparticles, and delivering tumor tissue is active drug targeting. These targeting molecules are binding with tumor tissue and sustained the release of drug molecules intracellularly to overcome the efflux (Jain et al. 2015).

8.5.2 Passive Drug Targeting

Directly delivering drugs to the tumor is passive targeting and both drugs and tumor characteristic are essential for efficient drug delivery. The passive targeting drug carrier complex is stable in metabolism, excretion, opsonization, phagocytosis, and longer circulation in blood. The pH and temperature play a vital role in regulating controlled drug delivery (Khanna 2012). Passive targeting of nanoparticles to cancer depends on enhanced permeability and retention effect. Nanoparticle size (10–100 nm), surface charges, drug-polymer coating are induced systemic circulation and recognition by the phagocytic system. However, the accumulation of drug-nanocarrier depends on drugs loaded and cancer type (Talekar et al. 2015).

8.5.3 *Novel Drug Delivery System (NDDS)*

Most of the anti-cancer drugs are of higher molecular weight, increased lipophilicity, and decreased solubility. Increasing solubility of anti-cancer drugs is altering their pharmacological properties and conjugating with nanoparticles (Yin et al. 2014). Nanotechnology extends its horizons to cancer research is using diagnosis and efficient drug delivery. The nanoparticle roles in tumor diagnosis are localization, marginal detection, adjacent structures, mapping of lymph nodes, and metastases detection. Nanodelivery enhances the identification of variable targets and avoids drug resistance (Talekar et al. 2015). Delivery systems like liposomes, polymeric micelles, dendrimers, nanocantilevers, carbon nanotubes, mesoporous silica nanoparticles, quantum dots, magnetic nanoparticles, gold nanoparticles (AuNP), poly(lactic-co-glycolic acid), multimodal nanoparticles were widely used in cancer targeting (Alam et al. 2014, 2015).

The physicochemical and biological characteristics of nanoparticles are able to optimize dosage concentration would provide a better understanding of nanotoxicity. Similarly, developing and explicit typical testing and characterization protocols can able to provide a synthesis of efficient nanotherapeutics (Babu et al. 2014). The analysis of nanoparticle dispersion, agglomeration, aggregation, and concentration would provide a better understanding of *in vitro* and *in vivo* behavior. Nanoparticles are induced by biological alteration such as oxidative damage leading to oxidative stress, genotoxicity, inflammation of the lung, macrophage activation, modification in the liver, clotting factors, and systemic thrombosis. Nanoparticles are the ability to travel the blood-brain barrier through the intravenous then induces neurotoxicity (Alam et al. 2014, 2015).

8.6 Mesoporous Silica Nanoparticles

The cationic surfactants like tetraethyl orthosilicate, tetramethylorthosilicate, tetramethoxyvinylsilane, sodium meta-silicate, and tetrakis (2-hydroxyethyl) orthosilicate are using as a template for mesoporous silica nanoparticles. Mesoporous silica structure can be easily tailored and produced a variety of structures like discord, wormhole like hexagonal, cubic, lamellar mesophases. Morphology of the mesoporous silica nanoparticles (MSNs) is spheres, hollow sphere, fibers, tubules, gyroid, helical fibers, and crystal structure. The dimension of mesophases is varied from nanometer to centimeters (Mann et al. 1997; Ying et al. 1999; Hoffmann et al. 2006).

8.6.1 Mesoporous Silica Nanoparticle Types

The mesoporous silica nanoparticles are the most widely using nanocarrier for drug delivery vehicles and are mainly divided into 5 major types. The Mobil Research and Development Corporation was first synthesized mesoporous silica particles are called mobil composition of matter (MCM-41). The MCM is generally hexagonal shape and varied pore size range from 2–50 nm, the pore arrangement is ordered structure (Huo et al. 1996; Trewyn et al. 2007). MCM-48 is a cubic arrangement, and MCM-50 is alamella-like arrangement (Oye et al. 2001). The polyethylene oxide and polyalkylene oxide block copolymers used as a template are designed as Santa Barbara Amorphous, or SBA-11 (cubic), SBA-12 and 15 (hexagonal), SBA-16 (cubic cage structure), and pore size is 4.6–30 nm (Zhao et al. 1998). FSM is a folded sheath mesoporous material using pharmaceutical applications, absorbent, and catalysis (Tozuka et al. 2005). Besides, Hiroshima Mesoporous Material-33, Centre for Research Chemistry and Catalysis, and Korea Advanced Institute of Science and Technology have been synthesized is varied pore symmetry and shape (Nandiyanto et al. 2009; Heikkilä et al. 2007).

8.6.2 Mesoporous Silica Nanoparticle Using as a Carrier

The MSNs have attracted researchers because of their high surface area tunable pore size and high drug loading capacity. Synthesis of mesoporous silica nanoparticles was achieved by two methods by (a) simultaneous grafting of surfactant micelles along with silica precursor, and (b) the silica precursor is allowed to accumulate over the pre-formed surfactant micelle. MSN is synthesized from silica precursor tetraethylorthosilate (TEOS), and silanol group appeared on the surface hydroxyl group (Ambrogio et al. 2011; Ekkapongpisit et al. 2012; Elmore 2007). Surface modification is performed by additional hydrolysis and condensation with the functional groups. The MSNs surface area, particle size, pore size, surface charge, and structures are tunable depending on the usage of precursor molecules. The shape of the MSN is varied from spherical to rod shape due to the alteration of nanoparticle synthesis. The shape of these nanoparticles is determining their biological activity, such as depending on the organelle distribution and subcellular localization of the cells (Anand 2007; Azim et al. 2011; Babu et al. 2014). MSN has wide applications, including purification of chemical and pharmaceutical, decolorization of food, improving solubility and surface affinity, sustained and controlled release.

Inorganic MSNs are used in a large number of applications. In the past decade, MSNs have increased attention to their potential biomedical applications. Mesoporous structure, with its tailored and high surface area, has significant advantages for drug delivery systems compared with traditional drug nanocarriers (Jadhav et al. 2015). The protamine coated MSN (MSN – PRM) is reported that selectively releases drugs in the presence of trypsin in the proximity of cancer cells. They are effectively blocking the outward growth of cancer due to drug molecules from the MSNs (Radhakrishnan et al. 2013). The hydrophobic drugs such as peptide and protein drugs like cyclosporine A. Silica nanoparticles improved the solubility of poorly water-soluble drugs and enhanced the absorption and bioavailability of these compounds (Lodha et al. 2012). The silica precursors are negatively charged, and they interact with the positively charged surfactants via electrostatic interactions. The silica species are polymerizing at the micelle surface and creating a negatively charged silica surface (Patil et al. 2011). The anti-cancer drugs are loaded in different types of MSNs are shown in Table 8.2.

8.6.3 Controlled Drug Delivery System

Improving the delivery efficiency of the conventional drug is existing demand in drug delivery system than investing in new drugs. Delivery of a drug in a controlled and slow mode at a target site is required to overcome the disadvantages in the prevailing scenario (Tiwari et al. 2012; Singh and Lillard 2009). The MSNs are the most widely researched drug carrier for an efficient drug delivery system because of their chemical inertness and high drug loading capacity. The efficacy of MSN is determined by its size, shape, and surface functionality. Besides, high surface area, and their modification, tunable pore size, enhances the efficient drug delivery system (Slowing et al. 2008; Yanes and Tamanoi 2012; Barbe et al. 2004). The shape of the nanoparticle influences its biological activity, phagocytosis, evasion, and

Table 8.2 Various anti-cancer drugs loaded in different types of Mesoporous silica nanoparticles

Anti-cancer drug	Types of Mesoporous Nanomaterial	Reference
Curcumin	MCM-41	Harini et al. (2015)
Curcumin	MSNAP	Harini et al. (2019)
Carbamazepine	Supercritical mesoporous nanoparticles	Patil et al. (2011)
Cyclosporin	MSN	Lodha et al. (2012)
Doxorubicin	CmMSN	Mishra et al. (2014)
Fluorescein, Texas Red, and rhodamine B	SBA-15	Slowing et al. (2008)
5-fluorouracil	MCM – 41	Kumar et al. (2017)
Flurbiprofen	FSM-16	Tozuka et al. (2005)
Ibuprofen	MCM-41	Regi et al. (2001)
Ibuprofen	TUD-1	Heikkilä et al. (2007)
Paclitaxel, telmisartan	Nanonetwork and Hierarchical Structure - MSN	Jadhav et al. (2015)
Protamine	Protamine-Capped Mesoporous Nanoparticles	Radhakrishnan et al. (2013)
Vancomyocin	SBA, MSU, and FSM	Trewyn et al. (2007)

increased circulatory time in the blood (Tang et al. 2012). Probably, biodistribution and bioaccumulation of MSN are increased with a high aspect ratio than spherical with a low aspect ratio (Zhou et al. 2011). Non-spherical MSN has a greater ability to escape from an immune response like cylindrical-shaped filamentous bacteria and viruses (Truong et al. 2015). Therefore, both drug-loaded spherical and non-spherical MSNs have induced cell death and among them, non-spherical induced more cell death when compared with spherical MSN. The surface functionalization of nano-drug carriers plays a vital role in binding with target biomolecules and easy to enter inside cells. Cationic polymers have an advantage in binding to the negatively charged plasma membrane than anionic counterparts (Frohlich 2012).

The conventional drug administration has been reformed with nano-drug delivery system for more target specificity, controlled drug release, high biodistribution, decreased toxicity, and dosage. The drug loading efficiency of nanocarrier plays a vital role in the theranostic application, and it varies from the composition of both drug and carrier (El-Sayed et al. 2009).

8.7 Drug Loaded Mesoporous Silica Nanoparticles Induced Cell Death

The MSNs are used as a drug delivery vehicle for the first time reported by Regi et al. (2001). The schematic diagram represents the curcumin loading on mesoporous silica nanoparticles (Fig. 8.1) and MSN is synthesized by three methods, and they are single micelle templating, vesicle templating, and micro-emulsion templating method. Drugs are loaded to the carrier through conjugate during the synthesis of MSN. Generally, functionalized mobil composition of matter has high loading capacity than functionalized KAUST Catalysis Center due to their surface area and pore volume. The hallow mesoporous silica nanoparticles are having a high capacity of loading drugs because of their core and porosity. Hyaluronic acid, folic acid,

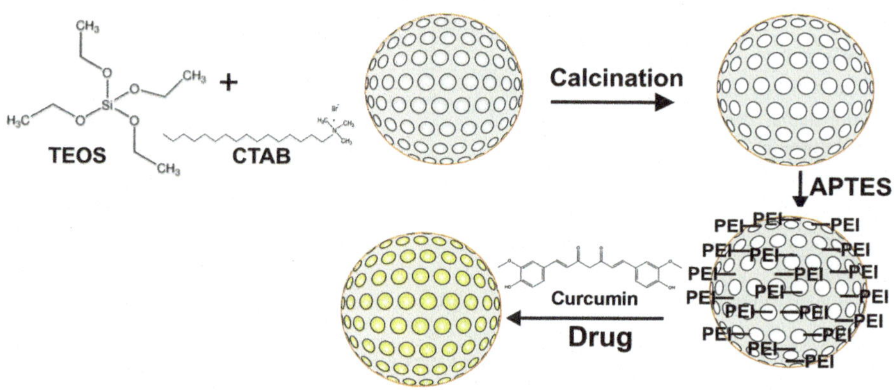

Fig. 8.1 The schematic diagram of curcumin-loaded in a mesoporous silica nanoparticle

3-aminopropyltriethoxysilane, polyethanolimine, polycaprolactone, polyethylene glycol is effectively conjugating with drugs and carriers then involved in the releasing process. Doxorubicin loaded in HMSNAP then induced cell death effectively in HeLa cells (Mishra et al. 2014). Topotecan, quercetin, and curcumin are loaded into MCM-41 then induced cell death in MD Anderson MB-232 (MDA-MB-232) cells (Murugan et al. 2016; Harini et al. 2015). Paclitaxel, 5-fluorouracil, etoposide loaded into MSNs and treated with Human colonic HT-29 cells and prostate cancer cells (Kumar et al. 2017).

8.7.1 Curcumin Loaded MCM-41 Induced Michigan Cancer Foundation-7 (MCF-7) Cell Death

Mobil composition of matter-41 was prepared in the silica precursor is allowed to accumulate over the pre-formed surfactant micelle in acid hydrolysis and condensation of silica in acid or base. Scanning electron microscope images of MCM-41 revealed that they are appeared in spherical shapes and sizes (Fig. 8.2). MCM-41 is smooth surfaced, and the presence of crystal structure appeared in Fig. 8.2e indicated its curcumin loading. MCM-41 synthesized from this method showed both micro and nanoparticles. Transmission electron microscope (TEM) images of MCM-41 (Fig. 8.3) showed it is non-homogenous in nature. MCM-41 coated with curcumin exposed that pores are saturated with curcumin loading, as shown in Fig. 8.3b.

These nanoparticles were characterized by high resolution transmission electron microscope, and the MCM-41 and curcumin-mobil composition of matter-41 (CUR-MCM-41) were analyzed by HR-TEM (Fig. 8.3) revealed that carrier size ranges from 100–200 nm and measured pore size from the instrument software ranges from 4–12 nm. The MCM-41 pores are wide open, however, CUR-MCM-41 pore closure occurred because of curcumin loaded on MCM-41.

DNA fragmentation assay indicated that curcumin-induced apoptosis of MCF-7 cells, as shown in Fig. 8.4. 4′, 6-diamidino-2-phenylindole stained cells were imaged in a fluorescence microscope and showed that cells with extended DNA revealed the fragmentation of DNA in the cell. Cells that have undergone curcumin treatment showed increased DNA fragmentation. Whereas cells treated with MCM-41 coated with polyethylenimine did not show the presence of fragmented DNA. But MCM-41 is coated with polyethylenimine and loaded with 50μM of curcumin-induced DNA fragmentation in cells which have taken up MCM-41 particle. The cells which didn't take up any MCM-41 particle showed intact DNA.

The curcumin-induced cell death was measured fragmented DNA through comet assay or single-cell gel electrophoresis. The cells are embedded in agarose in a microscopic slide run on electrophoresis. The curcumin-induced cell death DNA is formed as tail DNA, as shown in Fig. 8.5. Depending on the cell death, the tail of DNA is prolonging on comet assay.

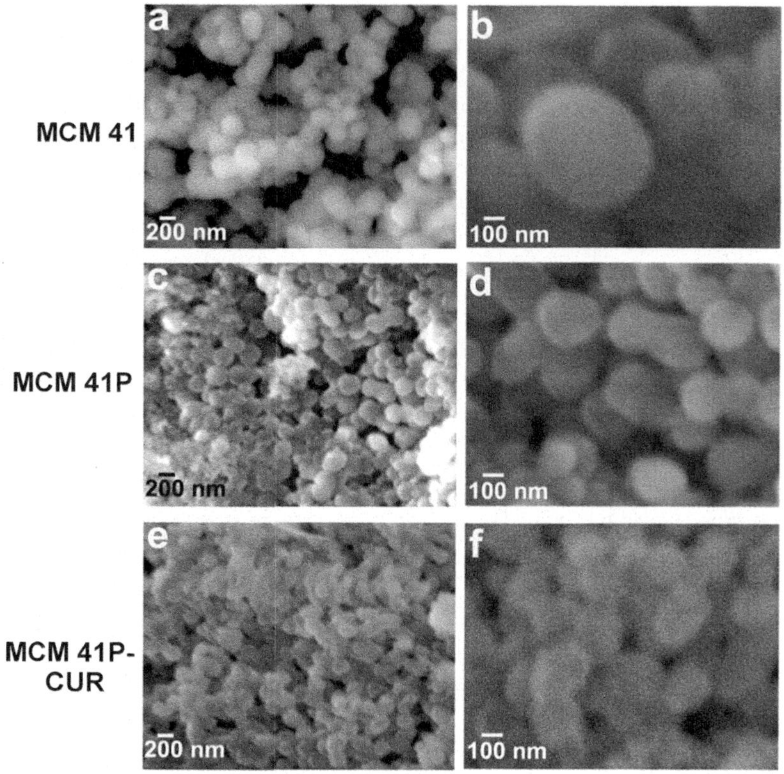

Fig. 8.2 SEM images of MCM-41 (**a**-**b**) without coating of polyethylenimine and (**c**-**d**) coating with polyethylenimine (**e**-**f**) curcumin-loaded MCM-41

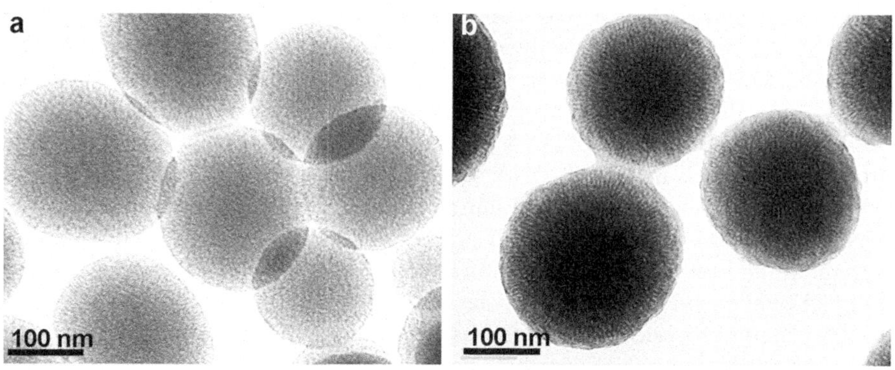

Fig. 8.3 Transmission electron microscopic analysis of MCM-41 (**a**) coating with polyethylenimine(**b**) curcumin-loaded MCM-41

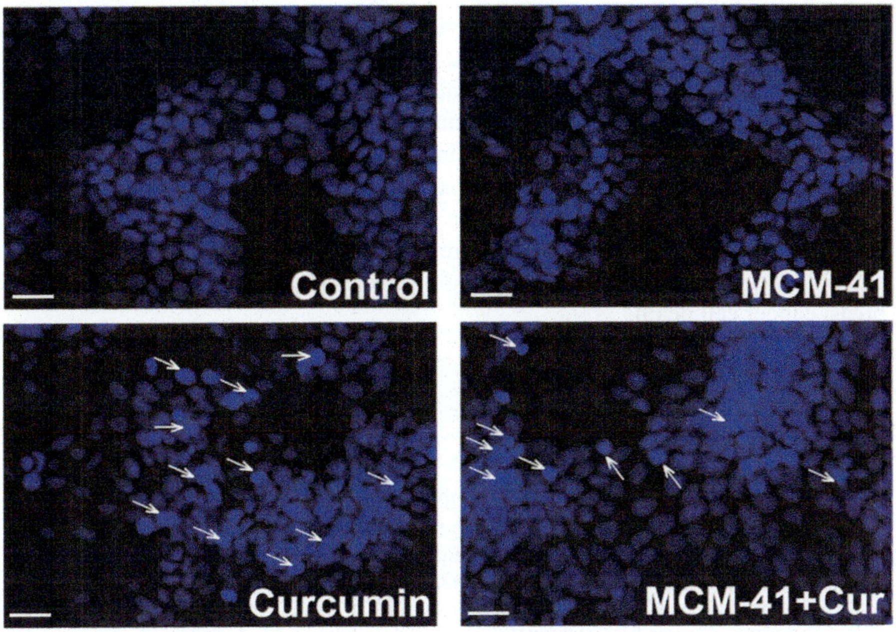

Fig. 8.4 Curcumin-loaded MCM-41 induced cell death in MCF-7 cells. The control and MCM-41 treated cells are healthy and curcumin alone and curcumin loaded MCM-41 treated cells induced cell death is indicated by an arrow. The scale bar represents 100μM

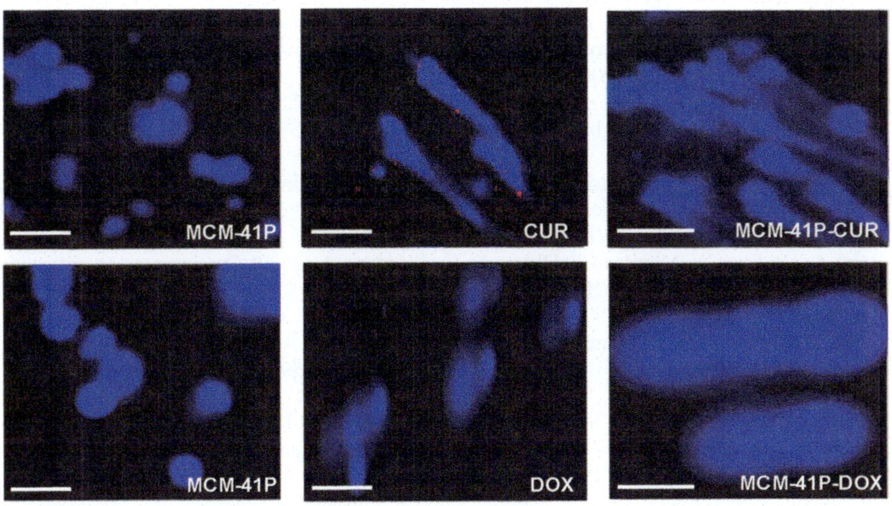

Fig. 8.5 Curcumin/doxorubicin-loaded MCM-41 induced cell death in MCF-7 cells is analyzed through comet assay. The curcumin loaded MCM-41 induced more DNA fragmented when compared with curcumin alone treated cells. The doxorubicin-loaded MCM-41 induced more cell death through DNA fragmentation compared with doxorubicin alone treated cells. The control does not have any DNA fragmentation. The scale bar represents 50μM

8.8 Conclusions

The therapeutic molecules such as natural products and chemotherapy loaded nanocarriers are targeted delivery and sustained release at the intracellular level then induced cell death effectively. Depending on the conjugate, the drug loading is effectively enhanced and sustained release from the carrier. The nanocarrier and drugs are associating with host proteins, then inducing cell death were identified through proteomics approaches (Kathiresan et al. 2009). Therefore, mesoporous silica nanocarriers significantly play a major role in targeted delivery and induce breast cancer cell death.

References

Abraham A, Kattoor AJ, Saldeen T, Mehta JL (2019) Vitamin E and its anticancer effects. Crit Rev Food Sci Nutr 59:2831–2838. https://doi.org/10.1080/10408398.2018.1474169

Alam F, Naim M, Aziz M, Yadav N (2014) Unique roles of nanotechnology in medicine and cancer. Indian J Cancer 51:506–510. https://doi.org/10.4103/0019-509x.175320

Alam F, Naim M, Aziz M, Yadav N (2015) Unique roles of nanotechnology in medicine and cancer-II. Indian J Cancer 52:1–9. https://doi.org/10.4103/0019-509x.175591

Ambrogio MW, Thomas CR, Zhao YL, Zink JI, Stoddart JF (2011) Mechanized silica nanoparticles: a new frontier in theranostic nanomedicine. Acc Chem Res 44:903–913. https://doi.org/10.1021/ar200018x

Anand P (2007) Bioavailability of curcumin: problems and promises. Mol Pharm 4:807–818. https://doi.org/10.1080/01926230701320337

Ananga A, Obuya J, Ochieng J, Tsolova V (2017) Grape seed nutraceuticals for disease prevention: current status and future prospects. In: Phenolic compounds–biological activity, pp 119–137. https://doi.org/10.5772/66908

Azim HA, de Azambuja Colozza EM, Bines J, Piccart MJ (2011) Long-term toxic effects of adjuvant chemotherapy in breast cancer. Ann Oncol 22:1939–1947. https://doi.org/10.1093/annonc/mdq683

Babu A, Templeton AK, Munshi A, Ramesh R (2014) Nanodrug delivery systems: a promising technology for detection, diagnosis, and treatment of cancer. AAPS Pharm Sci Tech 15:709–721. https://doi.org/10.1208/s12249-014-0089-8

Baldi A, De Luca A, Maiorano P, D'Angelo C, Giordano A (2020) Curcumin as an anticancer agent in malignant mesothelioma: a review. Int J Mol Sci 21:1839. https://doi.org/10.3390/ijms21051839

Barbe C, Bartlett J, Kong L, Finnie K, Lin HQ, Larkin M, Calleja S, Bush A, Calleja G (2004) Silica particles: a novel drug-delivery system. Adv Mater 21:1959–1966. https://doi.org/10.1002/adma.200400771

Bawa P, Pradeep P, Kumar P, Choonara YE, Modi G, Pillay V (2016) Multi-target therapeutics for neuropsychiatric and neurodegenerative disorders. Drug Discov Today 21:1886–1914. https://doi.org/10.1002/adma.200400771

Berger MF, Mardis ER (2018) The emerging clinical relevance of genomics in cancer medicine. Nat Rev Clin Oncol 15:353–365. https://doi.org/10.1038/s41571-018-0002-6

Bonta RK (2020) Dietary phenolic acids and flavonoids as potential anti-cancer agents: current state of the art and future perspectives. Anti Cancer Agents Med Chem 20:29–48. https://doi.org/10.2174/1871520619666191019112712

Bourdeanu L, Luu T (2014) Targeted therapies in breast cancer: implications for advanced oncology practice. J Adv Pract Oncol 5:246–260. https://doi.org/10.6004/jadpro.2014.5.4.2

Brambilla P, Milani M, Spada D, Pedrazzoli P, Martino M, Bottini A, Generali D (2015) Breast cancer circulating biomarkers: advantages, drawbacks, and new insights. Tumor Biol 36:6653–6665. https://doi.org/10.1007/s13277-015-3944-7

Breckenridge DG, Germain M, Mathai JP, Nguyen M, Shore GC (2003) Regulation of apoptosis by endoplasmic reticulum pathways. Oncogene 22:8608–8618. https://doi.org/10.1038/sj.onc.1207108

Bulusu KC, Guha R, Mason DJ, Lewis RPI, Muratov E, Motamedi YK, Cokol M, Bender A (2016) Modelling of compound combination effects and applications to efficacy and toxicity: state-of-the-art, challenges and perspectives. Drug Discov Today 21:225–238. https://doi.org/10.1016/j.drudis.2015.09.003

Cardoso F (2018) 4th ESO–ESMO international consensus guidelines for advanced breast cancer (ABC 4). Ann Oncol 29:1634–1657. https://doi.org/10.1093/annonc/mdy192

Chae HS, Xu R, Won JY, Chin YW, Yim H (2019) Molecular targets of genistein and its related flavonoids to exert anti-cancer effects. Int J Mol Sci 20:2420. https://doi.org/10.3390/ijms20102420

Cheng AL, Hsu CH, Lin JK (2001) Phase I clinical trial of curcumin, a chemo preventive agent, in patients with high-risk or pre-malignant lesions. Anticancer Res 21:2895–2900

Deb A, Andrews NG, Raghavan V (2018) Natural polymer functionalized graphene oxide for co-delivery of anti-cancer drugs: in-vitro and in-vivo. Int J Biol Macromol 113:515–525. https://doi.org/10.1016/j.ijbiomac.2018.02.153

Ekkapongpisit M, Giovia A, Follo C, Caputo G, Isidoro C (2012) Biocompatibility, endocytosis, and intracellular trafficking of mesoporous silica and polystyrene nanoparticles in ovarian cancer cells: effects of size and surface charge groups. Int J Nanomedicine 7:4147–4158. https://doi.org/10.2147/ijn.s33803

Elizabeth CC, Sergio CM, Kang J, Muthukaruppan A, Boersma MN, Stone A, Barraclough J, Lee CS, Black MA, Miller LD, Gee JM, Nicholson RI, Sutherland RL, Print CG, Musgrove EM (2012) Cyclin E2 overexpression is associated with endocrine resistance but not insensitivity to CDK2 inhibition in human breast cancer cells. Mol Cancer Ther 11:1488–1499. https://doi.org/10.1158/1535-7163.mct-11-0963

Elmore S (2007) Apoptosis: a review of programmed cell death. Toxicol Pathol 35:495–516. https://doi.org/10.1080/01926230701320337

El-Sayed AE, Futaki S, Harashima H (2009) Delivery of macromolecules using arginine-rich cell-penetrating peptides: ways to overcome endosomal entrapment. AAPS J 11:13–22. https://doi.org/10.1208/s12248-008-9071-2

Fresco P, Borges F, Diniz C, Marques MPM (2006) New insights on the anti-cancer properties of dietary polyphenols. Med Res Rev 26:747–766. https://doi.org/10.1002/med.20060

Frohlich E (2012) The role of surface charge in cellular uptake and cytotoxicity of medical nanoparticles. Int J Nanomedicine 7:5577–5591. https://doi.org/10.2147/ijn.s36111

Furlan V, Bren U (2020) Protective effects of [6]-gingerol against chemical carcinogens: mechanistic insights. Int J Mol Sci 21:695. https://doi.org/10.3390/ijms21030695

Gan RY, Li HB, Sui ZQ, Corke H (2018) Absorption, metabolism, anti-cancer effect and molecular targets of epigallocatechin gallate (EGCG): an updated review. Crit Rev Food Sci Nutr 58:924–941. https://doi.org/10.1080/10408398.2016.1231168

Golse N, Adam R (2017) Liver metastases from breast cancer: what role for surgery? Indications and results. Clin Breast Cancer 17:256–265. https://doi.org/10.1016/j.clbc.2016.12.012

Gupta SC, Patchva S, Aggarwal BB (2013) Therapeutic roles of curcumin: lessons learned from clinical trials. AAPS J 15:195–218. https://doi.org/10.1208/s12248-012-9432-8

Hafidh RR (2017) A comprehensive anti-cancer molecular study for genistein the promising anti-cancer drug. J Contemp Med Sci 3:264–269. https://doi.org/10.22317/jcms.09201704

Harini L, Karthikeyan B, Srivatsava S, Rajagopalan S, Kathiresan T (2015) Mesoporous silica nanoparticle loaded with curcumin reduces the cell survival of MCF-7 cells. In: Rajendran V, Prabu P, Geckeler KE(eds) Bio-Nanomaterials for Biomedical Technology, pp 243–248

Harini L, Karthikeyan B, Srivastava S, Suresh SB, Ross C, Gnanakumar G, Rajagopal S, Sundar K, Kathiresan T (2017) Polyethylenimine-modified curcumin-loaded mesoporus silica nanoparticle (MCM-41) induces cell death in MCF-7 cell line. IET Nanobiotechnol 11:57–61. https://doi.org/10.1049/iet-nbt.2016.0075

Harini L, Srivastava S, Gnanakumar GP, Karthikeyan B, Ross C, Krishnakumar V, Kannan VR, Sundar K, Kathiresan T (2019) An ingenious non-spherical mesoporous silica nanoparticle cargo with curcumin induces mitochondria-mediated apoptosis in breast cancer (MCF-7) cells. Oncotarget 10:1193–1208. https://doi.org/10.18632/oncotarget.26623

Heikkilä T, Salonen J, Tuura J, Hamdy MS, Mul G, Kumar N, Salmi T, Murzin DY, Laitinen L, Kaukonen AM (2007) Mesoporous silica material TUD-1 as a drug delivery system. Int J Pharm 331:133–138. https://doi.org/10.1016/j.ijpharm.2006.09.019

Hoffmann F, Cornelius M, Morell J, Fröba M (2006) Silica-based mesoporous organic–inorganic hybrid materials. Angew Chem Int Ed 45:3216–3251. https://doi.org/10.1002/anie.200503075

Housman G, Byler S, Heerboth S, Lapinska K, Longacre M, Snyder N, Sarkar S (2014) Drug resistance in cancer: an overview. Cancers (Basel) 6:1769–1792. https://doi.org/10.3390/cancers6031769

Hsia TC, Huang YP, Jiang YW, Chen HY, Cheng ZY, Hsiao YT, Chen CY, Peng SF, Chueh FS, Chou YC, Chung JG (2018) Phenethyl isothiocyanate induces apoptotic cell death through the mitochondria-dependent pathway in gefitinib-resistant NCI-H460 human lung cancer cells in vitro. Anticancer Res 38:2137–2147. https://doi.org/10.21873/anticanres.12454

Hudlikar R, Wu R, Cheng D, Kuo DHC, Wang L, Peter R, Yin R, Li S, Kong AN (2020) Anthocyanins and cancer prevention. In: Natural products for cancer chemoprevention, pp 351–373. https://doi.org/10.1007/978-3-030-39855-2_11

Huo Q, Margolese DI, Stucky GD (1996) Stucky surfactant control of phases in the synthesis of mesoporous silica-based materials. Chem Mater 8:1147–1160. https://doi.org/10.1021/cm960137h

Jadhav KS, Dumbare PS, Pande VV (2015) Mesoporous silica nanoparticles (MSN): a nanonetwork and hierarchical structure in drug delivery. J Nanomed Res 2:1–8. https://doi.org/10.15406/jnmr.2015.02.00043

Jain V, Jain S, Mahajan SC (2015) Nanomedicines based drug delivery systems for anti-cancer targeting and treatmen. Curr Drug Deliv 12:1–19. https://doi.org/10.2174/1567201811666140822112516

Kathiresan T, Harvey MC, Sokolowski BH (2009) The use of 2-D gels to identify novel protein–protein interactions in the cochlea. In: Auditory and vestibular research. Humana Press, pp 269–286. https://doi.org/10.1007/978-1-59745-523-7_16

Katiyar C, Gupta A, Kanjilal S, Katiyar S (2012) Drug discovery from plant sources: an integrated approach. AYU 33:10–19. https://doi.org/10.4103/0974-8520.100295

Khanna VK (2012) Targeted delivery of nanomedicines. ISRN Pharmacol 2012(571394):9. https://doi.org/10.5402/2012/571394

Kimbung S, Lomana N, Hedenfalk I (2015) Clinical and molecular complexity of breast cancer metastases. Semin Cancer Biol 35:85. https://doi.org/10.1016/j.semcancer.2015.08.009

Kocaadam B, Şanlier N (2017) Curcumin, an active component of turmeric (*Curcuma longa*), and its effects on health. Crit Rev Food Sci Nutr 57:2889–2895. https://doi.org/10.1080/10408398.2015.1077195

Kohno T, Hai-Bang T, Zhu Q, Amen Y, Sakamoto S, Tanaka H, Morimoto S, Shimizu K (2017) Tubulin polymerization-stimulating activity of Ganoderma triterpenoids. J Nat Med 71:457–462. https://doi.org/10.1007/s11418-017-1072-y

Kortenkamp A (2008) Breast cancer and environmental risk factors: an appraisal of the scientific evidence. Breast Cancer Res 10:45. https://doi.org/10.1186/bcr1929

Koul A, Bansal MP, Chaudhary H, Chugh NA (2019) Lycopene enriched tomato extract suppresses chemically induced skin tumorigenesis in mice. Int J Vitam Nutr Res 90:493–513. https://doi.org/10.1024/0300-9831/a000597

Kumar B, Kulanthaivel S, Mondal A, Mishra S, Banerjee B, Bhaumik A, Banerjee I, Giri S (2017) Mesoporous silica nanoparticle based enzyme responsive system for colon specific drug delivery through guar gum capping. Colloids Surf B: Biointerfaces 150:352–361. https://doi.org/10.1016/j.colsurfb.2016.10.049

Kweia KA, Kunga Y, Salaria K, Holcomba IN, Pollack JR (2010) Genomic instability in breast cancer: pathogenesis and clinical implications. Mol Oncol 4:255–266. https://doi.org/10.1016/j.molonc.2010.04.001

Liang J, Hänsch GM, Hübner K, Samstag Y (2019) Sulforaphane as anti-cancer agent: a double-edged sword? Tricky balance between effects on tumor cells and immune cells. Adv Biol Regul 71:79–87. https://doi.org/10.1016/j.jbior.2018.11.006

Lodha A, Lodha M, Patel A, Chaudhuri J, Dalal J, Edwards M, Douroumis D (2012) Synthesis of mesoporous silica nanoparticles and drug loading of poorly water soluble drug cyclosporin A. J Pharm Bioallied Sci 4:S92–S94. https://doi.org/10.4103/0975-7406.94153

Lustberg MB, Ramaswamy B (2011) Epigenetic therapy in breast cancer. Curr Breast Cancer Rep 3:34–43. https://doi.org/10.1007/s12609-010-0034-0

Mann S, Burkett SL, Davis SA, Fowler CE, Mendelson NH, Sims SD, Walsh D, Whilton NT (1997) Sol–gel synthesis of organized matter. Chem Mater 9:2300–2310. https://doi.org/10.1021/cm970274u

Min KJ, Nam JO, Kwon TK (2017) Fisetin induces apoptosis through p53-mediated up-regulation of DR5 expression in human renal carcinoma Caki cells. Molecules 22:1285. https://doi.org/10.3390/molecules22081285

Mishra AK, Pandey H, Agarwal V, Ramteke PW, Pandey AC (2014) Nanoengineered mesoporous silica nanoparticles for smart delivery of doxorubicin. J Nanopart Res 16:2515. https://doi.org/10.1007/s11051-014-2515-y

Mocanu MM, Nagy P, Szöllősi J (2015) Chemoprevention of breast cancer by dietary polyphenols. Molecules 20:22578–22620. https://doi.org/10.3390/molecules201219864

Moore J, Yousef M, Tsiani E (2016) Anticancer effects of rosemary (Rosmarinus officinalis L.) extract and rosemary extract polyphenols. Nutrients 8:731. https://doi.org/10.3390/nu8110731

Moorthi C, Manavalan R, Kathiresan K (2011) Nanotherapeutics to overcome conventional cancer chemotherapy limitations. J Pharm Pharm Sci 14:67–77. https://doi.org/10.18433/j30c7d

Moran MP, Fernandez JM, Tortosa CR, Tortosa MR (2016) Curcumin and health. Molecules 21:264. https://doi.org/10.3390/molecules21030264

Murugan C, Rayappan K, Thangam R, Bhanumathi R, Shanthi K, Vivek R, Thirumurugan R, Bhattacharyya A, Sivasubramanian S, Gunasekaran P (2016) Combinatorial nanocarrier based drug delivery approach for amalgamation of anti-tumor agents in breast cancer cells: an improved nanomedicine strategy. Sci Rep 6:34053. https://doi.org/10.1038/srep34053

Nagahama K, Utsumi T, Kumano T, Maekawa S, Oyama N, Kawakami J (2016) Discovery of a new function of curcumin which enhances its anti-cancer therapeutic potency. Sci Rep 6:1–14. https://doi.org/10.1038/srep30962

Nandiyanto ABD, Kim SG, Iskandar F, Okuyama K (2009) Synthesis of spherical mesoporous silica nanoparticles with nanometer-size controllable pores and outer diameters. Microporous Mesoporous Mater 120:447–453. https://doi.org/10.1016/j.micromeso.2008.12.019

Nik-Zainal S (2016) Landscape of somatic mutations in 560 breast cancer whole- genome sequences. Nature 534:47–54. https://doi.org/10.1038/nature17676

Olefsky JM (2001) Nuclear receptor minireview series. J Biol Chem 276:36863–36864. https://doi.org/10.1074/jbc.r100047200

Oye G, Sjöblom J, Stöcker M (2001) Synthesis, characterization and potential applications of new materials in the mesoporous range. Adv Colloid Interf Sci 89-90:439–466. https://doi.org/10.1016/S0001-8686(00)00066-X

Pandya U, Dhuldhaj U, Sahay NS (2019) Bioactive mushroom polysaccharides as antitumor: an overview. Nat Prod Res 33:2668–2680. https://doi.org/10.1080/14786419.2018.1466129

Parker JS (2009) Supervised risk predictor of breast cancer based on intrinsic subtypes. J Clin Oncol 27:1160–1167. https://doi.org/10.1200/jco.2008.18.1370

Patil A, Chirmade UN, Trivedi V, Lamprou DA, Urquart A, Douroumis D (2011) Encapsulation of water insoluble drugs in mesoporous silica nanoparticles using supercritical carbon dioxide. J Nanomed Nanotechnol 2. https://doi.org/10.4172/2157-7439.1000111

Priyadarsini IK (2013) Chemical and structural features influencing the biological activity of curcumin. Curr Pharm Des 19:2093–2100. https://doi.org/10.2174/138161213805289228

Priyadarsini IK (2014) The chemistry of curcumin: from extraction to therapeutic agent. Molecules 19:20091–20112. https://doi.org/10.3390/molecules191220091

Radhakrishnan K, Gupta S, Divya P, Praveen C, Dipshika C, Ashok M (2013) Protamine-capped mesoporous silica nanoparticles for biologically triggered drug release. Part Part Syst Charact 31:449–458. https://doi.org/10.1002/ppsc.201300219

Rahimi HR, Nedaeinia R, Shamloo AS, Nikdoust S, Oskuee RK (2016) Novel delivery system for natural products: nano-curcumin formulations. Avicenna J Phytomed 6:383–398. http://www.ncbi.nlm.nih.gov/pmc/articles/pmc4967834/

Rauf A, Imran M, Butt MS, Nadeem M, Peters DG, Mubarak MS (2018) Resveratrol as an anti-cancer agent: a review. Crit Rev Food Sci Nutr 58:1428–1447. https://doi.org/10.1080/10408398.2016.1263597

Regi VM, Rámila A, Del Real RP, Pérez-Pariente J (2001) A new property of MCM-41? Drug delivery system. Chem Mater 13:308–311. https://doi.org/10.1021/cm0011559

Rutqvist LE (1993) Radiation therapies for breast cancer: current knowledge on advantages and disadvantages. Recent Results Cancer Res 127:119–127. https://doi.org/10.1007/978-3-642-84745-5_17

Saraswathy M, Gong S (2013) Different strategies to overcome multidrug resistance in cancer. Biotechnol Adv 31:1397–1407. https://doi.org/10.1016/j.biotechadv.2013.06.004

Shafie NH, Ishak AH, Ismail A, Tohid SFM (2019) Induction of endoplasmic reticulum stress pathway by green tea Epigallocatechin-3-Gallate (EGCG) in colorectal Cancer cells: activation of PERK/p-eIF2α/ATF4 and IRE1α. Biomed Res Int:3480569. https://doi.org/10.1155/2019/3480569

Sharma GN, Dave R, Sanadya J, Sharma P, Sharma KK (2010) Various types and management of breast cancer: an overview. J Adv Pharm Technol Res 1:109–126

Shien T, Doihara H (2014) Resection of the primary tumor in stage IV breast cancer. World J Clin Oncol 5:82–85. https://doi.org/10.1016/j.ejso.2018.08.002

Singh R, Lillard JW Jr (2009) Nanoparticle-based targeted drug delivery. Exp Mol Pathol 86:215–223. https://doi.org/10.1016/j.yexmp.2008.12.004

Sinn HP, Kreipe HA (2013) Brief overview of the WHO classification of breast tumors, 4th edn. Focusing on issues and updates from the, 3rd edn. Breast Care (Basel) 8:149–154. https://doi.org/10.1159/000350774

Slowing II, Escoto JL, Wu CW, Lin VSY (2008) Mesoporous silica nanoparticles as controlled release drug delivery and gene transfection carriers. Adv Drug Deliv Rev 60:1278–1288. https://doi.org/10.1016/j.addr.2008.03.012

Sun J, Nawaz Z, Slingerland JM (2007) Long-range activation of GREB1 by estrogen receptor via three distal consensus estrogen-responsive elements in breast cancer cells. Mol Endocrinol 11:2651–2662. https://doi.org/10.1210/me.2007-0082

Takahashi S, Bhattacharjee S, Ghosh S, Sugimoto N, Bhowmik S (2020) Preferential targeting cancer-related i-motif DNAs by the plant flavonolfisetin for theranostics applications. Sci Rep 10:1–13. https://doi.org/10.1038/s41598-020-59343-2

Talekar M, Tran TH, Amiji M (2015) Translational nano-medicines: targeted therapeutic delivery for cancer and inflammatory diseases. AAPSJ 17:813–827. https://doi.org/10.1208/s12248-015-9772-2

Tang F, Linlin Li L, Chen D (2012) Mesoporous silica nanoparticles: synthesis, biocompatibility and drug delivery. Adv Mater 24:1504–1534. https://doi.org/10.1002/adma.201104763

Tiwari G, Tiwari R, Sriwastawa B, Bhati L, Pandey S, Pandey P, Bannerjee SK (2012) Drug delivery systems: an updated review. Int J Pharm Investig 2:2–11. https://doi.org/10.4103/2230-973x.96920

Tozuka Y, Wongmekiat A, Kimura K, Moribe K, Yamamura S, Yamamoto K (2005) Effect of pore size of FSM-16 on the entrapment of flurbiprofen in mesoporous structures. Chem Pharm Bull 53:974–977. https://doi.org/10.1248/cpb.53.974

Trewyn BG, Slowing II, Giri S, Chen HT, Lin VSY (2007) Synthesis and functionalization of a mesoporous silica nanoparticle based on the sol–gel process and applications in controlled release. Acc Chem Res 40:846–853. https://doi.org/10.1021/ar600032u

Truong NP, Whittaker MR, Catherine W, Mak CW, Davis PT (2015) The importance of nanoparticle shape in cancer drug delivery. Expert Opin Drug Deliv 12:129–142. https://doi.org/10.1517/17425247.2014.950564

Wang S, Tan M, Zhong Z, Chen M, Wang Y (2011) Nanotechnologies for curcumin: an ancient puzzler meets modern solutions. J Nanomater:1–8. https://doi.org/10.1155/2011/723178

Weigelt B, Peterse JL, van't Veer LJ (2005) Breast cancer metastasis: markers and models. Nat Rev Cancer 5:591–602. https://doi.org/10.1038/nrc1670

Yanes RE, Tamanoi F (2012) Development of mesoporous silica nanomaterials as a vehicle for anti-cancer drug delivery. Ther Deliv 3:389–404. https://doi.org/10.4155/tde.12.9

Yin H, Liao L, Fang J (2014) Enhanced permeability and retention (EPR) effect based tumor targeting: the concept, application and prospect. JSM Clin Oncol Res 2:1–5. https://doi.org/10.1007/978-1-60761-609-2_3

Ying JY, Mehnert CP, Wong MS (1999) Synthesis and applications of supramolecular templated mesoporous materials. Angew Chem Int Ed 38:56–77. https://doi.org/10.1002/(SICI)1521-3773

Zhang F, Zhang JG, Qu J, Zhang Q, Prasad C, Wei ZJ (2017) Assessment of anti-cancerous potential of 6-gingerol (Tongling White Ginger) and its synergy with drugs on human cervical adenocarcinoma cells. Food Chem Toxicol 109:910–922. https://doi.org/10.1016/j.fct.2017.02.038

Zhao D, Huo Q, Feng J, Chmelka BF, Stucky GD (1998) Nonionic triblock and star diblock copolymer and oligomeric surfactant syntheses of highly ordered, hydrothermally stable, mesoporous silica structures. J Am Chem Soc 120:6024–6036. https://doi.org/10.1021/ja974025i

Zhou H, Zink JI, Nel AE (2011) Aspect ratio determines the quantity of mesoporous silica nanoparticle uptake by a small GTPase-dependant macropinocytosis mechanism. ACS Nano 28:4434–4447. https://doi.org/10.1021/nn103344k

Chapter 9
Insights on the Biotechnological Applications of Marine Fungal Exopolysaccharides

A. M. V. N. Prathyusha, G. Triveni, G. Mohana Sheela, B. Anand Kumar, G. Bhargava Ram, T. Chandrasekhar, and Pallaval Veera Bramhachari

Abstract Microbial exopolysaccharides are the subject of research in many scientific areas, focusing on elucidating structure-function relationships, identifying novel biosynthetic pathways, and unraveling applications based on their distinct properties. Marine fungi signify a relatively untapped bio-resource for novel exopolysaccharides molecules; however, marine fungi are evidenced with great industrial significance during past decades. High molecular diversity, complexity and growth in unique and extreme habitats make the marine fungi produce unusual exopolysaccharides. However, due to their unique rheological properties, marine fungal exopolysaccharides have found several applications in the environment, food, cosmetic and pharmaceutical industries. The current review emphasizes untapped marine fungi focusing on exploring the fungal exopolysaccharides with novel structures and unique functional properties.

Keywords Marine Fungi · Exopolysaccharides · Industrial applications · Unique functional properties

A. M. V. N. Prathyusha · G. Triveni · G. M. Sheela · B. A. Kumar · G. B. Ram · P. V. Bramhachari (✉)
Department of Biotechnology, Krishna University, Machilipatnam, Andhra Pradesh, India

T. Chandrasekhar
Department of Environmental Science, YogiVemana University, Kadapa, Andhra Pradesh, India

K. M. Gothandam et al. (eds.), *Environmental Biotechnology Volume 4*, Environmental Chemistry for a Sustainable World 68,
https://doi.org/10.1007/978-3-030-77795-1_9

9.1 Introduction

Marine ecosystems are highly dynamic, which harbor enormous marine microbes' diversities and are presumed to produce chemically and biologically active compounds (Shearer et al. 2007). Marine microbes such as bacteria, fungi, viruses are ecologically diverse communities due to their role in nutrient cycling dynamics, biogeochemical processes, and importance in marine food chains. Marine fungi are considered an important source for ecological components and relatively untapped bioresources for novel natural products. Yet, scanty literature is available concerning their biodiversity and global distribution (Tisthammer et al. 2016; Raghukumar 2004). Most marine fungal diversity was reported in the coastlines such as riverine, mangroves and estuarine environments, signifying the influence of environmental catastrophes that inhabits terrestrial fungi in marine habitats. Several factors can affect fungal distribution, abundance in marine habitats. Therefore, fungi characterised from marine habitats exhibit few morphological traits analogous to terrestrial counterparts (Menezes et al. 2010).

Marine fungi are categorized into two groups obligate and facultative. Obligate marine fungi sporulate and germinate completely in marine habitats. In contrast, the facultative marine fungi inhabit terrestrial or freshwater ecosystems, growing and sporulating in marine environments (Raghukumar 2008). Marine fungi often produce exopolysaccharides with novel chemical and structural moieties, including their varied biological activities. Our recent report depicted that marine fungal exopolysaccharides are multifunctional, high molecular weight biopolymers produced by fungi and several other microbes during their metabolic process (Prathyusha et al. 2018). Exopolysaccharides have a wide range of molecular weight, many reactive groups with varied chemical compositions and exhibit diversity in their structural and functional properties. Interestingly, due to the presence of derivable groups on the molecular structures, these polysaccharides can be modified with much ease as per the demand and requirement. However, due to their enormous genetic and biochemical diversity, marine fungi are considered a potential source of exopolysaccharides with impending applications in medicine, food, pharmaceutical, and cosmetic industries (Mahapatra and Banerjee 2013a, b).

Most synthetic polymers are petroleum-derived hydrocarbons and are not biodegradable; for a matter of fact, their resistance to microbial degradation, lack of the required degree of purity and functional properties, accumulation in the environment. Therefore, this evidence has facilitated to promote interest in biodegradable polymers. Polymers from natural sources are available in huge quantities and are renewable resources, whereas synthetic polymers are synthesized from non-renewable petroleum sources (Ghanbarzadeh and Almasi 2013). Biological polymers have several advantages against synthetic polymers due to their accessibility from natural resources, chemically modifying, and inexpensive. Natural polymers are also ideal building blocks for generating systems that mimic the structural and biochemical properties of *in vivo* cellular environment. Exopolysaccharides being natural biomaterials are highly stable in the biological fluids, non-toxic, safe, and biodegradable (Hamid Akash et al. 2015). The following limitations for bacterial exopolysaccharides.

(a) Cost of production relative to their commercial value main constraints to full commercialization is their production costs, mostly related to substrate cost and downstream processing (Ates 2015).
(b) Different metabolic pathways might be followed because of the different nutrient composition and presence of contaminants. Thus, different polymers and/or (unwanted) byproducts might be synthesized by the bacteria.
(c) Non-reacted components might accumulate in the broth and eventually become inhibitors, which lowers product yield.
(d) Separation processes become especially challenging when polymers are considered for medical applications because a high-purity product is envisaged (Freitas et al. 2011).

The following are the advantages of fungal exopolysaccharides

(a) upstream and downstream processing of these exopolysaccharides is easier
(b) can produce a much larger amount in a shorter time than plant or algal polysaccharide production (Szewczuk-Karpisz et al. 2016).

However, exopolysaccharides research is currently a subject of great surge due to their interesting physicochemical and rheological properties with novel functionality of exopolysaccharides that act as a raw biomaterial with a wide range of applications in Microbial Enhanced Oil recovery (Sveistrup et al. 2016), wastewater treatment, cosmetic (Agrawal et al. 2018) and pharmaceutical industry etc. Salinity, wire ranges of mineral content, low temperature, pH extremes, high pressure, and special lighting conditions affect production and structural differences in exopolysaccharides among marine and terrestrial fungi (Bonugli-Santos et al. 2015). The current review particularly emphasizes untapped marine fungi, focusing on exploring the exopolysaccharides with novel structures and unique functional properties.

9.2 Characteristics of Exopolysaccharides from Marine-Derived Fungi

Exopolysaccharides produced by a marine fungus have different chemical structures and properties, viz., sugar composition, degree of polymerization, type of linkages, and molecular weight. Exopolysaccharides are classified based on sugar composition (homopolysaccharides and heteropolysaccharides) and nature (acidic, neutral and basic polysaccharides). In general, marine fungal exopolysaccharides are mostly heteropolysaccharides and acidic; however, few among them are homopolysaccharides with glucose as a dominant monomer. The saccharide content is varied among exopolysaccharides produced by diverse marine fungi, but glucose, galactose, xylose, mannose, rhamnose, and fucose sugars occur most frequently.

Several studies support the idea that polysaccharides are generally colorless, non-toxic, odorless, amorphous, mostly insoluble in water and no sweet taste (Mahapatra and Banerjee 2013a, b; Osińska-Jaroszuk et al. 2015). Several reports on the importance of exopolysaccharides molecular structure and their biological activity because bioavailability may be linked to their structural patterns. Notably, the biological responses of exopolysaccharides are closely related to their chemical structure, degrees of branching, glycosidic bonds, molecular weight, polymerization, backbone length and 3D conformation. Nonetheless, complete characterizations of exopolysaccharides are essential for a better understanding of the relation between structural and functional properties in addition to their nutritional and therapeutic applications. However, the molecular weight of exopolysaccharides also portrays an essential role in biological functions. The studies on exopolysaccharides revealed that a petite number of exopolysaccharides with low-molecular-weight demonstrated immunomodulatory, antioxidant and antitumor activities (Jing et al. 2014).

Aureobasidium pullulans yeast-like fungus were commonly found in terrestrial and aquatic habitats, widely present in temperate and tropical zones. These fungi are reported to produce a biodegradable, tasteless, odorless, non-toxic, non-ionic, edible exopolysaccharides "Pullulan," which is water-soluble (insoluble in organic solvents) with Poly α (1 $\rightarrow$ 6) maltotriose repeating Units. It is readily soluble in hot and cold water, which forms a clear, viscous solution with high adhesion and film forming capabilities. The main advantages of pullulans are blood compatible nature, non-immunogenic, non-carcinogenic and non-mutagenic (Ravella et al. 2010; Ma et al. 2014). This pullulan's molecular weight ranges from 4.5×10^4 to 6×10^5 Da, which is greatly affected by culture conditions. Hayashibara company (Okayama, Japan) commercially produced pullulan from 1976 onwards.

9.3 Role of Marine Fungal Exopolysaccharides in Medicine

Marine fungal exopolysaccharides are used in treatment, viz. antioxidant and antimicrobial agents help accelerate wound healing and fight against bacterial and viral infections, and antitumor agents in supporting chemotherapy (Osińska-Jaroszuk et al. 2015). Notably, based on their biomedical applications, microbial exopolysaccharides can be categorized into two:

- In gels' formation, with high osmotic pressure, viscosity and water absorption capacities, to synthesize pharmaceutical materials such as drug release agents and plasma substitutes.
- To synthesize vaccines or new drugs with biological properties such as antioxidant, antigenicity, antitumor activities.

However, biomedical applications of marine fungi were limited only to the synthesis of new drugs and vaccines. But fungal polysaccharides like pullulan,

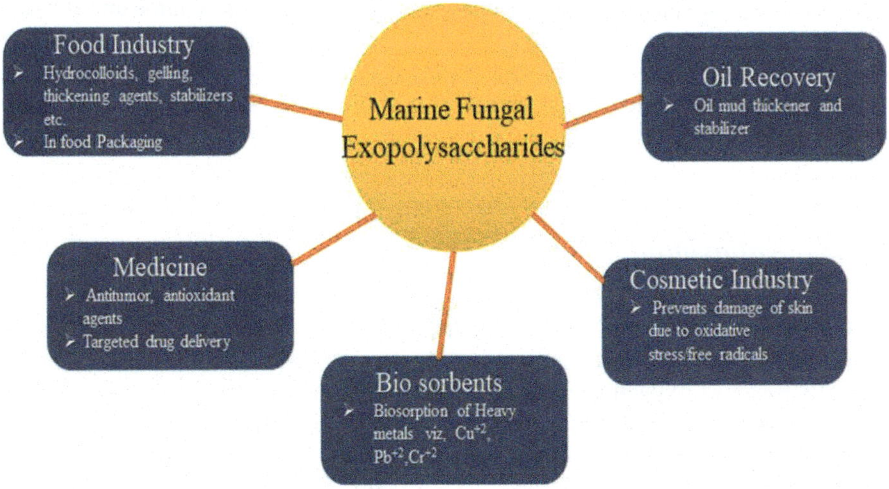

Fig. 9.1 An overview of Marine Fungal Biopolymers and their role

scleroglucan and dextran were previously reported for their biomedical applications in targeted drug tissue engineering and wound healing, respectively (Fig. 9.1).

9.3.1 *Antitumor Activity*

Cancer is one of the deadliest diseases globally, with high morbidity and mortality. Chemotherapy used in systemic therapy has significant side effects. The antitumor property of exopolysaccharides is closely associated with their immunostimulant and immunomodulatory activities, which can be used as alternatives for chemotherapy with almost no side effects (Osińska-Jaroszuk et al. 2015). Notably, polysaccharides extracted from natural sources, especially from the marine environment, have grabbed a great interest due to their unique bioactivities. Recently, an exopolysaccharide Yancheng polysaccharide (YCP) purified from *Phoma herbarum* YS4108, a filamentous fungus isolated from the Yellow Sea near Sheyang Port, China, reported having significant antitumor activity. Yancheng polysaccharide has a backbone with α-1,4-D-glucan and a lower proportion of α-1,6-linked glucopyranosyl and glucuronic acid residues as non-reducing terminal (Selbmann et al. 2002; Yang et al. 2005). Yancheng polysaccharide was reported to reduce tumor growth in xenografted cells (Heps, S180 and Lewis) without prompting any abnormalities in experimental mice's body weight and behaviour and great effectiveness and reduced toxicity. Yancheng polysaccharide has been with immunostimulating properties on B cells via a receptor-mediated pathway. This involves Toll-like receptors, TLR2 and TLR4, which results in stimulation of mitogen-activated protein kinase and Nuclear factor kappa B (NF-κB) signalling pathways, thus highlighting the efficiency of Yancheng polysaccharide as an immune modulator and in oncologic immunotherapy. Yancheng polysaccharide also exhibited good antioxidant activity. Strikingly these findings have surged great interest in Exopolysaccharides YCP against antitumor therapy and found appropriate for clinical trials (Zhang et al. 2013).

9.3.2 Antioxidant Activity

Several studies support the idea that some biopolymers can inhibit oxidative stress and, hence, diverse pharmaceuticals, cosmetic and food industries. Elevated free radicals/Reactive oxygen species in humans may lead to aging and T cell damage. As a result, diminished immune function while increasing risk for cardio and cerebrovascular diseases. Interestingly, in normal conditions, free radicals control cell growth and inhibit viruses and bacteria's growth. Nonetheless, increased levels of free radicals (H_2O_2, superoxide anion, OH radicals and NO) can also trigger some chronic human diseases, viz. arteriosclerosis, aging and cancer (Pahlavani and Harris 1998). Reactive oxygen species are generated during the electron transport chain in humans (Weydert and Cullen 2010). Notably, low Reactive oxygen species levels are essential in many biochemical processes, while elevated Reactive oxygen species levels can affect tissues and disrupt the redox homeostasis that causes oxidative stress (Georgieva 2005). Strikingly, the natural extracts, particularly the marine fungal polysaccharides, are high potential sources of antioxidants, with their ability to scavenge free radicals. Currently, Exopolysaccharides with antioxidant activity have been given more and more interest and are essentially significant in natural medicine research and production.

The antioxidant activity of exopolysaccharides from marine fungi has been most extensively studied among their numerous biology and medical applications. Many marine fungi belong to the genus *Penicillium, Aspergillus, Fusarium, Pseudomonas, Epicoccum, Arthrobacter, Cordyceps* (listed in Table 9.1) could produce exopolysaccharides with potential antioxidant activities. Ye et al. (2012) reported the *in vitro* antioxidant activity of an acidic heteropolysaccharide with molecular weight 8.83×10^5 Da produced by marine fungus *Pseudomonas* PF-6. Exopolysaccharides exhibited strong hydroxyl radical scavenging activity at a concentration of 0.6 mg/mL and superoxide radicals scavenging activity at 0.25 mg/mL. The exopolysaccharides obtained from *Pseudomonas* PF-6 exhibited stronger antioxidant activities than those from other microbial exopolysaccharides and Ascorbic acid (Ye et al. 2012). Sun et al. (2009) isolated three different exopolysaccharides from marine filamentous fungus *Penicillium* sp. F23–2, and assessed their *in vitro* antioxidant activity. The results indicated that the three exopolysaccharides possessed good antioxidant activity, particularly in scavenging superoxide and hydroxyl radicals (Sun et al. 2009). However, numerous studies report with antioxidant and radical scavenging activities of exopolysaccharides from *Cordyceps sinensis*, and related species had any preventive effect against H_2O_2 Fas-induced apoptosis of Jurkat T

Table 9.1 Marine Fungal Biopolymers and their role

[a]Marine Fungal Exopolysaccharides	Organism	Role	References
Pullulans	*Aureobasidium pullulans*, *Aureobasidium pullulans var. melanogenium* P16 (Isolated from Mangrove system)	Enhanced oil Recovery, Food additive, drug delivery, thickener, a viscosity stabilizer, preparation of non-toxic, biodegradable, edible plastic materials	Ma et al. (2014)
Scleroglucan	*Sclerotium* sp.	Enhanced oil recovery, food additive, drug delivery, cosmetic and pharmaceutical products, biocompatible materials	Sveistrup et al. (2016)
Pestan	*Pestalotiopsis* sp. KCTC 8637	Wastewater treatment as biosorbent	Moon et al. (2006)
EPS	*Aspergillus fumigatus*	Biosorbent	Yin et al. (2011)
As1–1	*Aspergillus* sp. Y16	Antioxidant	Chen et al. (2011)
AEPS-1	*Cordyceps sinensis*	Antioxidant, Anticancer activity, Immunomodulator	Leung et al. (2009) and Wang et al. (2011)
Fw-1	*Fusarium oxysporum*	Antioxidant	Chen et al. (2015)
CMP-1	*Cordyceps militaris*	Immunomodulator	Jing et al. (2014)
YCP	*Phoma herbarum* YS4108	Antioxidant, antitumor agent with additional biological activities including antioxidation, anticoagulation and antiplatelet aggregation	Zhou et al. (2011)
GW-12	*Penicillium solitum*	Antioxidant	Yan et al. (2014)
MF-1	*Fusarium equiseti ANP2*	Antioxidant	Prathyusha et al. (2018)
PS-I	*Fusarium solani SD5*	Antioxidant	Mahapatra and Banerjee (2012, 2013a, b)
PS1–1, PS1–2 and PS2–1	*Penicillium* sp. F23–2	Antioxidant	Sun et al. (2009)
EPS	*Pseudomonas* PF-6	Antioxidant	Ye et al. (2012)
ENP-1 and ENP-2	*Epicoccum nigrum* JJY-40	Antioxidant	Sun et al. (2011)

(continued)

Table 9.1 (continued)

[a]Marine Fungal Exopolysaccharides	Organism	Role	References
AVP	*Aspergillus versicolor* LCJ-5-4	Antioxidant	Chen et al. (2012)
YSS	*Aspergillus terreus*	Antioxidant	Wang et al. (2013)
EPS	*Arthrobacter* ps-5	Biosorption	Shuhong et al. (2014)

[a]Marine Fungal Exopolysaccharides mentioned in column 1 are named after purified EPS fractions

lymphocytes (T-cells) (Leung et al. 2009). Therefore, an in-depth exploration of exopolysaccharides' antioxidant activity will be required to determine if the exopolysaccharides will be useful in the food and pharmaceutical industry.

9.4 Role of Marine Fungal Exopolysaccharides in the Food Industry

In foods, polysaccharides are commonly used at very low levels (1–3%) based on formulation weight. Polysaccharides have been usually expended as food additives to modify the texture and rheology of aqueous suspensions, which gives the product desirable customer appeal. Biopolymers at low levels disperse in water to form extremely viscous dispersions or gels with varied thickening, gelling, emulsifying, binding and stabilizing effects. However, polysaccharides' addition slows down starch retrogradation, enhances fiber content, intensifies moisture retention, and increases foods' overall quality. Polysaccharides also regulate the formation of ice crystals in frozen foods and confer stability to products that go through successive freeze-thaw cycles (Pilnik and Rombouts 1985). Even though food products are emerging more and more diverse and complex, the necessity for new and versatile food additives are increasing.

Furthermore, polysaccharide gums form non-fat substitutes, functioning as a starting material for soluble dietary fiber with beneficial health effects. Currently, plants, seaweed and crustaceans are the sources of polysaccharide gums; still, the exploration for novel compounds continues (Gimeno et al. 2004). Antioxidants are generally added to processed foods, as they could reduce oxidative damage in tissues directly by scavenging the free radicals and indirectly by increasing cells' natural defences. However, almost all the antioxidants used in processed foods are synthetic and are assumed for liver damage and carcinogenesis. Therefore, it is indispensable to discover, develop and use effective natural antioxidants (Laurienzo 2010).

Packaging of food products is essential for the preservation and vending. In food packaging, several factors need to be considered: viz. gas and water vapour penetrability, mechanical properties, sealing ability, resistance to grease, acid, water and UV light, transparency, antifogging property, printing ability, accessibility, and cost.

For several years, petroleum-based products that are easily available, biologically resistant with excellent water barrier properties have been used in food packaging. However, synthetic packaging's main disadvantage has been recognized with severe environmental issues due to its non-biodegradable nature. Hence, research has been focused on increasing biodegradable, edible films from biological and renewable resources (Ferreira et al. 2016). Notably, fungal exopolysaccharides are the major alternatives in the Food industry due to their ability to form thin layers that can coat the foods, potential antioxidant abilities, stabilizer-texturizer properties.

Pullulan is one of the few biopolymers that have attracted much interest currently owing to its peculiar characteristics. Pullulan is widely used as a fiber and film-forming agent because of physiological properties such as oxygen impermeability, inhibition of fungal growth, non-hygroscopic and non-reducing. Pullulans can be used as dietary fiber and as prebiotic oligosaccharides, owing to their partial degradation by amylases. Pullulans can be used as an alternative to starch in bakeries, enhancing the texture of foods, viscosity and improving moisture retention. However, using pullulans in foods is limited globally except in Japan (Farris et al. 2014; Ma et al. 2017; Pınar et al. 2013; Singh et al. 2008).

Interestingly, scleroglucan is also a versatile, functional constituent in food, serves as a stabilizer-texturizer in desserts and sauces because of its stability at low pH and high temperature. Scleroglucan can form edible films and add to the formulation of low-calorie foods because it is not digestible. Several Japanese patents recently issued permission to synthesize and apply scleroglucan in food industries (Schmid et al. 2011).

9.5 Role of Marine Fungal Exopolysaccharides in Oil Recovery

Microbial enhanced oil recovery is one of the most feasible, eco-friendly and potentially cost-effective techniques compared to conventional enhanced oil recovery techniques. Microbial enhanced oil recovery technique has a great potential way to prolong the declining life of oil reservoirs. Microbial enhanced oil recovery involves microorganisms/microbe metabolites (biosurfactants, exopolysaccharides, biogases) and their byproducts (Safdel et al. 2017; Sen 2008). Using microbial enhanced oil recovery techniques, microbes and their metabolites cause alteration at oil/water/rock interfaces and fluctuations in the flow phase and behaviors. Remarkably, the exopolysaccharides secreted by marine fungi protect them from temporary desiccation and predation along with assistance in adhesion to surfaces. Microbial enhanced oil recovery using biopolymers is mostly selective plugging of high-permeability zones and modifying the reservoir's permeability to redirect water flood to the oil-rich channel. Another important use of biopolymers is their potential as mobility control agents by enhancing the viscosity of displacing water, henceforth improving mobility and sweep efficiency (Al-Sulaimani et al. 2011; Sen 2008).

Scleroglucan reported better stability in oil recovery than xanthan over a wide range of pH and temperature. In watered-out basins, where seawater's pressure is not adequate for oil recovery, the addition of scleroglucan can salvage oil significantly. Scleroglucan lubricates the drill and regulate backpressures generated during drilling. Furthermore, scleroglucan is a useful oil mud stabilizer and thickener; it also increases the viscosity of very thin oil muds, which cannot be drilled otherwise. Several reports describe the rheology of scleroglucan-oil muds that persists to be stable between 20–80 °C and have a higher resistance to field pollutants (Schmid et al. 2011; Sveistrup et al. 2016).

9.6 Role of Marine Fungal Exopolysaccharides in the Cosmetic Industry

Ageing of the skin is a diverse and complex biological process that happens in all living organisms at different rates. Intrinsic and extrinsic (premature cutaneous ageing) ways of skin ageing processes affect cell condition and functioning. Intrinsic ageing implies natural programmed cellular ageing. Cellular ageing is exemplified by reduced collagen synthesis, loss of hydration and degeneration of elastic fiber networks. However, in a most recent report, extrinsic ageing (also known as photo ageing) occurred due to external factors such as pollution, excessive alcohol consumption, poor nutrition, and smoking and predominantly by exposure to Sun. Therefore the solar exposure results in wrinkling of skin and pigmentation (Agrawal et al. 2018). Remarkably the oxidative stress/damage is also a principal reason for skin ageing. Topical and oral usage of antioxidants has shown the anticipatory effect on skin ageing resulted from free radicals.

Nonetheless, the marine fungi exopolysaccharides are an excellent source for natural antioxidant compounds. The polysaccharide exopolysaccharides two isolated from marine filamentous fungus *Keissleriella* sp. YS 4108 was reported to possess strong scavenging activity for superoxide radicals (Sun et al. 2004). Exopolysaccharides isolated from several marine fungi exhibited strong scavenging activity for superoxide radicals, which can be used effectively in the cosmetic industry.

9.7 Role of Marine Fungal Exopolysaccharides in Biosorption

Rapid growth in industrialization resulted in the accumulation of large quantities of effluent (toxic heavy metals, chemicals). Their existence sets environmental–disposal issues because of their non-degradable and persistent nature. Many heavy metals currently represent a comprehensive ecological hazard that is carcinogenic, toxic or mutagenic even at trace levels. Heavy metals like Cu, Pb and Cr are used in manufacturing, electroplating, automotive, mining etc. (Ogunbileje et al. 2013). Nonetheless, severe exposure to heavy metals causes severe health problems like anemia, liver and kidney dysfunction, brain damage and finally, death. Few traditional methods for removing heavy metals from diluted solutions include oxidation/reduction, precipitation, filtration, electrochemical processes, ionic exchange, evaporation and membrane separations. Traditional methods have several drawbacks, viz. incomplete removal, high cost and energy consumption, low selectivity, produce toxic slurries that are difficult for elimination. Therefore, the best alternative is biosorption, which sequesters heavy metal ions effectively in solution from ppm to ppb level. Notably, microbial Exopolyaccharides, especially marine fungi, are ideal candidates for biosorption and helps in the adsorption and binding of heavy metals. Exopolysaccharide aggregates metal cations by interaction with negatively charged functional groups on exopolysaccharides (Kim et al. 1996).

Pestan is a fungal exopolysaccharide synthesized by *Pestalotiopsis* sp. KCTC 8637 can be used in wastewater treatment as a biosorbent for heavy metals lead and zinc. One gram of pestan can adsorb 120 mg lead and 60 mg zinc, signifying that Pestan is more efficient in removing lead than zinc from aqueous solutions (Moon et al. 2006). Exopolysaccharides of *Aspergillus fumigatus* also showed biosorption efficiency for heavy metals copper, lead and cadmium. The efficiency of heavy metal sorption from exopolysaccharides in order was Cu(II) followed by Pb(II) and Cd(II) (Yin et al. 2011). In another study, a novel exopolysaccharide produced from *Arthrobacter* ps-5 was reported with biosorption efficiency for Cu^{2+}, $Pb^{2+}$and Cr^{6+} metals. Each gram of exopolysaccharides can adsorb up to 84.47 mg of Cr^{6+}, 169.15 mg of Cu^{2+}, and 216.09 mg of $Pb^{2+,}$ respectively, signifying that this exopolysaccharide can adsorb lead more efficiently than Pestan (Shuhong et al. 2014).

9.8 Conclusions and Future Perspectives

Many researchers reported several marine fungal exopolysaccharides with their potential application in biotechnology, medicine, and the cosmetic industry. Research related to the prospecting of marine fungal exopolysaccharides may result in discovering new exopolysaccharides that are different from their terrestrial counterparts. Marine habitats are thought to be poorly explored, and in light of the ongoing research associate with marine fungal diversity based on culture-dependent and independent approaches, it is reasonable to anticipate that new fungal taxa recovered from marine habitats will be reported in less time. However, with the additional studies on exopolysaccharides' structure and biological function, a better understanding of structure characterization and exopolysaccharides' physicochemical properties will be further illustrated. However, polysaccharides from the biological resource are promising constituents for several diseases; it is essential to review the polysaccharides studies to point out their application and exploitation. Several studies have been reported in the last decade about various functional properties of polysaccharides due to the lack of research on natural polysaccharide safeguard mechanisms and clinical trials, the application in medicine. It is limited to a small range, with the preparation of exopolysaccharides, quality standards, route of administration and other technical problems are resolved, and the future must have more significant applications. Our review might promote further understanding of exopolysaccharides' biological activities and use it as a valuable source for medicine, cosmetic and food industries.

Acknowledgements The authors gratefully acknowledge UGC, Government of India, for the financial support in the form of UGC-CSIR-JRF & SRF under the Grant No: F.No. 19-1/2015(SA-1).

Conflict of Interest The authors declare that the research was conducted in the absence of any commercial or financial relationships that could be construed as a potential conflict of interest.

References

Agrawal S, Adholeya A, Barrow CJ, Deshmukh SK (2018) Marine fungi: an untapped bio-resource for future cosmeceuticals. Phytochem Lett 23:15–20. https://doi.org/10.1016/j.phytol.2017.11.003

Al-Sulaimani H, Joshi S, Al-Wahaibi Y, Al-Bahry S, Elshafie A, Al-Bemani A (2011) Microbial biotechnology for enhancing oil recovery : current developments and future prospects. Biotechnol Bioinformatics Bioeng 1(2):147–158. https://d1wqtxts1xzle7.cloudfront.net/42515079

Ates O (2015) Systems biology of microbial exopolysaccharides production. Front Bioeng Biotechnol 3:200. https://doi.org/10.3389/fbioe.2015.00200

Bonugli-Santos RC, dos Santos Vasconcelos MR, Passarini MR, Vieira GA, Lopes VC, Mainardi PH (2015) Marine-derived fungi: diversity of enzymes and biotechnological applications. Front Microbiol 6:269. https://doi.org/10.3389/fmicb.2015.00269

Chen Y, Mao W, Tao H, Zhu W, Qi X (2011) Structural characterization and antioxidant properties of an exopolysaccharide produced by the mangrove endophytic fungus *Aspergillus sp.* Y16. Bioresour Technol 102(17):8179–8184. https://doi.org/10.1016/j.biortech.2011.06.048

Chen Y, Mao W, Yang Y, Teng X (2012) Structure and antioxidant activity of an extracellular polysaccharide from coral-associated fungus, *Aspergillus versicolor* LCJ-5-4. Carbohydr Polym 87(1):218–226. https://doi.org/10.1016/j.carbpol.2011.07.042

Chen YL, Mao WJ, Tao HW, Zhu WM (2015) Preparation and characterization of a novel extracellular polysaccharide with antioxidant activity, from the mangrove-associated fungus *Fusarium oxysporum*. Mar Biotechnol 17(2):219–228. https://doi.org/10.1007/s10126-015-9611-6

Farris S, Unalan IU, Introzzi L, Fuentes-Alventosa JM, Cozzolino CA (2014) Pullulan-based films and coatings for food packaging: present applications, emerging opportunities, and future challenges. J Appl Polym Sci 131(13):1–12. https://doi.org/10.1002/app.40539

Ferreira ARV, Alves VD, Coelhoso IM (2016) Polysaccharide-based membranes in food packaging applications. Membranes 6(2):1–17. https://doi.org/10.3390/membranes6020022

Freitas F, Alves VD, Reis MA (2011) Advances in bacterial exopolysaccharides: from production to biotechnological applications. Trends Biotechnol 29(8):388–398. https://doi.org/10.1016/j.tibtech.2011.03.008

Georgieva NV (2005) Oxidative stress as a factor of disrupted ecological oxidative balance in biological systems – a review. Bulg J Vet Med:1–12. http://tru.uni-sz.bg/bjvm/vol08-no01-01.pdf

Ghanbarzadeh B, Almasi H (2013) Biodegradable polymers. In: Biodegradation-life of science, pp 141–174. https://doi.org/10.5772/52777

Gimeno E, Moraru CI, Kokini JL (2004) Effect of xanthan gum and CMC on the structure and texture of corn flour pellets expanded by microwave heating. Cereal Chem 81(1):100–107. https://doi.org/10.1094/CCHEM.2004.81.1.100

Hamid Akash MS, Rehman K, Chen S (2015) Natural and synthetic polymers as drug carriers for delivery of therapeutic proteins. Pol Rev 55(3):371–406. https://doi.org/10.1080/15583724.2014.995806

Jing Y, Cui X, Chen Z, Huang L, Song L et al (2014) Elucidation and biological activities of a new polysaccharide from cultured *Cordyceps militaris*. Carbohydr Polym 102(1):288–296. https://doi.org/10.1016/j.carbpol.2013.11.061

Kim W, Choi K, Kim Y, Park H et al (1996) Purification and characterization of a fibrinolytic enzyme produced from *Bacillus sp.* strain CK 11-4 screened from Chungkook-Jang. Appl Environ Microbiol 62(7):2482–2488. https://aem.asm.org/content/aem/62/7/2482.full.pdf

Laurienzo P (2010) Marine polysaccharides in pharmaceutical applications: an overview. Mar Drugs 8(9):2435–2465. https://doi.org/10.3390/md8092435

Leung PH, Zhao S, Ho KP, Wu JY (2009) Chemical properties and antioxidant activity of exopolysaccharides from mycelial culture of *Cordyceps sinensis* fungus Cs-HK1. Food Chem 114(4):1251–1256. https://doi.org/10.1016/j.foodchem.2008.10.081

Ma ZC, Fu WJ, Liu GL, Wang ZP, Chi ZM (2014) High-level pullulan production by Aureobasidium pullulans var. melanogenium P16 isolated from mangrove system. Appl Microbiol Biotechnol 98(11):4865–4873. https://doi.org/10.1007/s00253-014-5554-5

Ma X, Zhang L, Fan D, Xue W, Zhu C, Li X, Wang Y (2017) Physicochemical properties and biological behavior of injectable crosslinked hydrogels composed of pullulan and recombinant human-like collagen. J Mater Sci 52(7):3771–3785. https://doi.org/10.1007/s10853-016-0577-1

Mahapatra S, Banerjee D (2012) Structural elucidation and bioactivity of a novel exopolysaccharide from endophytic *Fusarium solani* SD5. Carbohydr Polym 90(1):683–689. https://doi.org/10.1016/j.carbpol.2012.05.097

Mahapatra S, Banerjee D (2013a) Fungal exopolysaccharide: production, composition and applications. Microbiol Insight 6:MBI-S10957. https://doi.org/10.4137/MBI.S10957

Mahapatra S, Banerjee D (2013b) Evaluation of in vitro antioxidant potency of exopolysaccharide from endophytic *Fusarium solani* SD5. Int J Biol Macromol 53:62–66. https://doi.org/10.1016/j.ijbiomac.2012.11.003

Menezes CBA, Bonugli-Santos RC, Miqueletto PB, Passarini MRZ, Silva CHD, Justo MR, Sette LD (2010) Microbial diversity associated with algae, ascidians and sponges from the north coast of São Paulo state, Brazil. Microbiol Res 165(6):466–482. https://doi.org/10.1016/j.micres.2009.09.005

Moon SH, Park CS, Kim YJ, Park Y (2006) Biosorption isotherms of Pb (II) and Zn (II) on Pestan, an extracellular polysaccharide, of *Pestalotiopsis* sp. KCTC 8637P. Process Biochem 41(2):312–316. https://doi.org/10.1016/j.procbio.2005.07.013

Ogunbileje JO, Sadagoparamanujam VM, Anetor JI, Farombi EO, Akinosun OM, Okorodudu AO (2013) Lead, mercury, cadmium, chromium, nickel, copper, zinc, calcium, iron, manganese and chromium (VI) levels in Nigeria and United States of America cement dust. Chemosphere 90(11):2743–2749. https://doi.org/10.1016/j.chemosphere.2012.11.058

Osińska-Jaroszuk M, Jarosz-Wilkołazka A, Jaroszuk-Ściseł J, Szałapata K, Nowak A, Jaszek M, Majewska M (2015) Extracellular polysaccharides from Ascomycota and Basidiomycota: production conditions, biochemical characteristics, and biological properties. World J Microbiol Biotechnol 31(12):1823–1844. https://doi.org/10.1007/s11274-015-1937-8

Pahlavani MA, Harris MD (1998) Effect of in vitro generation of oxygen free radicals on T cell function in young and old rats. Free Radic Biol Med 25(8):903–913. https://doi.org/10.1016/S0891-5849(98)00124-5

Pilnik W, Rombouts FM (1985) Polysaccharides and food processing. Carbohydr Res 142(1):93–105. https://doi.org/10.1016/S0008-6215(00)90736-5

Pınar O, Yangılar F, Oğuzhan P, Yangılar F (2013) Pullulan: production and usage in food ındustry. Afr J Food Sci Technol 4(3):57–63. http://www.interesjournals.org/AJFST. ISSN:2141-5455

Prathyusha AMVN, Mohana Sheela G, Bramhachari PV (2018) Chemical characterization and antioxidant properties of exopolysaccharides from mangrove filamentous fungi *Fusarium equiseti* ANP2. Biotechnol Rep 19:e00277. https://doi.org/10.1016/j.btre.2018.e00277

Raghukumar C (2004) Marine fungi and their enzymes for decolorization of colored effluents, Marine Microbiology: Facets and Opportunities. National Institute of Oceanography, Goa, pp 145–158. http://drs.nio.org/drs/bitstream/handle/2264/80/RaghukumarC_chap15.pdf?sequence=1

Raghukumar C (2008) Marine fungal biotechnology: an ecological perspective. Fungal Divers 31:19–35. http://drs.nio.org/drs/handle/2264/1392

Ravella SR, Quiñones TS, Retter A, Heiermann M, Amon T, Hobbs PJ (2010) Extracellular polysaccharide (EPS) production by a novel strain of yeast-like fungus *Aureobasidium pullulans*. Carbohydr Polym 82(3):728–732. https://doi.org/10.1016/j.carbpol.2010.05.039

Safdel M, Anbaz MA, Daryasafar A, Jamialahmadi M (2017) Microbial enhanced oil recovery, a critical review on worldwide implemented field trials in different countries. Renew Sust Energ Rev 74:159–172. https://doi.org/10.1016/j.rser.2017.02.045

Schmid J, Meyer V, Sieber V (2011) Scleroglucan: biosynthesis, production and application of a versatile hydrocolloid. Appl Microbiol Biotechnol 91(4):937–947. https://doi.org/10.1007/s00253-011-3438-5

Selbmann L, Onofri S, Fenice M, Federici F, Petruccioli M (2002) Production and structural characterization of the exopolysaccharide of the Antarctic fungus *Phoma herbarum* CCFEE 5080. Res Microbiol 153(9):585–592. https://doi.org/10.1016/S0923-2508(02)01372-4

Sen R (2008) Biotechnology in petroleum recovery: the microbial EOR. Prog Energy Combust Sci 34(6):714–724. https://doi.org/10.1016/j.pecs.2008.05.001

Shearer CA, Descals E, Kohlmeyer B, Kohlmeyer J, Marvanová L, Padgett D, Voglymayr H (2007) Fungal biodiversity in aquatic habitats. Biodivers Conserv 16(1):49–67. https://doi.org/10.1007/s10531-006-9120-z

Shuhong Y, Meiping Z, Hong Y, Han W, Shan X, Yan L, Jihui W (2014) Biosorption of Cu^{2+}, Pb^{2+} and Cr^{6+} by a novel exopolysaccharide from *Arthrobacter* ps-5. Carbohydr Polym 101(1):50–56. https://doi.org/10.1016/j.carbpol.2013.09.021

Singh RS, Saini GK, Kennedy JF (2008) Pullulan: microbial sources, production and applications. Carbohydr Polym 73(4):515–531. https://doi.org/10.1016/j.carbpol.2008.01.003

Sun C, Wang JW, Fang L, Gao XD, Tan RX (2004) Free radical scavenging and antioxidant activities of EPS2, an exopolysaccharide produced by a marine filamentous fungus *Keissleriella* sp. YS 4108. Life Sci 75(9):1063–1073. https://doi.org/10.1016/j.lfs.2004.02.015

Sun HH, Mao WJ, Chen Y, Guo SD, Li HY, Qi XH, Xu J (2009) Isolation, chemical characteristics and antioxidant properties of the polysaccharides from marine fungus *Penicillium* sp. F23-2. Carbohydr Polym 78(1):117–124. https://doi.org/10.1016/j.carbpol.2009.04.017

Sun HH, Mao WJ, Jiao JY, Xu JC, Li HY, Chen Y, Yang YP (2011) Structural characterization of extracellular polysaccharides produced by the marine fungus *Epicoccum nigrum* JJY-40 and their antioxidant activities. Mar Biotechnol 13(5):1048–1055. https://doi.org/10.1007/s10126-011-9368-5

Sveistrup M, van Mastrigt F, Norrman J, Picchioni F, Paso K (2016) Viability of biopol enhanced oil recovery. J Dispers Sci Technol 37(8):1160–1169. https://doi.org/10.1080/01932691.2015.1088450

Szewczuk-Karpisz K, Wiśniewska M, Pac-Sosińska M, Choma A, Komaniecka I (2016) Stability mechanism of the silica suspension in the *Sinorhizobium meliloti* 1021 exopolysaccharide presence. J Ind Eng Chem 35:108–114. https://doi.org/10.1016/j.jiec.2015.12.018

Tisthammer KH, Cobian GM, Amend AS (2016) Global biogeography of marine fungi is shaped by the environment. Fungal Ecol 19:39–46. https://doi.org/10.1016/j.funeco.2015.09.003

Wang ZM, Peng X, Lee KLD, Tang JCO, Cheung PCK, Wu JY (2011) Structural characterisation and immunomodulatory property of an acidic polysaccharide from mycelial culture of *Cordyceps sinensis* fungus Cs-HK1. Food Chem 125(2):637–643. https://doi.org/10.1016/j.foodchem.2010.09.052

Wang C, Mao W, Chen Z, Zhu W, Chen Y, Zhao C, Guo T (2013) Purification, structural characterization and antioxidant property of an extracellular polysaccharide from *Aspergillus terreus*. Process Biochem 48(9):1395–1401. https://doi.org/10.1016/j.procbio.2013.06.029

Weydert CJ, Cullen JJ (2010) Measurement of superoxide dismutase, catalase and glutathione peroxidase in cultured cells and tissue. Nat Protoc 5(1):51–66. https://doi.org/10.1038/nprot.2009.197

Yan M, Mao W, Chen C, Kong X, Gu Q, Li N, Xiao B (2014) Structural elucidation of the exopolysaccharide produced by the mangrove fungus *Penicillium solitum*. Carbohydr Polym 111:485–491. https://doi.org/10.1016/j.carbpol.2014.05.013

Yang XB, Gao XD, Han F, Xu BS, Song YC, Tan RX (2005) Purification, characterization and enzymatic degradation of YCP, a polysaccharide from marine filamentous fungus *Phoma herbarum* YS4108. Biochimie 87(8):747–754. https://doi.org/10.1016/j.biochi.2005.03.004

Ye S, Liu F, Wang J, Wang H, Zhang M (2012) Antioxidant activities of an exopolysaccharide isolated and purified from marine *Pseudomonas* PF-6. Carbohydr Polym 87(1):764–770. https://doi.org/10.1016/j.carbpol.2011.08.057

Yin Y, Hu Y, Xiong F (2011) Sorption of Cu(II) and Cd(II) by extracellular polymeric substances (EPS) from *Aspergillus fumigatus*. Int Biodeterior Biodegrad 65(7):1012–1018. https://doi.org/10.1016/j.ibiod.2011.08.001

Zhang X, Ding R, Zhou Y, Zhu R, Liu W, Jin L, Gao X (2013) Toll-like receptor 2 and toll-like receptor 4-dependent activation of B cells by a polysaccharide from marine fungus *Phoma herbarum* YS4108. PLoS One 8(3). https://doi.org/10.1371/journal.pone.0060781

Zhou T, Song YC, Feng MQ, Tan RX (2011) Cloning, expression and biochemical characterization of UDP-glucose 6-dehydrogenase, a key enzyme in the biosynthesis of an antitumor polysaccharide from the marine fungus *Phoma herbarum* YS4108. Process Biochem 46(12):2263–2268. https://doi.org/10.1016/j.procbio.2011.09.006

Printed by Printforce, the Netherlands